Nanomaterials:
From Research to Applications

Nanomaterials: From Research to Applications

H. Hosono, Y. Mishima, H. Takezoe, and K.J.D. MacKenzie
Tokyo Institute of Technology, JAPAN

Amsterdam • Boston • Heidelberg • London • New York • Oxford
Paris • San Diego • San Francisco • Singapore • Sydney • Tokyo

Elsevier
The Boulevard, Langford Lane, Kidlington, Oxford OX5 1GB, UK
Radarweg 29, PO Box 211, 1000 AE Amsterdam, The Netherlands

First edition 2006
Reprinted 2007

Copyright © 2006 Elsevier Ltd. All rights reserved

No part of this publication may be reproduced, stored in a retrieval system
or transmitted in any form or by any means electronic, mechanical, photocopying,
recording or otherwise without the prior written permission of the publisher

Permissions may be sought directly from Elsevier's Science & Technology Rights
Department in Oxford, UK: phone (+44) (0) 1865 843830; fax (+44) (0) 1865 853333;
email: permissions@elsevier.com. Alternatively you can submit your request online by
visiting the Elsevier web site at http://elsevier.com/locate/permissions, and selecting
Obtaining permission to use Elsevier material

Notice
No responsibility is assumed by the publisher for any injury and/or damage to persons
or property as a matter of products liability, negligence or otherwise, or from any use
or operation of any methods, products, instructions or ideas contained in the material
herein. Because of rapid advances in the medical sciences, in particular, independent
verification of diagnoses and drug dosages should be made

British Library Cataloguing in Publication Data
A catalogue record for this book is available from the British Library

Library of Congress Cataloging-in-Publication Data
A catalog record for this book is available from the Library of Congress

ISBN–13: 978-0-08-044964-7
ISBN–10: 0-08-044964-6

For information on all Elsevier publications
visit our website at books.elsevier.com

Printed and bound in *Great Britain*

07 08 09 10   10 9 8 7 6 5 4 3 2

Working together to grow
libraries in developing countries

www.elsevier.com | www.bookaid.org | www.sabre.org

ELSEVIER   BOOK AID International   Sabre Foundation

Contents

Preface	x
Materials Research at the Tokyo Institute of Technology *Seizo Miyata*	xi
List of Contributors	xiv

Part I   Revolutional Oxides

1   Function Cultivation in Transparent Oxides Utilizing Natural and Artificial Nanostructures   3

Hideo Hosono and Masahiro Hirano

1.1	General Introduction	4
1.2	Transparent Oxide Semiconductors	7
1.3	Transparent Nanoporous Crystal $12CaO \cdot 7Al_2O_3$	24
1.4	Encoding of Periodic Nanostructures with Interfering Femtosecond Pulses	40
	References	57

2   The Role of Lattice Defects in Oxides   62

Mitsuru Itoh

2.1	Introduction	62
2.2	Magnetic Materials (Spin Crossover in Oxides)	63
2.3	Ferroelectric Materials	72
2.4	Lithium Ion Conductivity in Oxides	79

2.5	Concluding Remarks	91
2.6	Clue to the Design of New Functional Oxide Materials	93
	References	95

3 Size Effect of Ferroelectric and High Permittivity Thin Films 99

Hiroshi Funakubo

3.1	Introduction	99
3.2	Size Effect of Ferroelectricity in PZT Thin Films	101
3.3	Size-Effect-Free Characteristics of Bismuth Layer Structured Dielectrics	107
3.4	Summary and Future Aspects	131
	References	131

Part II State-of-the-Art Polymers

4 Photonic Devices using Liquid Crystal Nanostructures 137

Hideo Takezoe

4.1	Photonic Effect	137
4.2	Lasing from CLCs	148
4.3	Optical Diode	160
4.4	Concluding Remarks and Future Problems for Practical Applications	167
	References	168

5 Nanocylinder Array Structures in Block Copolymer Thin Films 171

Kaori Kamata and Tomokazu Iyoda

5.1	General Introduction	171
5.2	Synthesis of Block Copolymers	172
5.3	Self-organization and Phase Behavior of Block Copolymer Microdomains	173
5.4	Phase-segregated Nanostructures in Block Copolymer Thin Films	191
5.5	Phase-segregated Nanostructures in Thin Films Effective for Practical Use Cylindrical Phases in Block Copolymer Thin Films	194

5.6	Nanocylindrical-structured Block Copolymer Templates	210
5.7	Summary and Future Directions	215
	References	216

6 Nano-Size Charge Inhomogeneity in Organic Metals 224

Takehiko Mori

6.1	Introduction	224
6.2	Universal Phase Diagram of the θ-Phase	226
6.3	Charge Order	231
6.4	Theoretical Background and Estimation of V	240
6.5	Discussion	255
6.6	Summary	257
	References	258

Part III Nanostructure Design for New Functions

7 Size Control of Nanostructures by Quantum Confinement **265**

Hiroyuki Hirayama

7.1	Introductory Remarks	265
7.2	Quantum Well States in Surface Nanostructures	267
7.3	Size Control of Nanostructures via Quantum Confinement	278
7.4	Practical and Future Applications	291
7.5	Summary	295
	References	295

8 Grain Boundary Dynamics in Ceramics Superplasticity **297**

Fumihiro Wakai and Arturo Domínguez-Rodríguez

8.1	Introduction	297
8.2	Motion and Topological Evolution of Grains	298
8.3	Physical Characteristics of Ceramics Superplasticity	304
8.4	Grain Refinement and Suppression of Grain Growth	305
8.5	Diffusion Enhancement	306
8.6	Superplastic Forming	309
8.7	Future Prospects	310
	References	310

9 Nanostructure Control for High-strength and High-ductility Aluminum Alloys — 315

Tatsuo Sato

- 9.1 Introduction — 315
- 9.2 History of High-strength and High-ductility Aluminum Alloys — 316
- 9.3 Discovery of GP Zones — 319
- 9.4 Clusters in the Early Stage of Phase Decomposition — 320
- 9.5 Ductility and PFZ Control by Nanoclusters — 342
- 9.6 Summary — 344
- References — 344

Part IV Nanostructure Architecture for Engineering Applications

10 Nanoporous Materials from Mineral and Organic Templates — 349

Kiyoshi Okada and Kenneth J.D. MacKenzie

- 10.1 Historical Background and Development — 349
- 10.2 Review of the Porous Properties of Nanoporous Materials Produced Using Mineral Templates — 357
- 10.3 Practical/Future Applications Related to Various Properties — 373
- 10.4 Summary — 378
- References — 379

11 Enhancement of Thermoelectric Figure of Merit through Nanostructural Control on Intermetallic Semiconductors toward High-temperature Applications — 383

Yoshinao Mishima, Yoshisato Kimura and Sung Wng Kim

- 11.1 Background and Principles — 384
- 11.2 Bulk Intermetallic Semiconductors for Future High-temperature Applications — 390
- 11.3 A Breakthrough in *ZT*: Nanostructured Materials — 405
- References — 414

| 12 | **Smart Coatings – Multilayered and Multifunctional in-situ Ultrahigh-temperature Coatings** | **419** |

Hideki Hosoda

	12.1	Introduction	420
	12.2	Oxidation Resistance of Ir	421
	12.3	Design of Multifunctional and Multilayered Coating Based on IrAl	423
	12.4	Physical Properties of IrAl Alloys	425
	12.5	Oxidation Behavior of Ir-rich IrAl	433
	12.6	Oxidation Behavior of Al-rich IrAl Alloys	437
	12.7	Oxidation Behavior of Co-added IrAl	440
	12.8	Summary	444
		References	445

Index **447**

Preface

A proposal submitted by the professors working in the area of Materials Science at the Tokyo Institute of Technology was selected as one of the twenty-first Century COE (Center of Excellence) programs entitled "Nanomaterial Frontier Cultivation for Industrial Collaboration". The objective of the project is to foster innovation in the field of nanomaterials, building on the strong tradition of Materials Science at the Tokyo Institute of Technology, typified by the success of the ferrite and polyacetylene materials developed there.

This book, which summarizes the achievements of this COE program, is divided into four parts: (1) Revolutional Oxides, (2) State-of-the-Art Polymers, (3) Nanostructure Design for New Functions, and (4) Nanostructure Architecture for Engineering Applications. Each part consists of three or four chapters related to inorganic, organic, and metallic nanomaterials.

This book is published with the support of the COE program. All the contributors in this program are grateful for the continuous support by the JSPS (Japan Society for the Promotion of Science). Acknowledgment is also given to the Tokyo Institute of Technology for its encouragement of this activity and its financial support. Finally, sincere thanks are due to Elsevier Science Ltd. for assistance with editing the manuscripts and for publishing this book.

Hideo Hosono
Kenneth MacKenzie
Yoshinao Mishima
Hideo Takezoe

Materials Research at the Tokyo Institute of Technology

Seizo Miyata

Professor Shirakawa is presented with the Nobel Prize by King Gustov
(Courtesy of Professor Kenneth J. Wynne)

The Tokyo Institute of Technology was originally founded as the Tokyo Vocational School in May 1881. The School was renamed as the Tokyo Technical School in 1890 and later became the Tokyo Higher Technical School in 1901. In 1929, the Tokyo Technical School was promoted to the status of a degree-conferring university and was renamed as the Tokyo Institute of Technology. Its new mission was to impart higher education to professional engineers and develop their capabilities for research and development, in order to contribute to the modernization of Japanese industries.

Materials research has been actively carried out throughout the entire history of the University. Currently, the Tokyo Institute of Technology is one of the best educational and research centers for materials science, not only in Japan but also by world standards, as shown in the following chapters.

In its history spanning more than a century, two Professors of the Tokyo Institute of Technology, Professor Yogoro Kato and Professor Hideki Shirakawa, stand out from the many excellent Professors of Materials Science.

Professor Yogoro Kato (1872–1967) was invited to join the Tokyo Higher Technical School as a Professor in 1907 and began the study of metal oxides with his former student Professor Takeshi Takei in 1929. Soon they discovered the existence of strong magnetization in ferrite even though it was an insulator. Thus began the application of ferrites to modern electronics and the era of ferrite technology was launched in 1932. Three years later, the TDK Corporation was founded to industrialize these inventions, and it is currently one of the world's leading electronics components manufacturers, with 31 000 employees.

In 1939, Professor Kato donated all his patent royalties to the Tokyo Institute of Technology to establish the Chemical Research Laboratory where Professor Shirakawa was to begin his career later as an Assistant Professor. Professor Hideki Shirakawa (1936–present), who was the Nobel Laureate for Chemistry in 2000, graduated from the Tokyo Institute of Technology with a degree in chemical engineering in 1961 and enrolled in the graduate program there, receiving his doctorate in engineering in 1966. Immediately on receiving his PhD, he was hired as an Assistant Professor at the Chemical Resources Laboratory of the Tokyo Institute of Technology and began working on the polymerization of polyacetylene, the work for which he received the Nobel Prize. On this occasion, he wrote:

> In the fall of 1967, only a short time after I started, I discovered polyacetylene film through an unforeseeable experimental failure. With the conventional method of polymerization, chemists had obtained the compound in the form of a black powder; however, one day, when a visiting scientist tried to make polyacetylene in the usual way, he only produced some ragged pieces of a film. In order to clarify the reason for the failure, I inspected the various polymerization conditions again and again. I finally found that the concentration of the catalyst was the decisive factor for making the film. In any chemical reaction, a very small quantity of the catalyst, of the order of m·mol, would be sufficient, but the result I got was for a quantity of mol, a thousand times higher than I had intended. It was an extraordinary unit for a catalyst. I might have missed the "m" for "m·mol" in my experimental instructions, or the visitor might have misread it. For whatever reason, he had added the catalyst in molar quantities to the reaction vessel. The catalyst concentration a thousand-fold higher than I had planned had apparently accelerated the rate of the polymerization reaction about a thousand times. Roughly speaking, as soon as acetylene gas was put into the catalyst, the reaction occurred so quickly that the gas was just polymerized on the surface of the catalyst as a thin film.

The film, which was shiny as an aluminum foil, was more intriguing than the rather uninteresting black powder that was normally synthesized, and gave very simple IR absorption spectra. Incidentally during 1975, Professor Alan MacDiarmid of the University Pennsylvania (UPenn) was visiting Japan to lecture on the electrically conducting inorganic polymer sulphur nitride, $(SN)x$. He met Professor Shirakawa on learning that a gleaming polymer film had been invented. As soon as he went back to UPenn, he called Dr. Kenneth Wynne, the Program Manager of the Office of Naval Research (presently a Professor at the Virginia Commonwealth University) who promptly decided to support the new project because he felt that it fitted well with his funding policies, which were "to try to find innovative projects... to try to place funding in focus areas which would have "impact"... something like picking good stocks."

In his capacity as a Visiting Scholar, Dr. Shirakawa met Professor Heeger, who was working on one-dimensional electrical conducting organic materials in which the conductivity upon doping. The separate threads soon came together at UPenn. when a novel type of electrical conducting polymer was discovered. On November 23, 1976, when Shirakawa and a post-doctoral research Fellow of Professor Heeger were measuring the electrical conductivity of polyacetylene, there was a sudden surge in the conductivity over seven orders of magnitude when it was doped with bromine.

Currently, electrically conducting polymers show promise for display technology applications such as polymer light emitting diodes, transparent electrodes for switching LC displays, and so on.

Other Professors of the Tokyo Institute of Technology who have made significant contributions to society as materials scientists include Professor Issaku Koga (1891–1982), who developed high precision clocks using quartz crystals and Professor Shu Kambara (1906–2000), who in 1941 discovered a novel method for the production of polyacrylonitrile fiber.

List of Contributors

Seizo Miyata
(21st COE Professor)
Senior Program manager,
Fuel cell and Hydrogen,
Technology Development Dept.
New Energy and Industrial Tchnology
Development Organization
20F Muza Kawasaki Building, 1310,
Omiya-cho, Saiwai-ku, Kawasaki
Kanagawa 212-8554, Japan
E-mail: miyatasiz@nedo.go.jp
Tel: & Fax: +81-44-520-5262

Hideo Hosono
Professor, Frontier Collaborative
Research Center,
Tokyo Institute of Technology,
Nagatsuta 4259, Mail Box R3-1,
Midori-ku, Yokohama 226-8503
E-mail: hosono@msl.titech.ac.jp
Tel: +81-45-924-5359
Fax: +81-45-924-5339

Masahiro Hirano
21st COE professor,
Frontier Collaborative Research Center,
Tokyo Institute of Technology,
Nagatsuta 4259, Mail Box S2-13,
Midori-ku, Yokohama 226-8503
E-mail: m-hirano@lucid.msl.titech.ac.jp
Tel: & Fax: +81-45-924-5127

Mitsuru Itoh
Professor, Materials and Structures
Laboratory
Tokyo Institute of Technology,
Nagatsuta 4259, Mail Box J2-19,
Midori-ku, Yokohama 226-8503
E-mail: Mitsuru_Itoh@msl.titech.ac.jp
Tel: & Fax: +81-45-924-5354

Hiroshi Funakubo
Associate Professor, Department
of Innovative and Engineered
Materials,
Tokyo Institute of Technology,
Nagatsuta 4259, Mail Box J2-43,
Midori-ku,
Yokohama 226-8503
E-mail: funakubo@iem.titech.ac.jp
Tel: & Fax: +81-45-924-5446

Hideo Takezoe
Professor, Department of Organic and
Polymeric Materials
Tokyo Institute of Technology,
O-okayama, 2-12-1, Mail Box S8-42,
Meguro-ku,
Tokyo 152-8552
E-mail: htakezoe@o.cc.titech.ac.jp
Tel: +81-3-5734-2436
Fax: +81-3-5734-2876

List of Contributors

Tomokazu Iyoda
Professor, Chemical Resources Laboratory
Tokyo Institute of Technology,
Nagatsuta 4259, Mail Box R1-25,
Midori-ku, Yokohama 226-8503
E-mail: iyoda@res.titech.ac.jp
Tel: +81-45-924-5266
Fax: +81-45-924-5247

Takehiko Mori
Professor, Department of Chemistry and Materials Science
S8-31 Tokyo Institute of Technology,
O-okayama, 2-12-1, Mail Box S8-31,
Meguro-ku, Tokyo 152-8552
E-mail: takehiko@o.cc.titech.ac.jp
Tel: +81-3-5734-2427
Fax: +81-3-5734-2876

Hiroyuki Hirayama
Professor, Department of Materials Science and Engineering
Interdisciplinary Graduate School of Science and Engineering
Tokyo Institute of Technology,
Nagatsuta 4259, Mail Box J1-13,
Midori-ku, Yokohama 226-8503
E-mail: hirayama@materia.titech.ac.jp
Tel: +81-45-924-5637
Fax: +81-45-924-5685

Fumihiro Wakai
Professor, Materials and Structures Laboratory, Tokyo Institute of Technology,
Nagatsuta 4259, Mail Box R3-23,
Midori-ku, Yokohama 226-8503
E-mail: wakai@msl.titech.ac.jp
Tel: +81-45-924-5361
Fax: +81-45-924-5390

Arturo Dominguez-Rodrigues
Professor, Department of Solid State Physics, University of Sevilla
Apto. 1065, 41080 Sevilla, Spain
E-mail: adorod@us.es
Tel: +34-954-557-849
Fax: +34-954-612-097

Tatsuo Sato
Professor, Department of Metallurgy and Ceramics Science, Tokyo Institute of Technology,
O-okayama, 2-12-1, Mail Box S8-13,
Meguro-ku, Tokyo 152-8552
E-mail: sato@mtl.titech.ac.jp
Tel: & Fax: +81-3-5734-3139

Kiyoshi Okada
Professor, Department of Metallurgy and Ceramics Science, Tokyo Institute of Technology,
O-okayama, 2-12-1, Mail Box S7-7,
Meguro-ku, Tokyo 152-8552
E-mail: kokada@ceram.titech.ac.jp
Tel: +81-3-5734-2524
Fax: +81-3-5734-3355

Kenneth J.D. MacKenzie
Professor, MacDiarmid Institute for Advanced Materials and Nanotechnology,
Victoria University of Wellington,
P.O. Box 600, Wellington, New Zealand
& Visiting COE Professor, Department of Metallurgy and Ceramics Science,
Tokyo Institute of Technology,
O-okayama, Meguro, Tokyo 152-8552, Japan
E-mail: Kenneth.MacKenzie@vuw.ac.nz
Tel: +64-4-463-5885
Fax: +64-4-463-5237

Yoshinao Mishima
Professor, Department of Materials
Science and Engineering
Interdisciplinary Graduate School of
Science and Engineering
Tokyo Institute of Technology,
Nagatsuta 4259, Mail Box G3-23,
Midori-ku, Yokohama 226-8502
E-mail: mishima@materia.titech.ac.jp
Tel: & Fax: +81-45-924-5612,

Yoshisato Kimura
Associate, Department of Materials
Science and Engineering
Interdisciplinary Graduate School of
Science and Engineering
Tokyo Institute of Technology,
Nagatsuta 4259, Mail Box G3-23,
Midori-ku, Yokohama 226-8502
E-mail: kimurays@materia.titech.ac.jp
Tel: & Fax: +81-45-924-5495

Sung Wng Kim
Research Fellow, Frontier Collaborative
Research Center,
Tokyo Institute of Technology,
Nagatsuta, 4259, Mail Box S2-13,
Midori-ku, Yokohama 226-8503,
Japan
E-mail: sw-kim@lucid.msl.titech.ac.jp
Tel: & Fax: +81-45-924-5127

Hideki Hosoda
Associate Professor,
Advanced Materials Division,
Precision and Intelligence
Laboratory, Tokyo Institute of
Technology,
Nagatsuta 4259, Mail Box R2-27,
Midori-ku, Yokohama 226-8503
E-mail: hosoda@pi.titech.ac.jp
Tel: & Fax: +81-45-924-5057

PART I

Revolutional Oxides

CHAPTER 1

Function Cultivation in Transparent Oxides Utilizing Natural and Artificial Nanostructures

Hideo Hosono and Masahiro Hirano

Abstract

In this chapter, we review the recent progress in optoelectronic applications of transparent wide band gap oxides. We concentrate especially on creating new functions in transparent oxides by forming or utilizing nanostructures embedded in the materials. First, our material design concept is introduced in relation to the electronic structures of the oxides. Then optoelectronic properties, electronic structures, and device applications are reviewed for (1) layered oxychalcogenides LnCuOCh (Ln = lanthanide, Ch = chalcogen), and (2) nanoporous crystal $12CaO.7Al_2O_3$ (C12A7). Finally, fabrication of periodic nanostructures in transparent materials by interfering femtosecond (fs) laser pulses is reviewed. Sharp blue-to-UV light emission was observed in LnCuOCh originating from room-temperature stable exciton and the stability of exciton is discussed in relation to their two-dimensional electronic structures. C12A7 has free oxygen ions clathrated in its subnanometer-sized cages, and new functions may be added to C12A7 by replacing the free oxygen ions with active anions. Quantum calculations indicate that cages trapping electrons in C12A7 can be regarded as quantum dots. Micro/nano-processing techniques using fs pulses is an emerging approach to adding new functions to transparent materials. A distributed feedback laser structure was fabricated solely using the laser pulses and its oscillation at room temperature was demonstrated.

Keywords: **transparent oxide semiconductor, transparent conductive oxide, oxide electronics, electride, defect reengineering, nano-fabrication, laser-processing, femtosecond laser, nano-porous materials.**

1.1 General Introduction

1.1.1 Research background

The Clerk number shows the order in terms of natural abundance of the elements in the earth's crust. The top ten on the list are oxygen, silicon, aluminum, iron, calcium, sodium, potassium, magnesium, hydrogen, and titanium. Human beings have created civilization utilizing materials consisting of these elements, i.e., oxides of the light or main group metals (Al_2O_3, SiO_2, CaO, and MgO) in the Stone Age, and Fe in the Iron Age, and the current Information Age is supported by Si-based semiconductor integrated circuits and SiO_2-glass fibers. There is no doubt that the coming age will be created using these abundant elements intelligently.

Oxide ceramics are probably among the oldest man made materials and because of their excellent properties including abundance and easy availability of ingredients, mechanical strength, and excellent durability against severe chemical and thermal environments, they have been widely used mostly for structural components since ancient times. In addition, superior optical transparency of oxides makes it possible to realize various optically passive components in recent years: a representative example is 'optical fiber' used for optical communication systems, which is responsible for ushering in the current Information Age. However, it has been believed that active functions based on 'the control of the density and polarity of mobile carriers,' a control realized in semiconductor materials, are not possible in oxides. For example, alumina and glasses, which are representative oxides, are optically transparent but electrically insulating, and even the modification of electronic conductivity in these materials is rather difficult. However, In-doped SnO_2 (ITO) was discovered in 1953, and it exhibits both optical transparency and electrical conductivity. ITO is used as transparent electrodes, which are unavoidable for flat-panel displays and solar cells, and a category of oxides showing similar characteristics is called 'transparent conducting oxides (TCOs).' Therefore, we may expect from the discovery of TCOs that active electronic functions will be realized in novel oxides or novel forms of conventional oxides such as multilayer structures and nanoparticles.

During the last two decades techniques for purifying oxides have advanced considerably. It is now possible to obtain highly purified materials, resulting from studies focused on fine ceramics in the 1980s. For instance, metal oxides containing impurities at the sub-ppm ($\sim 10^{-7}$) level are now commercially available. Furthermore, thin film deposition techniques for oxides have advanced through intensive studies in the last decade on high-Tc superconductors for electronic applications. Taking advantage of this situation, we have explored new types of TCOs while following a working hypothesis established on the basis of a consideration of chemical bonding and point defects, resulting in the discovery of more than ten new TCOs [1]. Important among them are p-type TCOs, such as $CuAlO_2$, reported in Nature (1997) [2], and a series of amorphous TCOs in which the Fermi-level is controllable by intentional doping (1996) [3]. The former is particularly important because most active functions in semiconductors originate from pn-junctions. The absence of practical applications of transparent oxide semiconductors as electronic active devices has been primarily due to the lack of a p-type TCO. The discovery of the p-type TCO opens a new frontier for TCOs, which should be called as 'transparent oxide semiconductors (TOSs)' (Fig. 1.1).

Fig. 1.1. Impact of discovery of P-type transparent conductive oxides.

Furthermore, recent progress in 'nanotechnology' provides a new approach for the cultivation of the active functions in oxides: The approaches include the use of 'natural nanostructures' embedded in oxides and the fabrication of 'artificial nanostructures' with assisitance of thin-film deposition, nanolithography techniques, and emerging femtosecond laser processing.

1.1.2 Research concepts and strategies

Transparent oxides are the most abundant and stable materials on earth and they are environment friendly. Although they have been used as ingredients in traditional materials such as cement, glass, and porcelain since the early stages of human history, only few active functions have been found in them. In fact, these materials are described as typical insulators in college textbooks. However, a widely accepted view that 'a transparent oxide cannot be a platform for electroactive materials' comes from only phenomenological observations. We think it possible to realize a variety of active functionalities in transparent oxides by appropriate approaches based on a deep insight into the electronic structure of these compounds and the use of modern concepts of nanotechnology.

There are two characteristic features of oxides: One is that a various kinds of constituting metal elements exist and correspondingly a wide variety of crystal structures exist, while no such vast variety is seen in elementary semiconductors (Si, Ge, diamond) or compound semiconductors (GaAs, GaN, CdS, and ZnS). The crystal structure of the semiconductors is limited to the diamond type. These varieties suggest that novel active functionalities, which are useful for novel devices, still remain undiscovered in oxides.

The other factor is the ionic nature of chemical bonds in oxides, which should be compared with the covalent nature of bonds in semiconductors. Thus, local structures in oxides are governed both by the preferential coordination number of the metal ions and the ionic radii ratio between the metal and oxygen ions, while the local coordination configurations in semiconductors are almost exclusively tetrahedral. Such a striking difference in bonding nature differentiates the energy band structure of oxides from that of semiconductors. That is, the valence band is composed of oxygen p orbitals and the conduction band of metal s orbitals in oxides. On the other hand, both the valence and conduction bands are formed by the s-p hybrid orbitals in semiconductors. This energy structure in oxides is intimately connected with the difficulty in achieving p-type conductivity because of the localization of the p-orbital. It also relates to the slight degradation of n-type carrier mobility in amorphous oxides from that in the crystalline state because the degree of the overlap of the spherically distributed s-orbitals among adjacent metal ions is insensitive to changes in the local coordination configurations from crystalline to amorphous states.

The ionic bonding in oxides also gives rise to a stronger 'electron lattice interaction' than in semiconductors. When electrons are in the ground state, the interaction may cause lattice distortions due to the 'Jahn Teller effect.' The interaction also stabilizes 'polarons' composed of charge carriers and their associated polarization due to the host lattice deformation field [4]. On the other hand, the interaction leads to the formation of a variety of excited states of electrons: a typical example is a self-trapped exciton which works as an energy localization center, leading to persistent defect formation.

In addition to the basic insight into electronic states in the bulk crystals, our intentions on goals to cultivate novel active functions are focused on the nanostructures of oxides. Among various types of the nanostructures, we evaluated, natural nanostructures embedded in transparent oxides such as quasi-multi-quantum-well structures realized in layered compounds and quasi-quantum dot structures in nanoporous compounds (Fig. 1.2). Novel optical and electrical properties are expected to emerge from the intrinsic nanostructures because of their unique crystal structures if we scrutinize them from the nanotechnological point of view.

We are also interested in fabricating artificial periodic nanostructures in transparent oxides using the interference of femtosecond (fs) laser pulses, which involves a self-aligned process. Extremely high-energy density pulses are now available from a table-top fs-laser through regenerative amplification. In general, transparent dielectrics are unfavorable

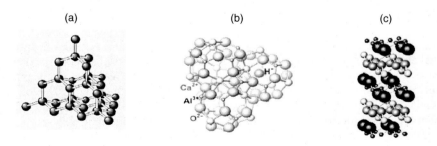

Fig. 1.2. Crystal structures: (a) diamond-type, (b) nanocage-type, and (c) layer-type.

for laser machining because most photons pass through the sample. Fs laser pulses overcome this difficulty due to large nonlinear effects arising from extremely high peak power, and thus they provide an opportunity to write three-dimensional nanostructures inside a transparent oxide, which is useful especially for optical integrated circuits in transparent oxides.

In summary, we chose transparent oxides with unique crystal and/or electronic structures and explored the novel active functions using state-of-the-art knowledge and techniques. The primary purpose is to explore the intrinsic potential of transparent oxides as functional materials toward the cultivation of new material frontiers. Emphases are placed on establishing new materials views and new methodologies, both of which may lead to effective and powerful tools for future study in this area.

This study is composed of three major subjects: transparent oxide semiconductors, nanoporous crystalline oxides, and the fabrication of periodic nanostructures in transparent dielectrics using fs-laser pulses.

1.2 Transparent Oxide Semiconductors

1.2.1 Introduction

It is generally believed that high-optical transparency is incompatible with high-electronic conduction, since optical transparency requires band gaps larger than 3.3 eV and such a large gap makes carrier doping very difficult. In this sense, transparent conductive oxides (TCOs) are exceptional materials. The first TCO developed was In_2O_3:Sn (ITO) in 1954, followed by other TCOs:SnO_2 and ZnO.

Although TCOs, featured as a transparent metal, have been commercialized intensively in transparent window electrodes and interconnections, their applications were limited to a narrow area because of the absence of p-type TCO; no active electronic devices such as bipolar transistors and diodes can be fabricated without pn-junctions. The breakthrough was the finding of the first p-type TCO, $CuAlO_2$ in 1997 by our group [2], which triggered the development of a series of p-type TCOs and transparent pn-junction devices such as UV light emitting diodes (LEDs). The achievement has significantly changed our conception of TCOs and has opened a new frontier called 'transparent oxide semiconductors (TOSs).' Therefore, we now consider that TOSs have the potential to develop new functionalities useful for novel optoelectronic devices that are hard to realize by current Si-based semiconductor technology.

The discovery of the p-type TOSs resulted from rational considerations regarding to the design of new TOSs based on knowledge about electronic structures that has been accumulated experimentally and theoretically. Our material design concept has been proven to be valid by the development of new TOSs including p-type TOSs. These new TOSs led to transparent electronic devices such as UV LEDs and transparent thin film transistors (TFTs). In addition, we have proposed that the use of natural nanostructures embedded in crystal structures of TOSs is very effective in the cultivation of new functions in oxides.

Such structures exist in layered and nanoporous compounds. From a processing point of view, techniques for growing high-quality single crystals or epitaxial thin films of the compounds are necessary to fabricate the devices. They are also essentially important for clarifying intrinsic properties associated with the structures. We have developed a unique epitaxial film growth technique, 'reactive solid phase epitaxy (R-SPE),' which is particularly suited for growing the layered compounds.

In this section, we first discuss the guiding principles for the development of new TOSs and then briefly review recent achievements we have made, which have opened new the frontier of TOSs. Finally, we introduce distinct optoelectronic properties associated with low-dimensional electronic structures, taking R-SPE-grown oxychalcogenide films as an example.

1.2.2 Guiding principles for developing new TOSs

The conduction band minimum (CBM) of most metal oxides is made of spatially spread spherical metal s orbital. Therefore, electrons in the metal oxides have small effective masses, and high electronic conduction is possible if high-density electron doping is achieved. This is the reason why several n-type TOSs have been found to date. In contrast, the valence band maximum (VBM) is made of oxygen 2p orbitals, which are rather localized, leading to small hole effective masses. Furthermore, the dispersion of the valence band tends to be small, and thus the VBM level is so deep that hole doping is difficult. Therefore, p-type TOS was not discovered before 1997. We proposed an idea that the use of metal d orbitals with energy levels close to those of O 2p orbitals may form highly hybridized orbitals with O 2p. We expected that it might raise the VBM level and make hole doping easier. We noticed that the $3d^{10}$ configuration of Cu^+ was a candidate because the Cu 3d energy level is just above the O 2p level. Further, the closed shell configuration of Cu^+ allows for large band gaps and optical transparency. This idea actually led to the discovery of $CuAlO_2$ [2]. This was followed by the subsequent discovery of new Cu^+-based p-type TOSs such as $CuGaO_2$ and $SrCu_2O_2$.

However, neither high-concentration hole doping nor large-hole mobility are achieved in these Cu^+-based p-type TOSs. Therefore, the material design concept was extended to use the chalcogen (S, Se, and Te) p orbitals instead of those of oxygen. That is, what we intended was to increase the valence band dispersion by forming hybridized orbitals between Cu 3d orbitals and chalcogen p orbitals that are more delocalized than O 2p. We preferred layered oxychalcogenides because they are optically transparent in the visible light region, although simple chalcogenides are transparent only in the IR region.

Electron mobility in amorphous TOSs is expected to maintain a large value (e.g., >20 $cm^2 \cdot V^{-1} \cdot s^{-1}$), comparable to those of corresponding crystalline materials, because electron transport paths (i.e., CBMs) are made of spherically spreading metal s orbitals. Such expectations are distinctly different from those in covalent amorphous semiconductors such as amorphous hydrogenated silicon (a-Si:H), where mobility in amorphous states is largely reduced from that of the crystalline state due to CBM and VBM being formed by the sp^3 hybrid orbitals (Fig. 1.3).

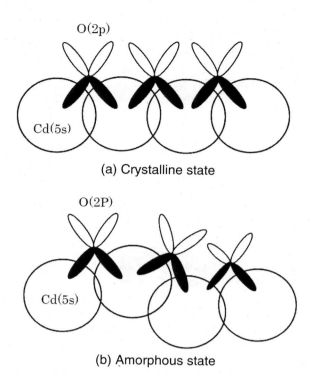

Fig. 1.3. Schematic illustration of electron transport paths in (a) crystalline and (b) amorphous $2CdO \cdot GeO_2$.

1.2.3 Research frontiers in TOS

TOS materials.

1) **p-type TOSs: Cu^+-bearing oxides**

   In 1997, we reported $CuAlO_2$ thin films as the first p-type TOS along with a chemical design concept for exploring p-type TOSs [1,2]. After that, a series of p-type TOSs based on Cu^+-bearing oxides such as $CuGaO_2$ [5] and $SrCu_2O_2$ [6] were found.

   In order to clarify the origin of p-type conduction, the electronic structure of $SrCu_2O_2$ was examined by photoelectron spectroscopy and band structure calculations using LDA [7]. The electronic structure around the band gap was found to be similar to that of Cu_2O despite a large difference in the band gap energies. That is, the admixed orbitals of 3d, 4s, and 4p of the Cu^+ ion are hybridized with 2p orbitals of the O^{2-} ligands, which constitute the VBM.

2) **Alternative p-type TOSs: $ZnRh_2O_4$**

   It is known that transition metal ions with $4d^6$ configurations located in an octahedral crystal field have a low-spin configuration in the ground state, which may be regarded as a 'quasi-closed shell' configuration. On the basis of the idea that such ions are expected to behave similarly to Cu^+ ions with $3d^{10}$ closed shell configurations and to enhance the dispersion of the valence band, we have found that normal spinel $ZnRh_2O_4$ is a p-type wide-gap semiconductor with a band gap of ~ 2.1 eV [8]. The electrical conductivity of the sputtered film was 0.7 S·cm^{-1} at 300 K without intentional doping. Magnetic susceptibility, photoelectron spectroscopy, and optical measurements revealed that the band gap originated from the ligand-field split of Rh^{3+} d orbitals in octahedral symmetry, while the valence bands were made of fully occupied t_{2g}^6 and the conduction band of empty e_g^0.

3) Deep-UV (DUV) TOS: β-Ga$_2$O$_3$

Conventional TCOs such as ITO and ZnO are opaque for DUV light (<300 nm) due to their small band gap (~ 3 eV), although the DUV region will be important for future biotechnologies such as DNA detection. DNA detection may be possible by electrical sensing or DUV optical absorption measurements. It is necessary to improve molecular selectivity to realize the DNA detection function. Our idea is to control the selectivity by applying voltages to the adsorption surface. Therefore, DUV-transparent TCOs are needed for these applications.

β-Ga$_2$O$_3$ is considered to be a good candidate because this material has a large band gap of 5 eV and good electronic conduction by bulk single-crystal β-Ga$_2$O$_3$ was reported [9]. We successfully fabricated conductive β-Ga$_2$O$_3$ thin films by high-temperature pulsed-laser deposition at 880°C [10] and subsequently succeeded in fabricating conductive films at 300°C by fine tuning the deposition conditions [11]. The optical band gap estimated from the $(\alpha h\nu)^2$-$h\nu$ plot was 4.9 eV.

4) Transparent Electrochromic Material: NbO$_2$F

We demonstrated [12] that oxyfluoride NbO$_2$F with a ReO$_3$-type structure was a promising electrochromic material with large band gap energy. Diffuse reflectance spectra revealed that the optical band gap of NbO$_2$F (3.1 eV) was larger than that of the well-known electrochromic oxide WO$_3$ (2.6 eV). Electronic conduction is rendered by heating the material at 500°C in H$_2$ atmosphere with Pt powders. The sintered sample was blue, and its electrical conductivity was 6×10^{-3} S·cm^{-1} at room temperature. The electrical conductivity increased with temperature, exhibiting semiconductor behavior. A reversible electrochromism between pale blue and deep blue was confirmed in H$_2$SO$_4$ and Na$_2$SO$_4$ aqueous solutions.

5) Amorphous TOSs

Amorphous TOSs (a-TOSs) have potential as transparent electrodes for flat-panel displays such as plastic or film LCDs and OLEDs, provided that a reasonably low electrical resistivity can be obtained. The vacant ns orbitals of metal ions with an electronic configuration of $(n-1)d^{10}ns^0$ ($n \geq 5$) are expected to form mobile carrier transport paths even in amorphous structures. The recent discovery of new a-TOSs supports this expectation [3]. These a-TOSs are characterized by a high electron mobility (~ 10 cm^2·V^{-1}·s^{-1}), which is remarkably large compared with that of a-Si (<1 cm^2·V^{-1}·s^{-1}) [13]. The origin of the carrier transport properties was theoretically clarified using the amorphous 2CdO-GeO$_2$ system as an example [14].

The CBM is mainly composed of Cd 5s orbitals, which are overlapped as illustrated in Fig. 1.3. Thus, high-electron mobility originates from continuous electron conduction paths formed by direct overlap of the Cd 5s orbitals.

However, amorphous TOSs containing Cd cannot be used for practical applications because of the toxicity of the Cd ion. We employed the In_2O_3-Ga_2O_3-$(ZnO)_m$ system instead of ZnO to clarify whether the 4s orbital has the ability to form a conduction path in an amorphous phase, since amorphous ZnO cannot be formed by a conventional film deposition process. In this system, In and/or Ga ions are expected to act as network formers. As a result, it was confirmed that the resultant films with m = 1 − 4 were amorphous and exhibited electrical conductivity on the order of 10^2 S·cm^{-1} and transparency in the visible light region [15].

Further, we found that amorphous films of Zn-Rh-O exhibit p-type conductivity. This was the first demonstration of p-type amorphous TOS. An all-amorphous pn-junction diode with good rectifying performance was successfully fabricated using the amorphous oxide films of Zn-Rh-O and In-Ga-Zn-O [16].

TOS epitaxial films.

1) Super-flat ITO epitaxial films prepared by pulsed-laser-deposition

High-quality epitaxial films are necessary for studying the intrinsic properties of electronic materials. We have developed several techniques for growing high-quality epitaxial films of TOSs using pulsed-laser deposition (PLD). For example, very low resistivity (7.8×10^{-5} Ωcm, the lowest to date) ITO epitaxial films were reproducibly grown on an atomically flattened (100)-YSZ single-crystal substrate at 600°C (Fig. 1.4) [17]. We also fabricated single-crystalline ITO films having

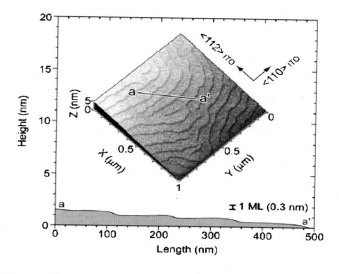

Fig. 1.4. AFM image of superflat ITO thin films. A step corresponding to a monolayer of ITO is clearly seen.

an atomically flat surface ($R_{rms} \sim 0.2$ nm @ 1×1 cm^2) on the (111) surface of YSZ at 900°C [18]. This atomically flat ITO worked as the base for successful fabrication of transparent pn-junctions [19,20], near-UV light emitting diode [21] UV detector [22], and transparent organic TFTs [23,24].

2) Lateral epitaxial growth of vanadyl-phthalocyanine (VOPc) on atomically flattened ITO film

We obtained a laterally grown VOPc layer on an epitaxial ITO film surface composed of atomically flat terraces and 0.29-nm-high steps using molecular beam epitaxy (MBE) method [23,24]. The VOPc (phase II) layer was heteroepitaxially grown on the (111) ITO surface with a relationship of (010){21-2} VOPc ∥ (111){110} ITO. The crystallographic orientation of the film differs distinctly from that of the films grown on the other substrates such as alkali halides. AFM images revealed that six kinds of two-dimensional VOPc domains were heteroepitaxially grown laterally on the ITO surface and contacted each other, forming domains and domain boundary structures. These results demonstrate, by taking VOPc as an example, that a transparent conductive epitaxial ITO film with the atomically flat and stepped surface is effective in growing organic molecules laterally. The laterally grown organic molecules on transparent conductive substrates are important for emerging molecular electronics technology. Further, epitaxial layers of VOPc on ITO films provide information for clarifying the mechanism of improved hole injection in organic LEDs.

TOS device.

1) UV-LED: p-type SrCu$_2$O$_2$/n-type ZnO [21]

We realized that the near-UV-LED are composed of pn heterojunction of TOSs, p-type SrCu$_2$O$_2$, and n-type ZnO. ZnO is an n-type TOS ($E_g = 3.38$ eV), which can emit UV ($\lambda = 380$ nm) due to a room temperature exciton ($E_x = 59$ meV). Efficient electroluminescence centered at 382 nm was observed when a forward current was injected into the pn heterojunction diode (Fig. 1.5). The threshold voltage for electroluminescence was ~ 3 V, which suggests that the origin of the electroluminescence was electron-hole recombination in the ZnO layer.

2) Transparent pn-Diode: p-ZnRh$_2$O$_4$/n-ZnO [22]

The pn heterojunction diodes composed of wide-gap oxide semiconductors of p-ZnRh$_2$O$_4$ and n-ZnO were successfully fabricated by R-SPE. The pn heterojunction diodes obtained have an abrupt interface and exhibit rectifying I–V characteristics with a threshold voltage of ~ 2 V, which is in good agreement with the band gap energy of ZnRh$_2$O$_4$. It verifies that the heterojunction formed by the narrow bandgap ZnRh$_2$O$_4$ and the wide bandgap ZnO works as a good carrier blocking contact. This behavior is similar to that of conventional pn junctions, not specific to the d-electron system. On the other hand, with irradiation of UV light, the fundamental absorption edge of ZnO produces photovoltages more effectively than that of ~ 2 eV light, which may result from the intrinsic nature of the d-electron bands, such as small absorption coefficients and less mobile carriers. We have demonstrated herein that R-SPE is suitable for fabricating oxide heterojunctions. This is an advantage of fabricating optoelectronic devices using TOSs.

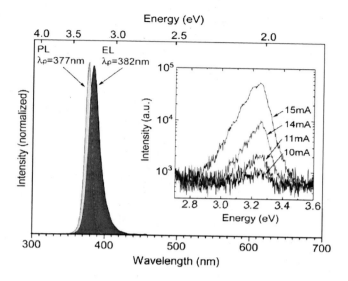

Fig. 1.5. Photoluminescence and current-injected luminescence spectra of p-SrCu$_2$O$_2$/n-ZnO heterojunction LED. The inset shows EL spectra for several forward currents injected to the junction.

3) Transparent UV detector [22]

The UV light radiation that reaches the earth's surface is 280 ∼ 400 nm in wavelengths (UV-A and UV-B); it plays a harmful role in that it may cause skin cancer. A transparent UV light detector was fabricated using a high-quality pn heterojunction diode composed of transparent oxide semiconductors, p-type NiO and n-type ZnO, and its UV light response was characterized at room temperature. The diode exhibited clear rectifying I–V characteristics with a forward threshold voltage of ∼ 1 V. Efficient UV-light response was observed up to ∼ 0.3 AW^{-1} at 360 nm (−6 V biased) (Fig. 1.6), which is comparable to that of a commercial GaN detector.

4) Organic TFTs [23, 24]

Transparent organic TFTs (OTFTs) were fabricated using a vanadyl-phthalocyanine (VOPc) film as an active p-channel, a lattice matched (Sc$_{0.7}$Y$_{0.3}$)$_2$O$_3$ film as a high-k gate dielectric, and an atomically flat ITO film as a bottom contact. Lateral growth of the VOPc epitaxial channel layer on the epitaxial (Sc$_{0.7}$Y$_{0.3}$)$_2$O$_3$ gate dielectric with a root-mean-square roughness (R$_{rms}$) of ∼ 1 nm was achieved by MBE. Laterally grown VOPc thin film contributes dominantly to obtaining a reasonably large field-effect mobility μ_{eff} of ∼ 5 × 10^{-3} cm$^2\cdot$V$^{-1}\cdot$s^{-1}, which provides significant improvement compared with reported values of OTFT based on a nonplanar phthalocyanine. These results also suggest that the bilayered film composed of the lattice-matched high-k gate dielectric of the Y-Sc-O system and the atomically flat ITO transparent electrode is applicable to other organic materials with larger mobility, leading to further improvements of transparent OTFTs.

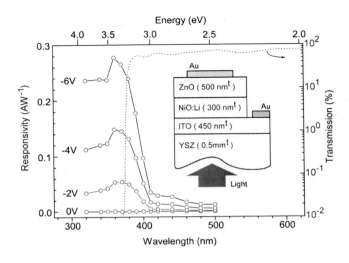

Fig. 1.6. Optical response of p-NiO/n-ZnO hetrojunction UV detector.

5) High-performance transparent TFTs [25]

TTFTs, using TOSs as a channel layer have merits compared with conventional TETs when applied to flat-panel displays. The merits include an efficient use of back light in LCD and emitted light in OLED, and insensitiveness of device performance to visible light irradiation. In addition, TFTs based on oxides have potential advantages over conventional FETs in terms of their high voltage and high temperature tolerances. Although several reports have appeared on the fabrication of TTFTs using conventional TOSs such as SnO_2 and ZnO [26], their performances were not satisfactory for practical applications. The large off-current and the unintentional normally-on characteristics originate from the fact that these conventional TOSs contain many carriers in the as-prepared state due to a relatively large nonstoichiometric chemical composition, i.e., oxygen vacancies, and it is hard to control the carrier density down to less than 10^{17} cm^{-1} without impurity counter doping. Their on/off-current ratio and field-effect mobility (μ_{eff}) are as large as ~ 5 cm$^2 \cdot$V$^{-1} \cdot$s^{-1}, and the TFT characteristics exhibit 'normally-on' characteristics unintentionally. In addition, as these TTFTs were fabricated in polycrystalline thin films, defects and grain boundaries in the active channel deteriorate the device performance.

The TOS-TTFT fabricated using a single-crystalline $InGaO_3(ZnO)_5$ film [25,27] displayed reasonable normally-off characteristics with good performance, such as the large $\mu_{eff} \sim 80$ cm$^2 \cdot$V$^{-1} \cdot$s^{-1}, the low off current $\sim 10^{-9}$ A, and on/off-current ratios larger than 10^5, distinguishing these materials from conventional TOS-TFETs reported to date. Figure 1.7(a) shows the device structure of the TTFT. An 80-nm-thick amorphous HfO_x layer was used for the gate insulator, and a transparent electrode of ITO was used for source, drain, and gate electrodes. The optical transmittance is almost 100% in the visible region (b). The output characteristics (c) show that source-to-drain (I_{DS}) current increases markedly as source-to-drain voltage (V_{DS}) increases at a positive gate bias V_{GS}, indicating

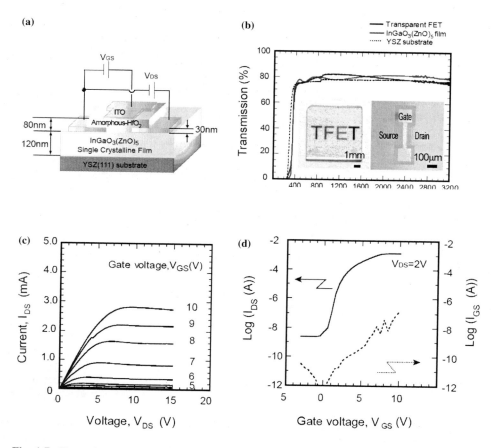

Fig. 1.7. Transparent TFT (a) device structure, (b) photo of device chip, (c) output characteristics, and (d) transfer characteristics.

that the channel is n-type and electron carriers are generated by a positive V_{GS}. A large $\mu_{eff} \sim 80$ cm$^2\cdot$V$^{-1}\cdot$s^{-1} is obtained, estimated both from the transconductance value and from the saturation current. The off-current is very low, on the order of 10^{-9} A, and the on/off current ratios larger than 10^5 are obtained (d). These characteristics are greatly improved over those reported for TTFTs fabricated using conventional TOSs. This achievement provides a practical method of fabricating TTFTs with reasonable performance, paving the way for realizing invisible circuits.

1.2.4 Natural quantum well structures in layered TOS compounds

In the layered compounds, layers A and B with different chemical compositions are regularly stacked along a certain crystal axis, mostly the c-axis. Figure 1.8 shows crystal

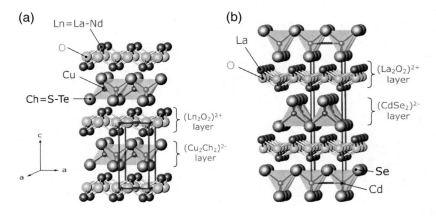

Fig. 1.8. Crystal structures of LnCuOCh and $La_2CdO_2Se_2$ as examples of layered compounds.

structures of LnCuOCh and $La_2CdO_2Se_2$ as examples of layered compounds [28,29]. In the case of LnCuOCh, $(Ln_2O_2)^{2+}$ and $(Cu_2S_2)^{2-}$ layers correspond to A and B layers, each being composed of three atomic layers. Judging from the band gap energies of La_2O_3 and Cu_2Ch, it is suggested that the $(Ln_2O_2)^{2+}$ layer has a much wider band gap than the $(Cu_2Ch_2)^{2-}$ layer, and thus large band offsets between two layers exist both at VBM and CBM [30]. Figure 1.9 illustrates a calculated energy band structure in k-space and the corresponding density of the states of each ion in LaCuOS, showing that both VBM and CBM are composed of a $(Cu_2S_2)^{2-}$ layer and La^{3+} and O^{2-} orbitals are located far above CBM and below VBM. That is, the calculated band structure provides validity to the assumed band structure with the large band offset as schematically shown on the right side of Fig. 1.9(a). This suggests that the energy structures of the layered compounds are analogous to those of multi-quantum-wells (MQWs) in artificial superlattice structures, which are composed of an alternative stack of thin films with lattice-matched compound semiconductors such as GaAs/AlGaAs. It is considered that the $(Ln_2O_2)^{2+}$ layer acts as

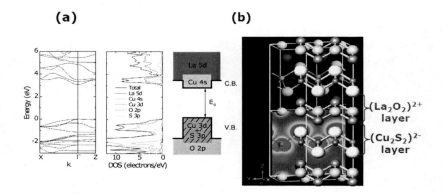

Fig. 1.9. Electronic band structure of LaCuOS: (a) Band structure in k-space, density of state of each constituent ion and schematics of the band structure, showing quasi-MQW structure.

Table 1.1. Features of layered compounds (natural MQW) compared with those of artificial MQW.

	Natural MQW	Artificial MQW
Fabrication process	Self-assembly (R-SPE)	Layer-by-layer (Vapor phase)
Layer dimension	Monoatomic or a few atomic layers	Several atomic layers
Flexibility of layer thickness	Inflexible	Flexible
Barrier height	Largely changeable	Less changeable
Layer charge	Charged	Noncharged

a barrier, while the $(Cu_2S_2)^{2+}$ layer acts as a quantum well in LaCuOS, and we call it the natural MQW. Table 1.1 shows comparison between natural and artificial MQWs.

Epitaxial film growth of layered TOS: reactive solid-phase epitaxy (R-SPE) [27]. It is absolutely necessary to prepare epitaxial films of layered compounds to investigate the intrinsic properties of the compounds to enable the fabrication of quantum devices fully utilizing the features of the natural MQW. However, it is extremely hard to grow single-crystalline thin films of the layered compounds, which contain several kinds of metal ions, by conventional vapor-phase epitaxy. We developed a novel method of fabricating single-crystalline thin films of layered complex oxides, called 'reactive solid-phase epitaxy (R-SPE)' [27]. In this process, the epitaxial template layer is grown on a substrate, followed by the deposition of a polycrystalline or amorphous film of complex oxides having a desired chemical composition. Single-crystalline films can be obtained if the bilayer film is annealed at high temperatures in an appropriate atmosphere through the solid-phase reaction. A detailed mechanism for R-SPE will be described in the following sections, taking LaCuOS as an example.

Layered oxychalcogenides LnCuOCh (Ln = lanthanide, Ch = chalcogen): novel transparent p-type Semiconductors. LnCuOCh is a layered compound, and its crystal structure consists of oxide $(Ln_2O_2)^{2+}$ with chalcogenide $(Cu_2Ch_2)^{2-}$ layers alternately stacked along the c-axis (Fig. 1.9) [30]. Each layer is doubly charged and contains two molecules or three atomic layers: Ln-O-Ln and Ch-Cu-Ch. As already described, the energy structure of LnCuOCh is regarded as a natural MQW with $(Ln_2O_2)^{2+}$ as a barrier and $(Cu_2Ch_2)^{2-}$ as well layers. It is expected that distinct characteristics inherent to the natural MQW will be realized similar to those of artificial MQW due to the confinement of electrons to the two-dimensional well layer.

1) **Growth of LnCuOCh by R-SPE [31–33]**

   The chemical composition of LnCuOCh is relatively complex, and the chalcogenide component, Cu_2Ch, evaporates easily from the film at high temperatures in vacuum. The evaporation of Cu_2Ch results in the deviation of the chemical composition from stoichiometry, resulting in the decomposition of LnCuOCh into Ln_2O_3, Ln_2O_2Ch, and Cu_2Ch. Although intensive efforts have been made to grow epitaxial LnCuOCh films at high temperatures by a PLD technique, the epitaxial films have not been obtained. Therefore, we employed the R-SPE technique to

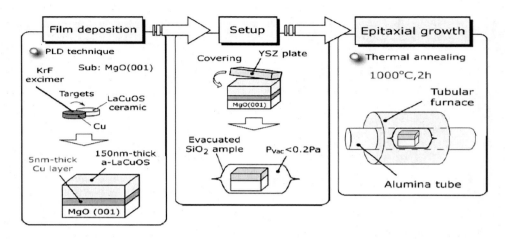

Fig. 1.10. Procedure for reactive solid-phase epitaxy (R-SPE).

epitaxial growth of LnCuOCh. As schematically shown in Fig. 1.10, a thin sacrificial Cu layer (~ 5 nm thick) was deposited first on a (001)-oriented MgO substrate by the PLD technique. Then, an amorphous LnCuOCh film was deposited on the Cu layer at room temperature. Finally, the a-LnCuOCh/Cu bi-layer film was thermally annealed at 1000°C in an evacuated SiO_2 glass ampoule. With thermal annealing, we obtained epitaxially grown LnCuOCh (Ln = La, Ce, Pr, Nd; Ch = $S_{1-x}Se_x$, $Se_{1-y}Te_y$) films. Figures 1.11 and 1.12 show SEM images and schematic illustrations of the as-deposit films and films that have been subjected to the thermal annealing at several temperatures. The Cu layer is composed of isolated islands in the as-deposited bilayered films. The epitaxial LaCuOS film starts to grow from a triple junction among three components: MgO substrate, Cu layer and amorphous LaCuOS layer at 500°C [36]. The epitaxial film area increases gradually with the annealing temperature and finally it occupies the entire area of the film at 1000°C, at which point the sacrificial thin Cu layer deposited before thermal annealing completely disappears in the final films. A cross-sectional, high-resolution electron microscopy (Fig. 1.13) image of the epitaxial LaCuOS film shows that layered patterns associated with the (001) planes of LaCuOS are stacked parallel to the substrate surface.

2) Electronic properties: Natural modulation doping and p-type degenerated conduction [34]

All the epitaxial LnCuOCh films exhibit p-type electrical conduction [35–37]. Hall mobility becomes larger with an increase in the Se content, reaching 8.0 $cm^2 \cdot V^{-1} \cdot s^{-1}$ in LaCuOSe; a value comparable to that of p-type GaN:Mg (Fig. 1.14). Further Mg doping increased the hole concentration up to 2×10^{20} cm^{-3}. The Hall mobility of the Mg-doped films is also enhanced from 0.2 to 4.0 $cm^2 \cdot V^{-1} \cdot s^{-1}$ by anion substitution from S to Se, resulting in an increase in the electrical conductivity from 5.9 to 140 $S \cdot cm^{-1}$ (Fig. 1.15).

It is noteworthy that the mobility of the Mg-doped films is reduced to only half that of the undoped films despite the heavy Mg ion doping. The coexistence of the

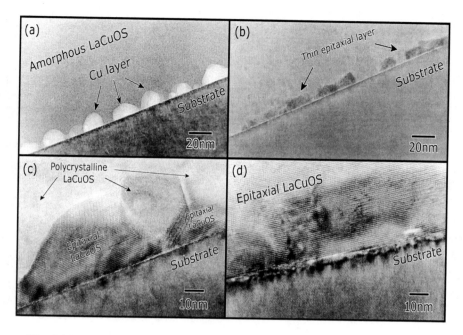

Fig. 1.11. SEM images of as-deposited and annealed films at several temperatures.

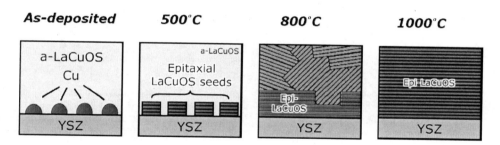

Fig. 1.12. Schematic illustrations of growth mechanism. (a) As deposited, (b) 500[Sup. 0]C, (c) 900[Sup. 0], and (d) 1000[Sup. 0]C.

high hole concentrations $>10^{20}$ cm^{-3} and the moderately large mobility is unusual in conventional semiconductors. It may be attributed to the modulation doping or delta doping in the natural MQW of LaCuOSe i.e., hole carriers generated by Mg^{2+} ion doping in the (La$_2$O$_2$)$^{2+}$ layer (carrier doping layer) are transferred to the (Cu$_2$Ch$_2$)$^{2-}$ layer (hole conduction layer) due to a large band offset at the VBM. As a result, the ionized dopants do not scatter mobile hole carriers, since the hole conduction layer is spatially separated from the doping layer, which realizes the modulation doping in the natural MQW, similar to that formed in the artificial MQW.

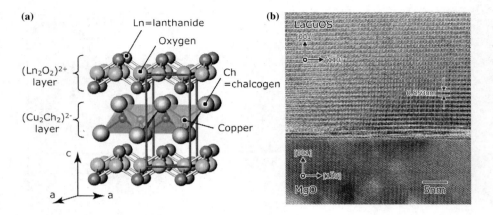

Fig. 1.13. (a) Crystal structure, and (b) HRTEM image of LnCuOCh epitaxial thin film fabricated by R-SPE.

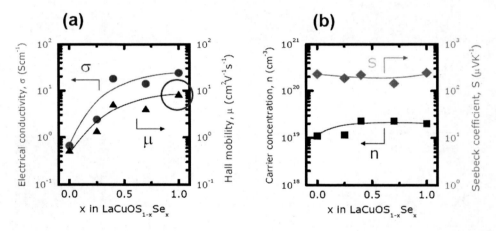

Fig. 1.14. Electronic properties of $LaCuOS_{1-x}S_x$: (a) Electrical conductivity and hole mobility. (b) Hole carrier concentration and Seebeck coefficient.

3) **Optical properties: room temperature excitons and two-dimensional electronic structures [38,39]**

The optical properties of LnCuOCh (Ln = La, Pr and Nd; Ch = S or Se) epitaxial films were examined and several unique optical properties likely associated with the two-dimensional electronic state were observed.

Absorption spectra of $LaCuOS_{1-x}Se_x$ and LnCuOS (Ln = La, Pr, and Nd) at 10 K are shown in Fig. 1.16. A sharp line exists around the fundamental absorption edge in LaCuOS, and it is assigned to as an associated exciton. The line splits into a doublet with increasing Se content, and the splitting is induced by the spin orbit interaction of Ch ions because the interaction is much larger for the Se ion than for the S ion. Further their replica structures are seen in the higher-energy side, and

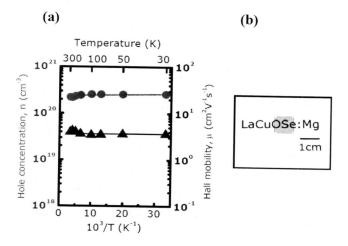

Fig. 1.15. (a) Temperature dependence of hole concentration and mobility of 10%-Mg doped LaCuOSe, showing p-type degenerated conduction, (b) photo of film on MgO substrate.

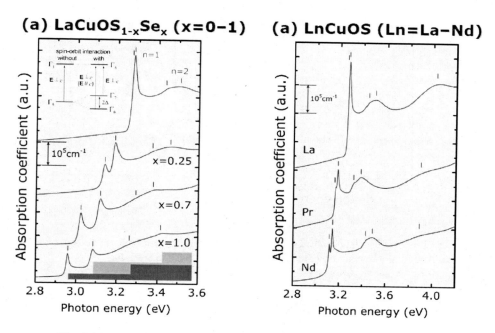

Fig. 1.16. Optical absorption spectra of LaCuO (a) and LnCuOS (b) at 10 K.

they are systematically tuned for a change in cationic and anionic ions. Figure 1.17 shows a schematic energy diagram of MQW compared with that of the bulk crystal. In the MQW structure, the density of states in both valence and conduction bands becomes a stepwise function due to the two-dimensional nature of the electron and the exciton states are generated at each step. Such behavior of the absorption

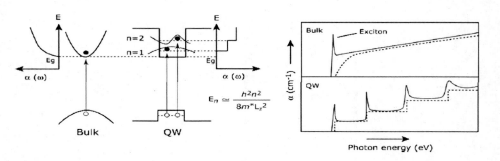

Fig. 1.17. Schematic representation of Energy diagram (a) and Optical absorption (b) of MQW and bulk.

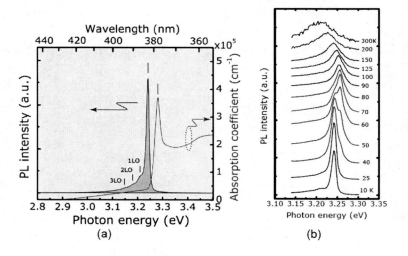

Fig. 1.18. Emission spectra of LaCuOS: (a) Comparison between emission and absorption spectra at 10 K. (b) Emission spectra at various temperatures.

spectrum in MQW is basically consistent with the observed spectra. The stepwise absorption spectra provide solid evidence that the layered structure of LnCuOCh is treated as a natural MQW.

A sharp line emission is observed corresponding to exciton absorptions below ~30 K (Fig. 1.18) and the line is assigned to a bound exciton because of the small Stokes shift from the intrinsic exciton absorption line. A shoulder structure starts to appear at 40 K and it increases with temperature and becomes a single line. With a further increase in temperature, the line becomes broadened accompanied with a large red-shift, indicating the line is due to an intrinsic exciton. From the decreasing manner of the emission intensity, the binding energy of the exciton is estimated to be ~ 50 meV for all LaCuOCh. Such a large binding energy is likely due to the two-dimensional nature of the exciton. Anionic and cationic substitutions tuned the emission energy from 3.21 to 2.89 eV ($\lambda = 386 \sim 429$ nm) at 300 K, which provides a way to engineer the electronic structure in light-emitting devices. These emission characteristics are quite favorable for use in the active layer of light-emitting devices.

4) Optical nonlinearity and exciton-exciton interaction [40,41]

The confined excitons in nanoparticles or nanolayered structures exhibit a large optical nonlinearity around the resonant energies. Accordingly, LaCuOCh shows potentially large optical nonlinearity. Thus, we have measured the third-order optical susceptibility, $\chi^{(3)}$, using a fs time-resolved degenerative four-wave mixing (DFWM) technique around the band gap energy. This method can also detect a small energy split as a quantum beat appearing in a time-evolutional profile of the diffracted beam. The beat is caused by the quantum interference between neighboring electronic levels.

The $\chi^{(3)}$ value for LaCuOS depends strongly on excitation energies, and it is enhanced to 4×10^{-9} esu at the absorption band peak (3.2 eV) (Fig. 1.19(a)). On the other hand, LaCuOSe has two absorption peaks (2.9, 3.1 eV) corresponding

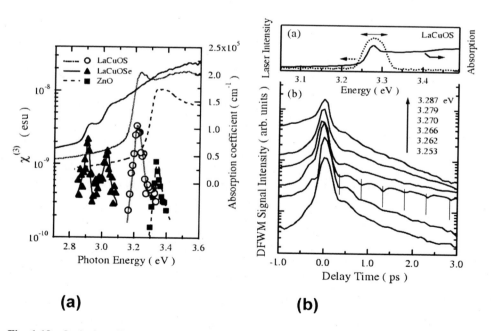

(a)   **(b)**

Fig. 1.19. Optical nonlinearity in LaCuO(S, Se): (a) third order optical nonlinearity and absorption spectra in LaCuO(S, Se) at RT. Those of ZnO are also shown for comparison. (b) DFWM signal intensity (dotted curve) and absorption (solid curve) spectra of LaCuOS around band-edge exciton; (a) laser scanning energy, and (b) DFWM traces for LaCuOS as a function of delay time for several excitation energies.

to a doublet exciton split due to the spin-orbit interactions of the Se ion. The $\chi^{(3)}$ values for LaCuOSe are enhanced at peak energies up to $\sim 2 \times 10^{-9}$ esu, which are comparable to that of LaCuOS and larger than that for the ZnO film (1×10^{-9} esu), which has a larger exciton binding energy (60 meV). The $\chi^{(3)}$ values are comparable to those for semiconductor CdS/Se nanoparticles dispersed in a glass matrix. These large $\chi^{(3)}$ values in LaCuOCh are attributable to the large density of exciton states caused by the confinement of excitons in the $(Cu_2Ch_2)^{2-}$ layer. The response time of the DFWM signals for LaCuOCh is 250 $\sim$ 300 fs, corresponding to a phase decay time of 1 $\sim$ 1.2 ps for excitons.

The DFWM signals for LaCuOS are also measured at 4 K by sweeping the excitation energies from 3.253 to 3.287 eV (Fig. 1.19(b)). The time evolution signals clearly exhibit beat structures for an excitation energy of 3.266 eV, which is induced by the interference between the phases of polarized excitons separated by a small energy gap. The experimental data provide a beat period of 480 fs, which corresponds to an energy separation of 9 meV. The energy coincides well with the splitting energy obtained from ab-initio band calculations incorporating spin-orbit interactions. We could, therefore, verify that the valence band maximum state of LaCuOCh is composed of a pair of states split by the spin-orbit interaction of Ch ions.

1.3 Transparent Nanoporous Crystal $12CaO \cdot 7Al_2O_3$

1.3.1 Background and approach

Porous materials, in which various kinds of molecules, ions and electrons can be trapped in pores or cages, have attracted growing attention because the cage can provide a closed microspace to store chemical species such as water and hydrogen molecules and to cause selective chemical reactions among encaged species: the cage surface is regarded as a clean surface having catalytic activity, or the encaged species itself may act as a catalyst. Furthermore, the encaged species are regarded as quantum dot if the cage size is small enough for the quantum effect to emerge. Thus, the porous materials are expected to realize novel functions owing to the low-dimensional nature of the caged state. These materials are often categorized in terms of the cage size and the connection topology of the cages, whether it is random or periodic. Intrinsic porous materials such as zeolites contain regularly distributed small pores with a diameter down to subnanometer. On the other hand, artificial porous materials including porous silicon, mesoporous silica, and anodic oxidized aluminum metal contain pores of relatively large size. Although the pores in most artificial materials are randomly distributed, pores in mesoporous silica are distributed periodically due to self-alignment during their preparation.

$12CaO \cdot 7Al_2O_3$ (C12A7) is a novel type of nanoporous crystal, having the distinct features in which the cage is positively charged and thus negatively charged anions and/or electrons are entrapped therein. Such features are distinctly different from those of porous materials, which have positively charged or neutral species entrapped in their cages [42]. In other words, C12A7 has exceptional features among porous compounds: positively charged nanometer-sized cages are closely packed in the lattice and share common planes as opening. Such a feature makes it possible to manipulate anions and to realize novel functions based on zero-dimensional electronic states of negative species which can interact with each other to some extent.

The exploration of active functions in C12A7 is in line with our material development policy: realization of active devices by materials constituted from only main group elements. The existing guiding principle for designing functional metal oxides such as semiconducting materials, magnetic materials, ionic conductors, high-temperature superconductors, and strongly correlated electron compounds depend mostly on the selection of transition metal or rare-earth ions. On the other hand, main-group metal oxides, which are the major ingredients of traditional ceramics that are important for passive use, have not been expected to exhibit such functionalities other than good transparency and electrical insulation. However, since these oxides are abundant in nature and are environmentally benign, it is desirable that the functional materials containing rare or harmful metal elements be replaced by the main group metal oxides in order to guarantee the sustainable growth of our civilization. An ultimate goal of the material research should be, therefore, to revive the main-group metal oxides as functional materials. To satisfy these contradicting requirements, new strategies are absolutely needed: a possible solution is an active use of the intrinsic nanostructures embedded inside the materials composed of main-group elements such as C12A7. This new strategy may provide novel functions for the environmental friendly oxides, which have not yet been realized using heavy metal ions.

C12A7 has been known as a constituent of alumina cements and is typically a transparent (or white powder) electric insulator. Despite the apparent commonplaceness of C12A7, it has a distinct feature in its crystal structure. The unit cell of C12A7 contains two molecules and may be expressed as $[Ca_{24}Al_{28}O_{64}]^{4+} + 2O^{2-}$ [47]. The first part, $[Ca_{24}Al_{28}O_{64}]^{4+}$, denotes a three-dimensional lattice framework which is illustrated in Fig. 1.20 [42]. The unit cell is composed of 12 cages with an inner free space ~ 0.4 nm in diameter. The cage has a mean effective charge of $+\frac{1}{3}$ (= +4 charges/12 cages) and is connected to eight other cages via 0.35 nm-wide-openings. Thus, the transport of encaged species among cages, which induces ionic or electrical conductivity, may take place through the openings. The latter part, $2O^{2-}$, is referred to as 'free oxygen' or

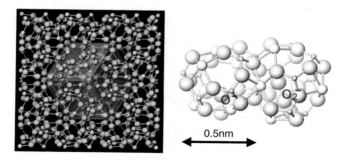

Fig. 1.20. Crystal structure of C12A7 (cubic lattice, viewed from ⟨111⟩ direction) and incorporation of anions (O^- and O_2^-) into cages. Since the lattice framework is positively charged (+1/3 per cage), the electroneutrality is preserved by the incorporation of the anions into the cages.

'extra-framework oxide ions,' and occupies 1/6 of the cage sites. The oxide ions can be partially or completely replaced by various monovalent anions such as OH^-, F^-, and Cl^- [43]. Such anion-exchange properties are complementary to conventional natural nanoporous crystals such as zeolites, whose lattice frameworks are principally formed with encaged cations.

Incorporation of unconventional anions in the cages was first reported by Hosono and Abe [44], who found superoxide ions (O_2^-) at a concentration of $\sim 1 \times 10^{19}$ cm^{-3} in C12A7 polycrystalline samples fabricated in air. This finding stimulated an examination of the ionic conduction properties of oxygen-related anions in C12A7, leading to the discovery of fast O^{2-} ion conduction by West's group [45]. The free oxygen is believed to be responsible for the fast ionic conduction. It is also noteworthy that the composition C12A7 easily forms a glass, which exhibits UV light-induced colorations due to the formation of point defects: ozonide (O_3^-) and aluminum-oxygen hole centers (Al-OHC) in the glass fabricated in an oxidizing atmosphere [46]. On the other hand, F^+-like centers are produced in the glass prepared in a reducing atmosphere, exhibiting phototropy (the color changes upon exposure to light and is restored near room temperature after illumination stops) [47]. Hence, an approach based on defect physics, anion chemistry, and thermodynamic analysis is effective for this material. Generally, our approach to render C12A7 electro-active is based on the incorporation of chemically unstable, i.e., 'active,' negative species into the positively charged cage structure inherent to this material by suitable thermal treatments or hot ion implantations. We employ a combination of conventional and state-of-the-art characterization techniques, including optical absorption, transient photoluminescence, Raman, continuous and pulsed EPR, NMR, TG mass spectroscopy, and muon spectroscopy. Such comprehensive characterization allows for better understanding of various novel phenomena discovered in C12A7.

In addition, one needs variable shapes of samples to proceed with materials research effectively. In particular, single crystals were indispensable for exploration of novel functions of C12A7 such as UV light-induced electrical conductivity, the synthesis of an electride and the clarification of the oxygen radical configuration in the cage. Furthermore, theoretical analysis provides a solid basis of understanding novel phenomena in C12A7, giving clear physical or chemical images to them. The calculation technique adopted the incorporation of quantum mechanical and fully relaxed embedded clusters, has such an advantage that it treats the C12A7 crystal as having a 'soft' framework structure that is deformed considerably when a specific ion or an electron is encaged. Good theoretical calculations are needed to appropriately understand electronic and optical properties when electrons are doped into such a complicated crystal structure [48].

1.3.2 Synthesis of various shapes of C12A7

Fully densified translucent C12A7 ceramics were obtained by sintering hydrated C12A7 powders in a *dry* oxygen atmosphere at 1300°C (the dry atmosphere is much better than a humid atmosphere). The average transmittance between 400–800 nm for 1-mm-thick samples was improved up to $\sim 70\%$ by sintering for 48 h. The elimination of water molecules existing in triple grain boundaries was considered to play a crucial role in improving the transparency [49].

We fabricated C12A7 single crystals by the floating zone method [50]. A conventional growth process resulted in the formation of a concave solid-liquid interface, leading to the formation of many bubbles and cracks in the crystals grown. We found that lowering the growth rate to 1 mm/h significantly reduced the bubble generation. Suppression of the bubbles is very important, because their formation frequently disturbs the stability of the molten zone. Further, by introducing an alumina tube as a heat reservoir at the heating zone, the shape of the solid-liquid interface could be controlled to a convex shape, allowing for the growth of high-quality crystals with higher growth rates. In addition, we have recently succeeded in growing a single crystal by Czochralski method in a 2.0% oxygen in nitrogen atmosphere using an iridium crucible. The crystal grown, which is orange because of the inclusion of a small amount of Ir ions, is of higher quality than those grown by the Fz method in terms of crystal size and inclusion of bubbles.

Polycrystalline C12A7 thin films were prepared by room-temperature deposition of amorphous C12A7 films on MgO substrates by PLD, followed by thermal annealing of the films in an oxygen atmosphere at temperatures above 800°C [51].

1.3.3 Incorporation of oxygen-related species [52–56]

1) Oxygen radical formation

It was found that extraordinarily high concentrations (on the order of 10^{20} cm^{-3}) of oxygen radicals, O^- and O_2^-, in C12A7 are formed in the cages by simply heating C12A7 ceramics in a dry oxygen atmosphere. Concentrations of O_2^- and O^-, were analyzed by a combination of electron paramagnetic spin resonance (EPR) and Raman spectroscopy. The resultant C12A7 had outstanding oxidative reactivity, such that even Pt metal was oxidized into the +4 state when the metal was placed on the material at high temperatures. The effect of oxygen partial pressure up to 400 atmosphere on the generation of active oxygen radicals was examined using a hot isostatic pressing furnace. The total concentration of the radicals increased with oxygen pressure and reached 1.7×10^{21} cm^{-3}, which is comparable to the maximum concentration of monovalent anions in the cages. The formation of oxygen radicals was markedly reduced as the pressure of water vapor increased during the heating.

The formation of oxygen radicals is attributed to the absorption of the oxygen molecules in the atmosphere into C12A7, which react with free oxygen to form an equal amount of O^- and O_2^-. That is,

$$O_2 \text{ (atmosphere)} + O_2^- \text{ (cage)} \rightarrow O^- \text{ (cage)} + O_2^- \text{ (cage)} \qquad (1.1)$$

The evaluation of the enthalpy and entropy of radical formation enables us to determine the temperature and oxygen partial pressure dependences of equilibrium oxygen radical concentration (Fig. 1.21). The rate-limiting processes for the radical formation are not the surface reactions, but the total ionic diffusion processes, in which the smallest diffusivity of O_2^- likely dominates the process. These results allow one to predict the total oxygen radical content for given annealing conditions.

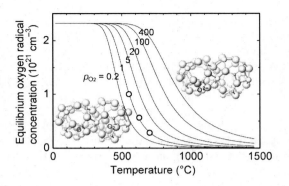

Fig. 1.21. Partial oxygen pressure and temperature dependences of equilibrium concentrations of oxygen radicals. Circles denote measured equilibrium concentrations.

2) **Configuration and dynamics of O_2^- radicals in cages [57]**

The O_2^- ion in C12A7 was studied by continuous-wave and pulsed EPR. A C12A7 single crystal heated in 40% ^{17}O-enriched gas showed the hyperfine splitting due to a $^{17}O_2^-$ nucleus. This fact clearly indicates that O_2^- is formed via the reaction of an O_2 molecule in the atmosphere with a free oxygen ion as given by Eq. (1.1). The angular variations of g-values and ^{17}O hyperfine splittings were measured for a single crystal at 20 K, and ^{27}Al-ESEEM powder patterns were measured at 4 K. These results verified that the O_2^- radical is located inside the cage, and they clarified that O_2^- is adsorbed on a Ca^{2+} ion of the lattice framework. Furthermore, the configuration of O_2^- is determined such that two constituent oxygen ions occupy crystallographically equivalent sites in the cage, indicating the O_2^- ion assumes a 'side-on' configuration, where the O–O bond is perpendicular to the two-fold rotation axis ($C_2 \parallel \langle 100 \rangle$) in the cage and points at the two calcium ions in the framework (Fig. 1.22).

Dynamic motion of the O_2^- was clarified from the temperature dependence of the g-values: O_2^- behaves like a solid $>\sim 20$ K; an anisotropic swinging rotation of the O–O bond along the C_2 axis is activated with temperature; and the anisotropy in the swinging motion disappears at >400 K (note the melting point of C12A7 is ~ 1700 K). These results were explained in terms of the electrostatic interaction between the Ca^{2+} ion and π-orbital of the O_2^- radicals.

3) **Partial oxidation of methane into syngas [58]**

Partial oxidation of methane into syngas (CO and H_2) was examined using C12A7 powders promoted by metals such as Ni, Co, Pt, Pd, and Ru. The effect of space velocity and metal loading was studied, and the catalytic ability of Ni/C12A7 was compared with that of a nickel catalyst supported on CaO, α-Al_2O_3, and other calcium aluminates with high CaO/Al_2O_3 ratios. On Ni, Pt, and Pd/C12A7, the reaction readily took place at temperatures as low as 500°C and reached thermodynamic equilibrium quickly. At 800°C and a space velocity of 240 000 ml·g^{-1}·h^{-1}, the activity of the five samples investigated decreased in the following order: 1% Pt/C12A7 > 5% Co/C12A7 > 5%

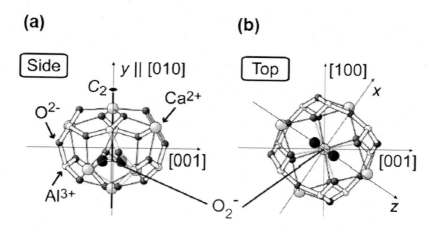

Fig. 1.22. Location of O_2^- inside cage in C12A7 determined by EPR study: (a) side view, and (b) top view.

Ni/C12A7 > 1% Ru/C12A7 > 1% Pd/C12A7. The activity and selectivity of Ni/C12A7 and Pt/C12A7 increased with metal loading, and the activity of 10% Ni/C12A7 was comparable to that of 1%Pt/C12A7·Ni/C12A7 exhibited a low coke formation rate, and it was more active than nickel supported on CaO, α-Al$_2$O$_3$, 3CaO·Al$_2$O$_3$, or CaO·Al$_2$O$_3$ because of the good dispersion of NiO on C12A7 and the existence of active oxygen ions incorporated in its nanosized cages (Fig. 1.23).

4) Generation of O$^-$ ion beam [59–62]

We succeeded in extracting intense O$^-$ ion beams by applying dc voltage to C12A7 incorporating a large amount of O$^-$ (1.3×10^{20} cm^{-3}) and O$_2^-$ (2.7×10^{20} cm^{-3}). The observed current was dominated by the O$^-$ ion. When the sample temperature was 800°C and the applied electric voltage exceeded 1 kV·cm^{-1}, the O$^-$ current density increased up to ~ 2 μA/cm^2, which is higher by three orders of magnitude than the maximum value obtained from yttrium-stabilized ZrO$_2$ (YSZ). A surface metal electrode, necessary for the YSZ to dissociate a drifting O^{2-} in the bulk into a pair of an O$^-$ radical and an electron on the surface, is not needed for C12A7. The presence of extremely large concentrations of O$^-$ and O$_2^-$ radical ions and the fast oxygen-ion conductive nature of C12A7 allow for the extraction of the O$^-$ radicals directly from the surface. The capability of generating high-density, pure O$^-$ ion current is useful for various novel oxidation processes such as the formation of gate dielectrics on semiconductors, organic chemical reactions, decomposition of pollutants, and sterilization. The development of continuous O$^-$ beam generation system using C12A7 ceramic membrane is now in progress (Fig. 1.24).

1.3.4 Incorporation of hydrogen-related species

1) Photoinduced conversion from insulator to conductor in C12A7:H$^-$

The first electronic conduction in the main-group light metal oxides was realized in C12A7. Hydride ions (H$^-$) were incorporated into the cages by thermal treatment in

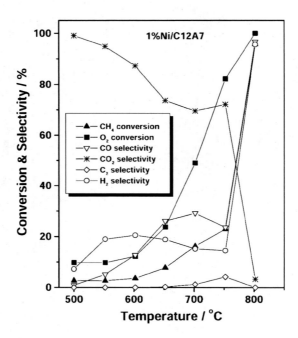

Fig. 1.23. CH_4 partial oxidation to syngas over 1% N1/C12A7 as a function of temperature. Gas space velocity = 30 000 ml·g^{-1}·h^{-1}.

a hydrogen atmosphere [63]. As-treated C12A7:H was colorless, transparent and a good insulator with an electronic conductivity of less than 10^{-10} S·cm^{-1}. However, upon irradiation with UV light, the C12A7:H exhibited a yellowish green color corresponding to optical absorptions at 2.8 and 0.4 eV with simultaneous conversion into an electronic conductor with a conductivity of 0.3 S·cm^{-1} at 300 K (Fig. 1.25). The conductive state continued even after the irradiation was stopped. The reverse process or reversion to the insulator occurred when the material was heated to above ~ 300°C accompanied by the decay of the optical absorption intensities. When the temperature rose above ~ 550°C, H_2 gas was released from the sample and the photosensitivity was completely lost. With UV light illumination, the H$^-$ ions emit electrons which are trapped in the cages, forming F$^+$-like centers. Further, a migration of the electrons at the F$^+$-like centers is responsible for conduction.

The photoinduced coloration and the transition from insulator to conductor are also observed in the H-incorporated C12A7 thin films. Hydrogen incorporation is achieved either by thermal treatment in hydrogen atmosphere at 1200 K or hot ion implantation of H$^+$ ions. The latter process is described later. The visible light absorption loss is estimated to be only 1% for 200 nm-thick-C12A7:H films with an electrical conductivity of 6.2×10 S·cm^{-1} at 300 K (Fig. 1.26). These properties enable novel applications such as the direct optical writing of conducting wires and areas on the insulating transparent media.

Generation of persistent carrier electrons in C12A7:H by electron-beam irradiation was also examined [64]. The surface layer of an insulating single-crystal C12A7:H

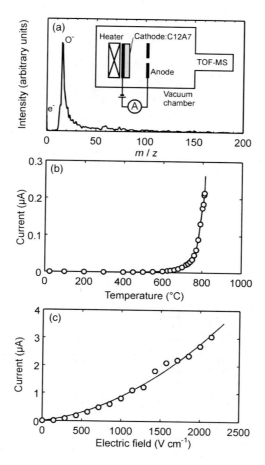

Fig. 1.24. Emission of O^- ions from C12A7. (a) Typical time-of-flight-type mass spectrum of ion current extracted from C12A7. (b) Ion current between two electrodes as a function of sample temperature. The extracting DC voltage was fixed at 375 V. The beam current came primarily from O^- ions. (c) Ion current from a C12A7 disk of 2 cm^{-2} area as a function of applied DC voltage at 810°C.

was directly converted to an electronic conductor by electron-beam irradiation, accompanied by a green coloration that is identical to the change induced by UV-light irradiation. Carrier electron formation at the maximum value of electron excitation was saturated at an electron beam dose of 10 $\mu C \cdot cm^{-2}$, which is comparable to the sensitivity of conventional photoresists for electron-beam lithography. The carrier electron formation yield per electron-hole pair was estimated to be as high as 0.05.

2) **Mechanism for photoinduced conversion [65–67]**

An embedded cluster approach was employed to calculate the energy state of the extra electron in C12A7. Our results indicate that the empty cage state forms a narrow conduction band called 'cage conduction band,' which is located higher

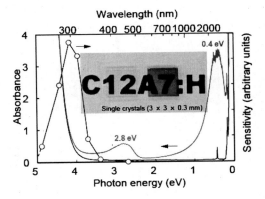

Fig. 1.25. Optical absorption spectra for H-loaded C12A7 single crystal before and after irradiation with UV light. Circles denote the photon energy dependence of the sensitivity on the coloration.

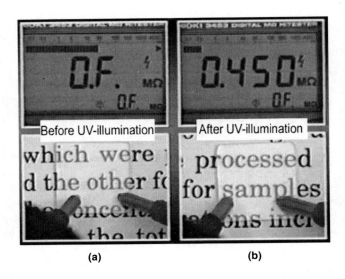

Fig. 1.26. Photoinduced conduction in H-loaded C12A7 thin film. The electrically conductive film looks transparent. (a) Before, and (b) After UV illumination.

than the top of the valence band (VB) by 5.2 ~ 5.7 eV and lower than the bottom of the framework conduction band by ~ 1.0 eV. When an electron is introduced in the band, its energy level becomes lowered by ~ 1 eV from the empty cage conduction band due to the electrostatic interaction with ions in the cage framework, resulting in the deformation of the cage structure. As a result, the electron is localized in a cage, forming an F^+-like center. Further, as the electron is loosely trapped in

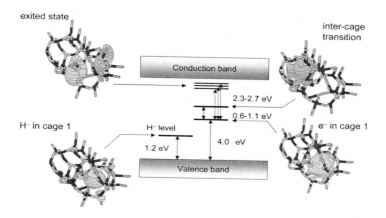

Fig. 1.27. Energy band diagram for H⁻ and electron in the cage of C12A7 obtained by first-principle calculation based on embedded cluster approach.

the cage, it can hop to the neighboring cages, inducing electrical conductivity with activation energy of ∼0.1 eV. The extra frame-work electron is regarded as a 'polaron.' The F^+-like center introduces two absorption bands with the peaks at ∼1 eV and ∼2.8 eV. These bands were assigned to the intercage charge transfer and the intracage s–p transitions of the electron, respectively (Fig. 1.27). These theoretical analyses are consistent with experimental observations, supporting the hypothesis that the photo-induced coloration and insulator–conductor conversion are induced by the photoionization of the H⁻ ion to form a pair of H^0 and an electron. However, no H^0 is detected by EPR measurements at room temperature, suggesting that H^0 is so unstable that it may further decompose into H^+ and an electron. The H^+ ions most likely react with the free oxygen ions to form OH^- ions. Thus, the total reaction is expressed as

$$H^- \text{ (cage)} \rightarrow H^0 \text{ (cage)} + e^- \text{ (cage)} \rightarrow H^+ \text{ (cage)} + 2e^- \text{ (cage)}$$
$$\rightarrow OH^- \text{ (cage)} + 2e^- \text{ (cage)} \quad (1.2)$$

The irreversible nature of the photo-induced process at room temperature is likely attributed to the decomposition of H^0.

ERP measurements at low temperature reveal that nearly equal amounts of H^0 and F^+-like centers were generated in C12A7:H when UV light was irradiated at low temperature as shown in Fig. 1.28 [68]. Further H^0 decreases abruptly with temperature and disappears completely above ∼100 K, while the F^+-like center decreases modestly and still survives above room temperature. This observation provides solid evidence for the existence of the H⁻ ion in the hydrogen-treated C12A7 and it clearly shows that reaction (1.2) is responsible for the photoinduced effects in C12A7:H⁻ (Fig. 1.29).

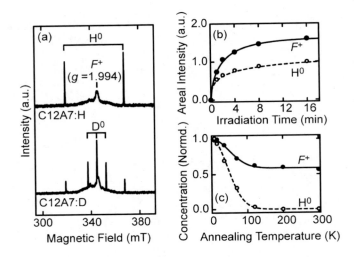

Fig. 1.28. (a) EPR spectra of C12A7:H and C12A7:D observed at 4 K after UV illumination. (b) Areal signal intensities of H^0 and F^+ as function of UV-illumination time: Almost 1:1 generation indicates a photodissociation of $H^- \longrightarrow H^0 + e^-$. (c) Isochronal annealing of H^0 and F^+. H^0 was completely annihilated above ~ 200 K while 60% of F^+ remains even at 300 K.

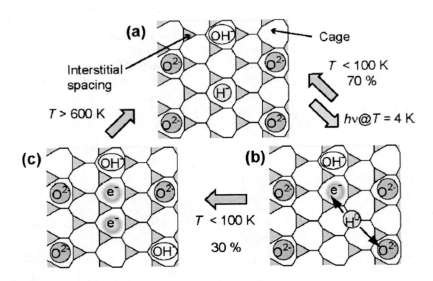

Fig. 1.29. Scheme of photo-induced insulator-conductor conversion: (a) Before UV illumination, H^- ions are incorporated in the cage of C12A7 in place of O^{2-}. (b) Upon UV illumination, H^- ions are photo-dissociated to H^0 and electron pairs ($H^- \rightarrow H^0 + e^-$). (c) The resulting H^0 is rather mobile and recombines with electrons ($H^0 + e^- \rightarrow H^-$) or reacts with free O^{2-} to form OH^- and e^- ($H^0 + O^{2-} \rightarrow OH^- + e^-$). By annealing at ~ 600 K, OH^- and electrons trapped in the cage return to the initial state via the reaction $OH^- + 2e^- \rightarrow O^{2-} + H^-$.

1.3.5 Formation of stable electride

1) C12A7 electride

An electride is a crystalline ionic salt in which electrons serve as anions. Typical examples are compounds of cryptands or crown-ethers as cations and the electrons as anions [69]. Although electrides are promising for novel applications including a field electron emitter due to their exotic characteristics such as small work functions, they are not stable at room temperature and/or in a moist atmosphere. Thus, syntheses of a stable electride have been sought for practical applications. The first thermally and chemically stable electride was realized in C12A7 [70], we eliminated the free O^{2-} ions from the cages by heating a single crystal of C12A7 in evacuated silica glass tubes together with calcium metal, leading to the formation of high-density (2×10^{21} cm^{-3}) electrons confined to the cages. The resultant C12A7 is described as $[Ca_{24}Al_{28}O_{64}]^{4+}(4e^-)$, i.e., C12A7 electride. The electrons localize in the cages, forming F^+-like centers, thereby behaving as substitution anions for the free O^{2-} ions. They can migrate by hopping to neighboring cages, giving electrical conductivity as high as ~ 100 S·cm^{-1}. Further, they are coupled antiferromagneticly, forming a diamagnetic pair or a singlet bipolaron.

2) Synthesis of C12A7 electride

We have developed three kinds of processes for preparing the C12A7 electride: thermal anneal of single-crystal C12A7 with calcium metal vapor [70], direct solidification of C12A7 in a carbon crucible [71], and hot ion implantation of inert gas ions [72].

Figure 1.30 shows changes in optical absorption spectra and the temperature dependence of electric conductivity of a C12A7 single crystal with annealing

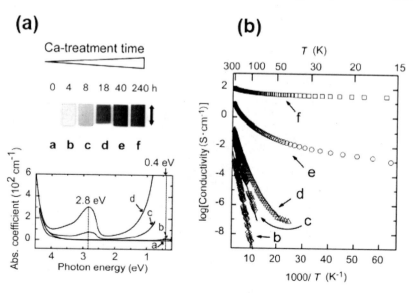

Fig. 1.30. (a) Coloration of single crystals processed by Ca-vapor annealing and their optical absorption spectra, and (b) electrical conductivity as a function of temperature.

time in calcium metal vapor at 700°C. The absorption bands peaked at 2.8 and 0.4 eV and the conductivity at room temperature is enhanced with annealing time. It indicates that the electron concentration increases with the annealing and reaches $\sim 2 \times 10^{21}$ cm^{-3} when the crystal is subjected to the chemical treatment for 240 h, corresponding to the almost complete replacement of the free oxygen with electrons. The C12A7 electrode is formed by the annealing of C12A7 single crystals in calcium vapor at 700°C. The electrical conductivity changes from semiconductor to degenerated conduction types with an increase in the electron concentration. Even metallic conduction is observed in the C12A7 electrode with an extremely high electron concentration.

When a mixture of C3A and CA is melted at $\sim 1600°$C in a carbon crucible with a cap and is cooled down to room temperature in an air, the solidified bulk changes to C12A7 and exhibits a green color and electrical conductivity (Fig. 1.31(a),(b)) [71]. Its optical absorption and ERP spectra indicate that electrons are incorporated in the solidified C12A7. Thus, direct solidification in a carbon crucible provides a simple mass-production process for the fabrication of the C12A7 electrode, although

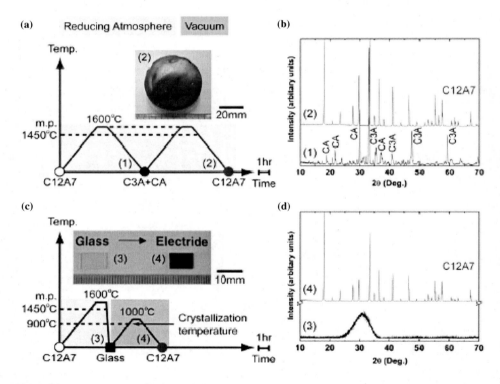

Fig. 1.31. Fabrication of C12A7 electrode via melting in carbon crucible with carbon cap in air. (a) Fabrication procedure of electrode by solidification of the reduced C12A7 melt. (b) Powder X-ray diffraction patterns of (1) C3A+CA and (2) Crystalline single phase C12A7. (c) Fabrication procedure for C12A7 electrode by crystallization of transparent reduced glass under vacuum. (d) Powder X-ray diffraction patterns of (3) Transparent glass and (4) Crystalline single phase C12A7.

complete displacement of free oxygen with electrons is very difficult. In this process, it is most probable that C_2^{2-} ions act as a template for the formation of the C12A7 crystallographic phase, and they are spontaneously released from the solidified C12A7, leaving electrons in the cages during the cooling process. When the reduced C12A7 glass, which is obtained by a rapid quenching of a melt with the stoichiometric composition of C12A7 in a carbon crucible, is crystallized in vacuum, the resultant polycrystalline C12A7 bulk becomes electrically conductive (Fig. 1.31(c),(d)). This supports our assumption that C_2^{2-} ions play a significant role in the incorporation of the electrons in C12A7.

The hot-ion implantation technique for the formation of the electride is described later.

3) Electron emission from C12A7 electride

We investigated electric-field electron emission from the C12A7 electride in vacuum [73]. Measured electron emission properties were explained by involving Fowler-Nordheim tunneling at a large electric field >200 kV·cm^{-1} and by thermionic emission at lower electric fields. The work function estimated from the emission characteristics was ~ 0.6 eV, which is comparable to values reported for conventional organic electrides. However, ultraviolet photoelectron spectroscopy gave a much larger value ~ 3.7 eV for the work function. The discrepancy may be attributed to the so-called band bending effect due to naturally formed n$^+$/n$^-$ layers on the surface (the surface is oxidized). Using the field emission of electrons from the electride at room temperature, a triode-structured field emission display (FED) device was constructed employing the electride as a cathode and a ZnO:Zn phosphor plate as an anode. Light emission was clearly observed under typical ambient light, demonstrating that the C12A7 electride is promising for FED devices (Fig. 1.32).

Recently, we have also examined the thermal field emission (TFE) from a flat surface of the C12A7 electride single crystal at temperatures up to 900°C and applied external voltages of 0 $\sim$ 6 kV in 10^{-5} Pa vacuum using the setup shown

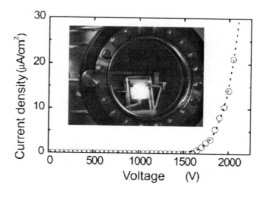

Fig. 1.32. Current-voltage characteristic for electron emission from single crystal C12A7 electride. The inset is a photograph of the operation of a field-emission-light-emitting device using a ZnO:Zn phosphor.

in Fig. 1.33 [74]. The TFE started to occur at $\sim 650°C$ and its current increased steeply at $\sim 900°C$, reaching ~ 80 μA (~ 1.5 A·cm^{-2}) (Fig. 1.34). The emission with a current of ~ 50 μA from a 80-μm-diameter area is sustained stably for more than 90 h at an extraction field of 10^5 V·cm^{-1}. The work function estimated using the Richardson-Dushman equation was ~ 2.1 eV, which is smaller than that of LaB$_6$ (~ 2.7 eV). These experiments strongly suggest that the C12A7 electrode has a high potential both for cold electron emission and TFE applications. The different characteristics in two modes for the electron emission are likely the result of the difference in the surface states of the C12A7 electrode.

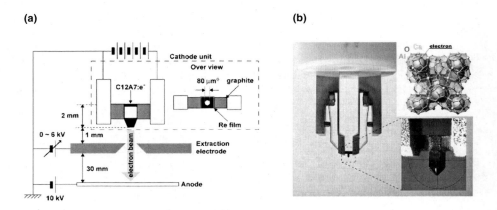

Fig. 1.33. (a) Experimental setup for TFE measurement, and (b) expanded pictures of TFE head using C12A7 single crystal.

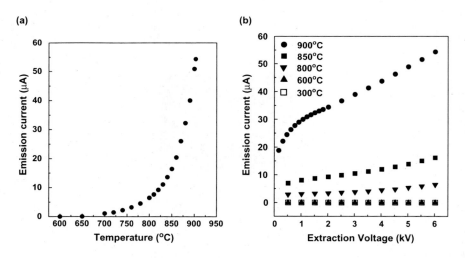

Fig. 1.34. (a) Electron emission current as a function of temperature. (b) Extraction voltage dependence of electron emission current for several temperatures.

1.3.6 Hot-ion implantation

We have succeeded in forming the H^- ions and electrons in the C12A7 films and single crystals by hot implantation of H^+ and inert gas ions [75]. In the H^+-ion-implanted films, the concentration of F^+-like centers and the electrical conductivity are induced by UV light illumination, indicating that H^- ions as photoactive donor are formed by proton implantation, and their concentration increases with increasing temperature of the C12A7 film. This fact indicates that the formation of H^- is accompanied by the thermal excitation processes. The optimum temperature for H^- formation was 600–700°C. Furthermore, the electrical conductivity after illumination was controllable by fluences of the implanted ions. For example, fluences of 1×10^{18} cm^{-2} gave electrical conductivity values as high as ~ 10 S·cm^{-1} at room temperature.

Upon inert gas ion (Ar^+, He^+) implantation into C12A7 films with fluences from 1×10^{16} to 1×10^{17} cm^{-2} at elevated temperatures of 600°C [72], the films were colored and exhibited electrical conductivity in the as-implanted state. This is caused by the extrusion of the free O^{2-} ions due to the 'kick-off effect' of the implanted inert gas ions, leaving electrons in the cages at concentrations up to $\sim 1.4 \times 10^{21}$ cm^{-3}. When the fluence was less than 1×10^{17} cm^{-2}, as-implanted films were transparent and insulating; they then exhibited coloration and electrical conductivity after being illuminated with UV light, suggesting that the H^- ions were formed by the inert gas implantation; probably preexisting OH^- ions act as precursors because the OH^- concentration is distinctly decreased. The electron concentration induced was proportional to the displacements per atom (dpa; 1 dpa means all atoms in the projected range are displaced once), suggesting that nuclear collision effects of the implanted ions play a dominant role in the extrusion of the O^{2-} ions (Fig. 1.35). The hot-ion implantation technique provides a novel physical process for preparing the electron-conductive C12A7 films.

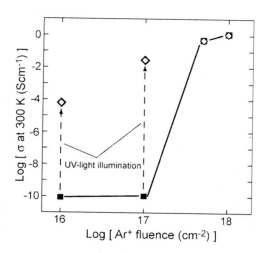

Fig. 1.35. Ar^+-dose dependence of electrical conductivity in C12A7 films. The symbols ■ and ◇ indicate electrical conductivities before and after UV-light irradiation.

1.4 Encoding of Periodic Nanostructures with Interfering Femtosecond Pulses

1.4.1 Background, history, and features of encoding

Background. Femtosecond (fs) pulses from a mode-locked laser, typically a Ti: sapphire laser, are characterized by an ultrashort time domain with good coherence over the entire pulse duration. These features provide additional advantages to conventional laser processing, including the capability of processing in transparent materials, nanosized patterning and clean processing nearly free from thermal effects. Those advantages originate from the inherent nature of the pulse. First, fs-laser pulses have extremely short durations, leading to very high peak powers due mostly to the temporal compression of the laser energy. Such high peak powers can process almost all kinds of materials typically by laser ablation [76–78]. As the laser energy is absorbed efficiently through multi-photon processes with such high power pulse, materials to be processed are not necessarily opaque to the laser wavelength. fs laser processing is particularly suited to transparent materials, which are difficult to process using conventional lasers with a continuous wave (cw) and a longer pulse width, such as CO_2 and YAG: Nd^{3+} or excimer lasers. Second, with the aid of nonlinear effects such as multi-photon absorption and self-focusing, we can concentrate the laser energy onto a very small spot. In addition, materials suffer ablation or structural changes only in limited areas where the accumulated energy exceeds a certain threshold value characteristic for each material. Nanoscale structures may be formed in this way. Finally, the pulse terminates before the completion of energy transfer from photoexcited electrons or electron-hole plasma to the lattice, which minimizes thermal effects during machining and makes the process clean [76–84], specifically when only a single pulse is involved. Using these features, a large number of excellent studies have been performed recently in fabricating various kinds of fine-scale structures on the surface and inside of transparent dielectrics, semiconductors, polymers, and metals.

Further, if we combine another distinct feature of the fs pulse, its excellent coherence [85–90], with the features mentioned, we can encode holograms with a very fine internal structure in various materials by splitting a single pulse into multiple beams and allowing them undergo interference. The encoded pattern is regarded as a 'periodic nanostructure' produced in a self-aligned process. This provides a new technique for fabricating nanostructures inside transparent materials, which are impossible to process by other technique such as lithography.

Progress in encoding. Phillips and Sauerbery at Rice University [83] performed initial studies, in which they used a fs KrF eximer laser to record holographic gratings in polyimide with the intention of realizing fine patterning resulting from the thermal-diffusion free process. The observation of a periodical fringe in diamond has been reported, which was accidentally encoded presumably due to the interference of the incident fs laser beam with a reflected beam [84]. In spite of these reports, material processing with interfering fs pulses had not attracted much attention before we succeeded in encoding holographic gratings in versatile materials including transparent dielectrics, semiconductors, polymers, and metals with interfering Ti: sapphire laser pulse [85–90]. Since then, several groups have reported the formation of periodic structures in polymers and glasses with the interfering fs pulse. Misawa's group at Tokushima (now at Hokkaido) University recorded three-dimensional structures in photoresist material with the interference of five beams split from a single fs pulse with the intention of forming a photonic crystal [91]. Hirao's group at Kyoto University [92] and Itoh's group at Osaka University [93] have succeeded in encoding embedded gratings in polymers and glasses. Research groups involved with this technique are increasing in number with the expectation that interfering fs pulses will open a new frontier in laser processing.

Features of encoding. Features of the encoding technique using interference fs laser pulses are summarized in Table 1.2.

In fs laser processing, almost all kinds of materials can be processed with interfering fs pulses, and the structures are formed in a designated position within materials. In particular, embedded structures can be formed in transparent materials. The process can be completed in a single shot; thus the throughput of the process is potentially very high. The overall size of gratings is as small as ~100 μm where the fringe spacing is in the range of 5–0.1 μm depending on the colliding angle of the beams and laser wavelength. Because the spacing is proportional to the laser wavelengths for a fixed crossing angle of the beams, the use of fs pulses with shorter wavelengths is a straightforward way to narrow the fringe spacing. On the other hand, as is common to fs laser processing, minimum structure sizes for line widths or dot diameters in the periodic structures can be reduced to the nanometer scale, which seems to be almost insensitive to the irradiated laser wavelength because of specific interactions of fs laser pulses with materials such as the generation of laser-induced plasma and Coulombic explosion [94]. An additional novel feature of the technique is that two- and three-dimensional periodic structures can be fabricated by a double exposure and multiple beam interfering methods, which provide a novel technique to prepare emerging photonic crystals.

Encoding mechanism. As fs laser pulses exhibit good coherence over the entire pulse duration, interference takes place when multiple fs pulse beams split from a single pulse and overlap each other both spatially and temporally. The interference pattern resulting from this overlap can immediately lead to the modulation of the electron-hole plasma density in the irradiated materials through multi-photon absorption and/or avalanche ionization, in which the plasma is excited more densely in enhanced regions of the interference than depressed regions. Density modulation of the plasma causes

Table 1.2. Features of encoding.

- Versatility of encoded Materials
  - Dielectric   : SiO_2 Glass, SiO_2 Thin Film, $LiNbO_3$, ZnO, ZrO_2, MgO, TiO_2, Diamond, α-C
  - Semiconductor: SiC, ZnSe, Si
  - Metal   : WC, Pt, Au, Al
  - Plastics   : Teflon
- Small size: ~100 μm diameter
- Single pulse encoding
- Surface and embedded types
- Nanoscale structure: ~15 nm
  - Self-focusing, Multiphoton process, Threshold limited process
- Multidimensional periodic structure

refractive-index modulation. A transient grating is immediately formed in the material right after the irradiation with the interfering fs pulse. The transient grating, which likely lasts for several picoseconds (ps), may induce reconstructed beams that interfere with the incident beams to form additional gratings. The plasma energy in the material likely relaxes to the lattice within ~ 1 ps and is accumulated locally in the lattice.

As the relaxation proceeds, the transient grating is converted into a permanent grating when the local energy in the lattice exceeds a threshold for laser ablation or for structural changes in the material. The permanent gratings can be formed when the laser energy in the enhanced region of the interference is over a threshold value.

1.4.2 Instrumentation

General features. A hologram encoding system using fs laser pulses does not differ significantly from conventional laser hologram exposure systems. The distinct difference between the two systems is that a fs laser pulse is used instead of continuous wave light typically from a gas laser having a large coherent length. A fs KrF excimer laser (248 nm, ~ 400 fs) [84], Ti:sapphire laser (800 nm, ~ 100 fs) [85–90, 92–97], second harmonics of Ti:sapphire (380 nm, ~ 80 fs) [91] and third harmonics of Ti:sapphire (290 nm, ~ 100 fs) [98,99] have been employed thus far as fs pulse sources. Because of the very small coherent length of the fs pulse, which is restricted by the spatial extent of the pulse, an optical delay line and a mechanism for detecting the temporal coincidence of the two beams with a spatial accuracy of ~ 1 μm are in general required to ensure spatial and temporal overlap. It is also desirable for the temporal coincidence measurement to be independent of the angle between interference beams. By varying this angle, we can control fringe intervals in gratings and shapes of unit elements in two-dimensional periodic structures. Sum frequency generation of the pulses using nonlinear optical crystals (colliding-angle dependent) [93] or third harmonic generation from air (colliding-angle independent) [95] is effectively used for most of the systems, but an alternative temporal coincidence detecting system without using the frequency up-conversion process is necessary when shorter wavelength fs pulses are employed. As a result, a distributed feedback dye laser (DFDL) technique was used for the fs KrF laser [83], and pump and prove techniques based on the optical Kerr effect and transient absorptions have been developed for the THG of the Ti:sapphire laser [98,99]. On the other hand, the use of a diffractive beam splitter in a multiple beam interfering system allows for temporal coincidence without the installation of any adjusting mechanism [91].

When comparing the encoding systems developed thus far, optical configurations are basically the same, but differences exist in the types of fs lasers and the mechanisms of detection for the temporal coincidence of the split beams, which are shown in Table 1.2.

Experimental setup. Figure 1.36 shows a schematic diagram of an experimental setup for grating encoding using near infrared (IR) fs laser pulses developed by our group [86]. A Ti:sapphire laser system consisting of a Ti:sapphire oscillator, a stretcher, a regenerative amplifier pumped by the second harmonic of an Nd:YVO$_4$ laser, and a compressor is used as a light source. It generated 800-nm light pulses at a repetition rate of 10 Hz, a pulse duration of ~ 100 fs (an effective coherence length of ~ 30 μm), and a maximum

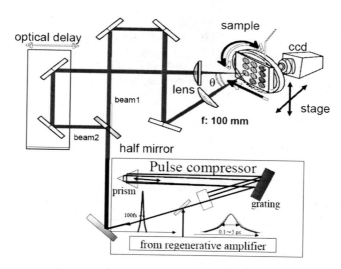

Fig. 1.36. Experimental setup for encoding of holographic grating by fs laser pulse.

energy of 3 mJ/pulse. A fs laser pulse from the laser system was divided into two beams by a half-silvered mirror. The separated beams propagated along different optical paths with one having a variable path length (optical delay line). The two beams were focused onto a single spot 50–100 μm in diameter by lenses with a focal length of 5∼10 cm.

The angle of intersection (θ) between these two beams was varied from 10 to 160°. The spatial length of each optical path was equalized with a precise movement of mirrors in the optical delay line, while the temporal coincidence signal was monitored. The third harmonic generation (THG) from air was used for the signal [95], as described in the following.

Once the temporal coincidence of the two beams was realized, samples were placed at the beam focus position. The encoding of the gratings was confirmed in situ by detecting diffracted light from an incident He–Ne laser. Figure 1.37 shows the grating is encoded only when the two beams coincide both spatially and temporally.

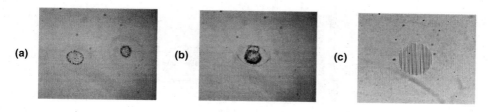

Fig. 1.37. (a) Two pulse beams are spatially noncollided. (b) Two pulse beams are collided spatially, but not temporally. (c) Two pulse beams are collided both spatially and temporally. The grating is encoded only in case (c).

As will be described later in this section, chirping of the fs pulse, which results from stretching the pulse width from ~100 fs to ~10 ps while keeping the total pulse energy unchanged, is very effective for encoding embedded gratings due mostly to the reduction of the surface damage [96,97]. In using the stretched pulse for encoding, the fs pulse is chirped by adjusting the compressor in the regenerated amplifier, while keeping other optical configurations unchanged.

Detecting mechanism for temporal coincidence. Representative examples of the detecting mechanism for the temporal coincidence of the beams are introduced below.

1) THG in air

It has been reported that THG occurs when an intense fs laser pulse is focused in various gases such as air [100] and argon [101]. The mechanism has been discussed theoretically [102]. This phenomenon has been applied to monitoring the temporal coincidence of two fs pulses [95].

When the energy of the first fs laser beam exceeds more than 1.5 mJ/pulse, a spot is observed at the beam center of the far-field pattern on a fluorescent screen, which is confirmed as the third harmonic of the laser fundamental (800 nm) at 266 nm. The THG was not observed when the pulse was focused in vacuum, indicating that air was responsible for the THG. Then, a second focused beam having a total energy of 50 µJ/pulse, which was too small to induce the THG by itself, was directed at the focal point to overlap spatially with the first beam. The far-field pattern of the second beam on the luminescence screen yielded only weak, white continuum light without a blue spot, probably due to the self-phase modulation in air when the relative time delay of the two pulses was large. While tuning the optical path length, the THG spot suddenly appeared at the center of the white continuum spot. Figure 1.38 shows the TH intensity in the second beam as a function of the delay time calculated from the change in the optical path length.

The intensity increases abruptly as the delay time approaches zero from below. The rise time for the TH enhancement estimated from the inset was ~200 fs,

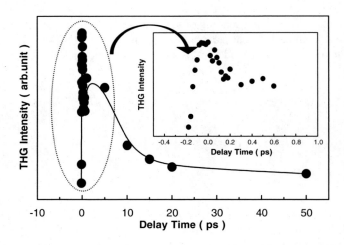

Fig. 1.38. THG intensity as a function of delay time between two colliding pulses.

which is in reasonable agreement with a pulse width of ~100 fs. With a further increase in the delay time, the intensity at first decreases sharply. The decrease then becomes gradual and is still observed at delay times above 50 ps. Despite the gradual decay on the long-delay-time side, the peak in Fig. 1.38 is sharp enough to determine the exact time coincidence of the two fs pulses. The trend in Fig. 1.3 was basically independent of the angle of intersection of the two beams, as expected from the inherent nature of THG in the gas phase. The use of THG in air makes it possible to arrange for the temporal coincidence of two fs pulses over a wide range of angles of intersection.

2) Pump and probe technique

The techniques using frequency up-conversion processes are not applicable for monitoring the temporal coincidence of UV fs pulses, because the resultant sum frequency light in the process has too short a wavelength to propagate in nonlinear crystals or even in air. To overcome this problem, alternative methods using a distributed feedback dye laser technique and two kinds of pump-probe techniques based on optical Kerr and transient absorption effects have been developed.

In the pump and probe techniques, the two beams to be interfered are used respectively as pump and probe pulses under low intensity conditions below the threshold for the ablation [98,99]. In the optical Kerr gating method, the pump pulse is polarized perpendicularly to the probe pulse using a $\lambda/2$ plate, which is inserted in the pump beam path before the quartz plate (Fig. 1.39(a)). On the other hand, the probe pulse goes into another $\lambda/2$ plate after passing through the quartz plate and then into a polarized beam splitter (PBS) placed at the front of a photodiode-detector (PD equipped with a phase-sensitive detection system).

Silica has third-order susceptibility with a very fast response, and thus the pump pulse induces a birefringence through the electronic Kerr effect, also known as the nonlinear refractive index change. When the $\lambda/2$ plates are rotated by 45° with respect to each other, the probe pulse can go through PBS only when the silica

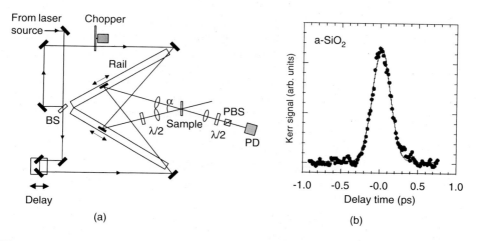

Fig. 1.39. Pump and probe technique to detect temporal coincidence of two fs pulses. (a) Experimental setup. (b) Kerr signal intensity as a function of delay time.

plate is irradiated simultaneously with the pump pulse (Fig. 1.39b). This geometry yields an autocorrelation measurement of the pulse. Although the observed pulse width stretches from that of the fundamental wavelength (870 nm), the exact reason for that has not been clarified and time zero is determined at the center of the band.

In the transient absorption method, the probe pulse is directly collected at the PD without either the $\lambda/2$ plates or the PBS device. The absorbance change of the probe pulse perturbed by pump pulse irradiation is detected by the same procedure as the Kerr measurement.

3) **Distributed feedback dye laser technique**
When the beams overlap each other spatially and temporally in a dye cell placed at the focal point of colliding fs KrF laser beams, a grating is induced in the cell, which in turn acts as a Bragg mirror, thereby forming a distributed feedback dye laser [83]. In this way, monitoring the output of the dye cell while tuning an optical delay line makes it possible to achieve temporal coincidence of the two KrF laser beams.

1.4.3 Example of encoded hologram

Several types of holographic structures have been encoded by these exposure techniques in various materials including dielectrics (SiO_2 glass, SiO_2 thin films, $LiNbO_3$, ZnO, ZrO_2, MgO, TiO_2, diamond, and carbon), semiconductors (SiC, ZnSe, and Si), Metals (WC, Pt, Au, and Al) and polymers (Polyimide, PMMA, and negative photoresist SU-8). Table 1.2 also summarizes types of periodic structures and encoded materials. Representative examples among them are discussed in detail in the next section.

Surface relief holograms.

1) **On silica glass using near-IR fs laser [87–90]**
When the intersecting beams were focused onto the surface, surface relief gratings were encoded due to material ablation. Figure 1.40 shows surface relief gratings encoded in silica glass for several θ values with a pulse width of ~ 100 fs and a total fluence of 0.3 mJ/pulse. Each grating forms a circle with a diameter of ~ 100 μm and is composed of parallel fringes with a constant spacing equal to $\lambda/[2 \sin(\theta /2)]$, where λ is the laser wavelength (800 nm). This observation provides clear evidence that the gratings were encoded as the result of interference between the two fs pulses at the fundamental wavelength (800 nm). The formation of a periodic valley structure ~ 1 μm depth is revealed by a cross-sectional FE-SEM image. This observation suggests that the grating structure results from laser ablation at this fluence level.

Infrared reflection spectra revealed that O–Si–O bond angles in the laser-irradiated area suffered photoinduced changes, leading to the densification of the silica glass. Such structural changes have been observed in silica glasses during high-energy ion implantation or neutron bombardment; densification typically saturates at $\sim 3\%$ [103]. As this compaction is accompanied by a refractive index increase of $\sim 0.7\%$, a refractive index modulation grating can be simultaneously formed beneath the surface relief grating.

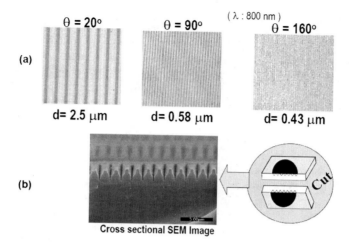

Fig. 1.40. Surface relief gratings encoded in silica glass for several θ values with pulse width of ~100 fs and total fluence of 0.3 mJ/pulse. (a) top view, and (b) cross-sectional SEM image.

2) On silica glass by UV fs laser [98,99]

When the interference UV fs pulse (290 nm) with a total energy of 20 µJ was irradiated at bulk silica glass, the micrograting structure is encoded at the surface through ablation as shown in Fig. 1.41. The colliding angle was set at 30°. The fringe spacing is in good agreement with the calculated value of 560 nm for $\lambda = 290$ nm.

The net fluence of the two laser pulses at the sample surface was 0.25 J/cm^2, which is similar to that of the 800 nm fs laser which encoded the surface grating on the silicon surface. The slight dependence of the threshold power on the laser wavelength implies that multi-photon absorption is not a major process, but avalanche ionization is responsible for the excitation of electron-hole pairs in silica glass.

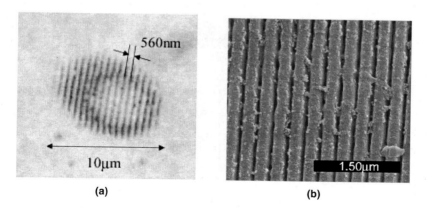

Fig. 1.41. Gratings recorded in ZnO using 260 nm fs laser: (a) total grating, and (b) expanded SEM image of grating.

The smaller spacing between grooves is expected when the colliding angle of the two pulses is increased. The scanning electron microscopic (SEM) image of the grating is shown in Fig. 1.41(b) for a colliding angle of 60°. The spacing agrees with the predicted value of 290 nm, which will be further narrowed to ∼ 150 nm with increasing angle. The deposition of debris or molten materials on the surface provides clear evidence that the grating is formed by laser ablation.

3) **On silica thin film [87,89,90]**
Surface relief gratings with valleys as shallow as 3.5 nm were recorded in thin-film silica on a silicon wafer by reducing the total fluence to 0.015 mJ. The surface profile of the fabricated grating was smooth, and neither small deposits nor macroscopic laser damage or cracking were observed at this fluence, suggesting that the grating formed resulted from the densification of the films, not from laser ablation. The valley depth of the grating (3.5 nm) relative to the film thickness (114 nm) supported this suggestion, as it agrees with the saturation level (∼ 3%) of radiation-induced densification in amorphous SiO_2. The difference in the properties of undensified and densified amorphous SiO_2 is manifested in the etching rates: the rate is markedly enhanced by stress-enhanced corrosion [104]. An AFM image and a cross section of the grating structure formed by etching with 1% HF solution (Fig. 1.42) shows that etching increases the depth of the valley by a factor of 5 ∼ 6 relative to that before etching. Chemical etching converted the refractive index modulation grating into a surface relief grating.

Embedded hologram.

1) **Full compressed pulse [86,93]**
When the focal point of the interfering fs laser pulses is positioned inside the target materials, an embedded grating is recorded through a refractive index modulation,

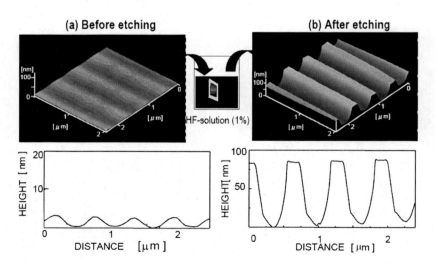

Fig. 1.42. AFM images of surface grating in silica thin films (a) Before and (b) After chemical etching using HF solution.

either due to structural alternations such as densification in silica glass, crystallographic phase changes from crystalline to amorphous states, or the formation of micropores due to the evaporation of a small amount of the material.

Such an embedded grating was recorded inside a diamond crystal [86]. Confocal microscopic images reveal the grating lies 0.45–1.05 μm beneath the surface. Raman spectra of the grating-encoded area suggest that the refractive index modulation was caused by a structural change from diamond to diamond-like carbon or amorphous carbon [105]. Similar embedded gratings in a multicomponent glass plate [93] and polymer [92] have been reported using fully compressed fs laser pulse.

2) Chirped pulse [96,97]

In spite of the successes in encoding the embedded gratings by fully compressed fs pulses (~100 fs), it is still difficult to encode embedded gratings deep inside versatile materials. The major reason for such shallow processing depth is likely that the absorption coefficient during the fs laser pulse irradiation is larger at the surface than inside the bulk of the material; therefore most of the pulse energy is lost around the surface, and the penetrating pulse is not intense enough for encoding. The deterioration of the coherence of the pulse due to the interaction between the intense pulse and material during propagation may be an additional cause.

One possible approach to overcome this difficulty is to use focal lenses with a large numerical aperture, which makes it possible to reduce the laser power density at the surface below that at the focal area inside the material, thereby keeping accumulated energy at the surface below the threshold, while enhancing that at the focal point above the threshold. This approach is successful for encoding embedded gratings for specific materials such as diamond [95] multicomponent glass [93], and polymers [92] as mentioned previously. However, it is unlikely that this approach alone provides enough controllability and flexibility to the encoding process of the embedded gratings.

An essential solution for overcoming this difficulty is to reduce the peak energy of the pulse while keeping the total energy of the pulse unchanged by expanding the pulse width. A chirped pulse, which is a partially compressed fs pulse, may meet these requirements. Before encoding embedded gratings with chirped pulses, surface relief gratings were encoded to confirm their effectiveness [96,97]. The diameters of the surface relief type gratings at the surface decrease slightly with an increase in the pulse width from 100 to 5000 fs, each pulse having a constant total energy. As the energy density at the periphery of the recorded area corresponds to the threshold for encoding, this observation implies that the threshold is governed more dominantly by the total energy than by the peak energy of the pulse in this range of pulse widths.

Figure 1.43 shows FE-SEM images of cross sections of surface relief gratings, which were recorded using 100 fs and 400 fs pulses with a total energy of 50 μJ/pulse. The 100 fs pulse encodes the grating with very shallow grooves ~1 μm deep located at the surface. On the other hand, the groove depth increases for the 400 fs pulse irradiation, indicating the pulse energy penetrates deeper with

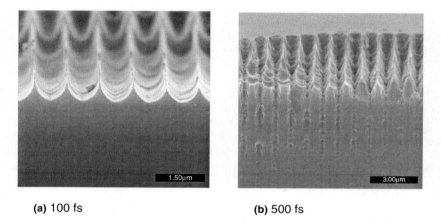

(a) 100 fs **(b)** 500 fs

Fig. 1.43. FE-SEM images of cross-sections of the surface relief gratings with different pulse: (a) 100 fs, and (b) 500 fs.

an increase in the pulse width or a decrease in the peak power. The shallow processing depth of the 100 fs pulse may be explained as follows. When the 100 fs pulse with an energy of ~ 5 TW/cm^3 is irradiated on the silica glass, it generates electron-hole pairs with a density of $\sim 10^{21}$ cm^{-3} around the surface within the pulse duration via multi photon absorption and avalanche ionization as predicated by Stuart et al. [106]. The resultant free electron absorption or dense electron plasma strongly absorbs IR light through a one-photon process, thereby dramatically enhancing the plasma density and thus preventing the penetration of the pulse into the silica by more than ~ 1 μm. On the other hand, the front part of a stretched pulse may penetrate into materials before the electron-hole pair density near the surface region becomes larger than $\sim 10^{21}$ cm^{-3}, at which point the tail of absorption due to laser-induced free electrons effectively controls the optical transmission at the wavelength of ~ 800 nm. These observations encourage the use of stretched pulses to form embedded gratings.

The cross area where the two stretched pulse beams cross was shifted stepwise inside the material in four depth layers from 30 to 1000 μm to encode the embedded grating in each layer, where a beam colliding angle of 45 degree, laser pulse width of 500 fs, and total pulse energy of 50 μJ/pulse were employed. No gratings were observed in the crossing area in the as-cut sample, but faint gratings are seen in the areas where beam enters on the surface. After chemical etching, embedded gratings appear in each layer due to stress-enhanced etching as described above (Fig. 1.44(a)). An expanded SEM image shown in Fig. 1.44(b) for the crossing area reveals periodic line grooves with a constant spacing (d) of ~ 1 μm in the grating. The spacing agrees with that given by the well known equation of $d = \lambda/[2\sin(\theta/2)]$, where λ is wavelength of the laser (800 nm) and θ is the colliding angle between the two incident pulses (45°). The faint gratings in the beam entrance areas on the surface, whose line spacing is ~ 1 μm, corresponding to that of the embedded grating, are most likely created as a result of the interference between an incident beam and a conjugated beam reconstructed from the transient grating, and formed

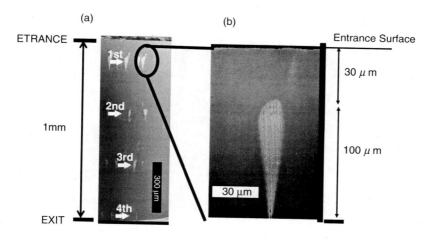

Fig. 1.44. Cross-sectional image of embedded gratings encoded in 4 layers. (a) Overall view. (b) Magnified view.

by a periodic density distribution of laser-induced excited states or electron-hole plasmas.

No noticeable difference is observed between the gratings encoded using a 'down- or up-chirped' pulse, suggesting a nonlinear optical interaction, which may degrade the coherence of the pulse, does not play a major role at least in the stretched pulse.

These results clearly indicate that the use of chirped pulses instead of fully compressed fs pulses, which results in stretching of pulse width from ∼100 to ∼5000 fs, provides controllability and flexibility to the encoding system, particularly giving rise to the capability of encoding embedded gratings in versatile materials.

Multidimensional periodic structure.

1) Double exposure technique [94]

Two-dimensional periodic structures can be formed by a double exposure technique, which is schematically shown in Fig. 1.45 [94]. First, a grating is encoded in a sample. Then, the sample is rotated 90° and the second grating is recorded exactly overlapping the first grating. Photos (a) and (b) in Fig. 1.45 show gratings with diameters of ∼15 μm and ∼50 μm encoded at laser pulse energies of 40 μJ and 100 μJ, respectively.

The encoded structure resulting from the overlap of two gratings is not always a simple superposition of two gratings, but a complicated structure as a result of interactions between the two encoding processes as shown in photo (c). Figure 1.46 shows an SEM image of a crossed grating encoded when the energy of the second pulse (80 mJ) was larger than that of the first (40 mJ) with a 45° angle between the beams for both exposures. The periodic vertical lines represent fringes of the grating encoded by the first exposure, since a periodicity of 1.0 μm equals $\lambda/[2\sin(\theta/2)]$.

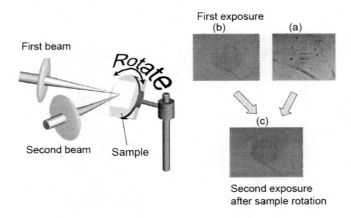

Fig. 1.45. Schematic illustration of double exposure technique.

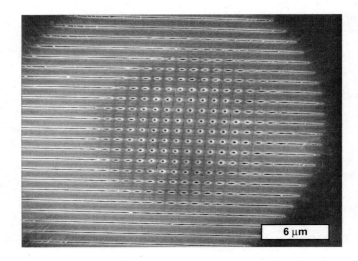

Fig. 1.46. SEM image of crossed grating encoded when energy of second pulse (80 mJ) was larger than in the first pulse (40 mJ) with a 45° between beams for both exposures.

It is noted that an array of round dots with a diameter of ∼140 nm is observed at the center of the crossed grating. The dots become ellipsoidal as one moves toward the edge of the grating and finally connect with each other to form periodic horizontal lines. Since the first grating was not encoded in the outer area and the spacing is consistent with the second exposure, the horizontal lines here are fringes of the second grating. The formation of the dot array at the central likely results from the interaction between the first grating and the incident beams in the second exposure.

When the energy of the first pulse (80 μJ/pulse) is larger than that of the second pulse (40 μJ/pulse), the resultant structure is composed of periodic narrow valleys recorded by the first exposure and a two-dimensional dot array with exfoliations and cracks around the dots (Fig. 1.47). The diameters of the dots are ∼100 nm, and

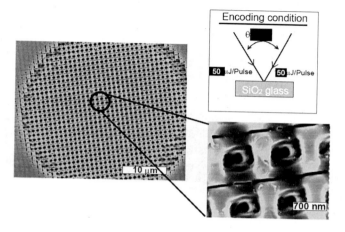

Fig. 1.47. SEM image of crossed grating encoded when energy of second pulse (50 mJ) equaled to in the first pulse with a 45° between beams for both exposures.

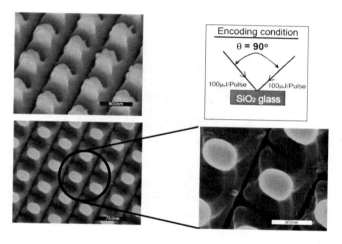

Fig. 1.48. SEM image of grating recoded with cross angle of 90°.

each dot seems to have a smaller dot inside, whose diameter becomes smaller as the pulse energy decrease. The dot array may result from the interaction between the periodic valleys and the incident beams in the second exposure due to the interference among two incident beams and the reconstructed beams from the first grating. If this were the case, encoded structures would be equivalent with those due to the interference among four beams divided from one pulse [107]. By increasing θ from 45 to 90°, an array of trapezoidal structures was produced; this array looks like a simple superposition of two orthogonal gratings (Fig. 1.48).

These results indicate that the double exposure technique using fs laser pulses offers a new tool for the formation of two-dimensional periodic nanostructures in various materials.

1.4.4 Application to optical device fabrication

As sizes of the recorded gratings (50 ~ 100 μm) are slightly smaller than that of optical devices, gratings could be used as building blocks for the optical devices. These gratings are applicable to diffracting mirrors used for optical waveguides, fiber gratings, and distributed feedback laser cavities.

Optical coupling device in waveguide. Two gratings capable of coupling and decoupling light to a waveguide were encoded on a waveguide fabricated on a $LiNbO_3$ substrate. Figure 1.49 shows that He–Ne laser light is coupled and decoupled to the waveguide through these gratings. If the grating acting as a diffraction mirror is encoded in the core region of the optical fibers, the fiber functions as a fiber grating.

Distributed feedback color center laser in LiF single crystal [108–111]. A distributed feedback (DFB) laser structure is encoded by the irradiation of a chirped fs pulse laser, and the DFB mode oscillation is achieved at room temperature by photoexcitation. An array of embedded microgratings with a diameter of ~40 μm is encoded in LiF single crystals at a depth of ~100 μm by irradiation with an interference chirped fs

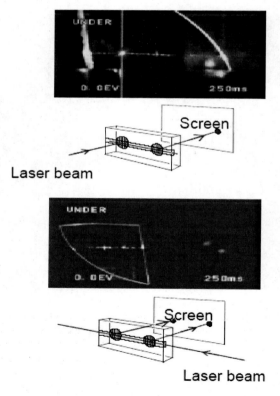

Fig. 1.49. Grating encoded in waveguide acting as optical coupler and decoupler.

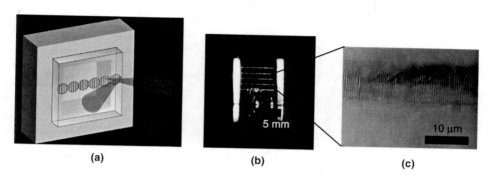

Fig. 1.50. Fabrication of DFB laser structure in LiF single crystal. (a) Schematic of encoding process (b) Encoded grating arrays, which are encoded in five layers inside LiF (c) Magnified image of portion of grating array.

laser pulse with an energy of ∼100 μJ/pulse and a pulse width of ∼500 fs (Fig. 1.50). Since the refractive index of the grating-encoded area becomes larger than that of the nonirradiated area, the grating array acts as an optical waveguide. Optical absorption and emission spectra for the fs-laser-irradiated region indicate that the irradiation of the intense fs laser pulse induces color centers F_2, F_2^+, and F_3^+. The emissions due to these centers are efficient enough for lasing. In this way, a DFB laser structure, which is an optical waveguide containing optical active centers and diffraction grating mirrors, is fabricated by simply irradiating an interfering fs pulse on a LiF crystal. The colliding angle of the laser beams is set to 102°, resulting in the fringe spacing of 510 nm, which should generate the DFB mode oscillation of 710 nm as the second mode.

To secure enough gain for DFB mode lasing, X-rays were generated from a Cu tube, operating at 30 kV and 30 mA, and were additionally irradiated for 8 hours at room temperature. A transverse pumping configuration was employed for the lasing, in which 450 nm light from an optical parametric oscillator (repetition frequency of 1 kHz, pulse duration of 10 ns, and energy of 0.8 mJ/pulse) was focused at the grating line with the aid of a cylindrical lens. The outputs of the DFB laser were measured using a conventional spectrometer with a CCD detector. Figure 1.51 shows an emission spectrum: a very narrow intense line at 707 nm is observed. In this figure, a broad emission band from X-ray-irradiated LiF without a grating is also shown for comparison. The observed oscillation wavelength agrees well with the theoretically expected value of 710 nm. The line width is less than the resolution of our measurement system (1 nm). These results clearly indicate the LiF color center laser oscillations in the DFB mode. Figure 1.52 shows a photograph that demonstrates the DFB laser oscillation. This is a miniature laser written only using a photon inside a salt crystal!

Acknowledgments

The results described here were obtained by cooperative research with many collaborators in the 'Hosono Transparent Electro-Active Materials (TEAM) Project' Exploratory

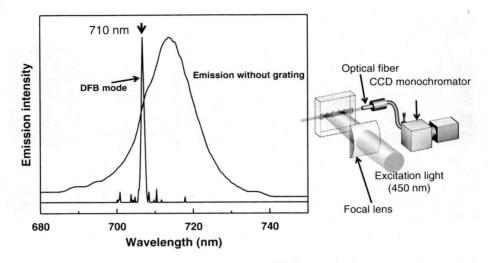

Fig. 1.51. Laser oscillation of LiF DFB laser and measuring setup.

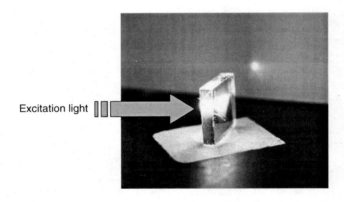

Fig. 1.52. Photograph of LiF DFB laser oscillation. Excitation conditions are: wavelength 450 nm, repetition 10 Hz, pulse width 10 ns, and excitation Intensity 0.8 mJ/pulse.

Research for Advanced Science and Technology (ERATO, Period: October, 1999–September, 2004) and ERATO–SORST (October, 2004–September, 2009) sponsored by the Japan Science and Technology Agency, Japan. Part of the research on C12A7 was supported by a Grant-in-Aid for Scientific Research (Creative Research, No.16GS0205) from MEXT, Japanese Government. The authors express their sincere thanks to Drs. Hiromichi OHTA (now with Nagoya University, Japan), Toshio KAMIYA, Masahiro ORITA (now with Hoya Corp.), Hidenori HIRAMATSU, Kenji NOMURA, Kazushige UEDA (now with Kyushu Inst. Tech. Japan), Hiroshi YANAGI, Katsuro HAYASHI, Masashi MIYAKAWA, Satoru MATSUISHI, Hayato KAMIOKA, Satoru NARUSHIMA (now with Asahi Glass Corp.), Kenichi KAWAMURA, Taisuke MIURA (now with Giga Photon Corp.), Nobuhiko SARUKURA (Institute for Molecular Research, Japan),

Peter SUSHKO (UCL, UK), Alexander SHLUGER (UCL, UK), and many graduate students of our laboratory.

References

[1] H. Kawazoe, H. Yanagi, K. Ueda and H. Hosono, MRS Bulletin **25** (2000) 28.

[2] H. Kawazoe, M. Yasukawa, H. Hyodo, M. Kurita, H. Yanagi and H. Hosono, Nature **389** (1997) 939.

[3] H. Hosono, M. Yasukawa and H. Kawazoe, J. Non-Cryst. Sol. **203** (1996) 334.

[4] N. Itoh and M. Stoneham, Materials Modification by Electronic Excitation, Cambridge: Cambridge Univ. Press, (2001).

[5] K. Ueda, T. Hase, H. Yanagi, H. Kawazoe, H. Hosono, H. Ohta, M. Orita and M. Hirano, J. Appl. Phys. **89** (2001) 1790.

[6] A. Kudo, H. Yanagi, H. Hosono and H. Kawazoe, Appl. Phys. Lett. **73** (1998) 220.

[7] H. Ohta, M. Orita, M. Hirano, I. Yagi, K. Ueda and H. Hosono, J. Appl. Phys. **91** (2002) 3074.

[8] H. Mizoguchi, M. Hirano, S. Fujitsu, T. Takeuchi, K. Ueda and H. Hosono, Appl. Phys. Lett. **80** (2002) 1207.

[9] N. Ueda, H. Hosono, R. Waseda and H. Kawazoe, Appl. Phys. Lett. **70** (1997) 3561.

[10] M. Orita, H. Ohta, M. Hirano and H. Hosono, Appl. Phys. Lett. **77** (2001) 4166.

[11] M. Orita, H. Hiramatsu, H. Ohta, M. Hirano and H. Hosono, Thin Solid Films **411** (2002) 134.

[12] H. Mizoguchi, M. Orita, M. Hirano, S. Fujitsu, T. Takeuchi and H. Hosono, Appl. Phys. Lett. **80** (2002) 4732.

[13] K. Shimakawa, S. Narushima, H. Hosono and H. Kawazoe, Philos. Mag. Lett. **79** (1999) 755.

[14] S. Narushima, M. Orita, M. Hirano and H. Hosono, Phys. Rev. B **66** (2002) 035203.

[15] M. Orita, H. Ohta, M. Hirano, S. Narushima and H. Hosono, Philos. Mag. B **81** (2001) 501.

[16] S. Narushima, H. Mizoguchi, K. Simizu, K. Ueda, H. Ohta, M. Hirano, T. Kamiya and H. Hosono, Adv. Mater. **15** (2003) 1409.

[17] H. Ohta, M. Orita, M. Hirano, H. Tanji, H. Kawazoe and H. Hosono, Appl. Phys. Lett. **76** (2000) 2740.

[18] H. Ohta, M. Orita, M. Hirano and H. Hosono, Appl. Phys. Lett. **91** (2002) 3547.

[19] H. Ohta, H. Mizoguchi, M. Hirano, S. Narushima, T. Kamiya and H. Hosono, Appl. Phys. Lett. **82** (2003) 823.

[20] H. Yanagi, K. Ueda, H. Hosono, H. Ohta, M. Orita and M. Hirano, Solid State Commun. **121** (2002) 15.

[21] H. Ohta, K. Kawamura, M. Orita, M. Hirano, N. Sarukura and H. Hosono, Appl. Phys. Lett. **77** (2000) 475.

[22] H. Ohta, M. Hirano, K. Nakahara, H. Maruta, T. Tanabe, M. Kamiya, T. Kamiya and H. Hosono, Appl. Phys. Lett. **83** (2003) 1029.

[23] H. Ohta, T. Kambayashi, M. Hirano, H. Hoshi, K. Ishikawa, H. Takezoe and H. Hosono, Adv. Mater. **15** (2003) 1258.

[24] H. Ohta, T. Kambayashi, K. Nomura, M. Hirano, K. Ishikawa, H. Takezoe and H. Hosono, Adv. Mater. **16** (2004) 312.

[25] K. Nomura, H. Ohta, K. Ueda, T. Kamiya, M. Hirano and H. Hosono, Science **300** (2003) 1269.

[26] M.W.J. Prins, S.E. Zinnemers, J.F.M. Cillessen and J.B. Giesbers, Appl. Phys. Lett. **70** (1997) 458.

[27] H. Ohta, K. Nomura, M. Orita, M. Hirano, K. Ueda, T. Suzuki, Y. Ikuhara and H. Hosono, Adv. Funct. Mater. **13** (2003) 139.

[28] D.O. Charkin, A.V. Akopyan and V.A. Dolgikh, Russ. J. Inorg. Chem. **44** (1999) 833 and references therein.

[29] H. Hiramatsu, K. Ueda, T. Kamiya, H. Ohta, M. Hirano and H. Hosono, J. Phys. Chem. B **108** (2004) 17344.

[30] K. Ueda, H. Hiramatsu, H. Ohta, M. Hirano, T. Kamiya and H. Hosono, Phys. Rev. B **69** (2004) 155305.

[31] H. Hiramatsu, K. Ueda, H. Ohta, M. Orita, M. Hirano and H. Hosono, Appl. Phys. Lett. **81** (2002) 598.

[32] H. Hiramatsu, K. Ueda, K. Takafuji, H. Ohta, M. Hirano, T. Kamiya and H. Hosono, J. Mater. Res. **19** (2004) 2137.

[33] H. Hiramatsu, H. Ohta, T. Suzuki, C. Honjo, Y. Ikuhara, K. Ueda, T. Kamiya, M. Hirano and H. Hosono, Cryst. Growth Des. **4** (2004) 301. K. Nomura, H. Ohta, T. Suzuki, C. Honjo, K. Ueda, T. Kamiya, M. Orita, Y. Ikuhara, M. Hirano and H. Hosono, J. Appl. Phys. **95** (2004) 5532.

[34] H. Hiramatsu, K. Ueda, H. Ohta, M. Hirano, T. Kamiya and H. Hosono, Appl. Phys. Lett. **82** (2003) 1048.

[35] K. Ueda, S. Inoue, S. Hirose, H. Kawazoe and H. Hosono, Appl. Phys. Lett. **77** (2000) 2701.

[36] K. Ueda, S. Inoue, H. Hosono, N. Sarukura and M. Hirano, Appl. Phys. Lett. **78** (2001) 2333.

[37] K. Ueda and H. Hosono, J. Appl. Phys. **91**, 4768 (2002). K. Ueda, K. Takafuji, H. Hiramatsu, H. Ohta, T. Kamiya, M. Hirano and H. Hosono, Chem. Mater. **15** (2003) 3692.

[38] H. Hiramatsu, K. Ueda, K. Takafuji, H. Ohta, M. Hirano, T. Kamiya and H. Hosono, J. Appl. Phys. **94** (2003) 5805.

[39] H. Hiramatsu, K. Ueda, K. Talafuji, H. Ohta, M. Hirano, T. Kamiya and H. Hosono, Appl. Phys. **A79** (2004) 1521.

[40] H. Kamioka, H. Hiramatsu, H. Ohta, K. Ueda, M. Hirano, T. Kamiya and H. Hosono, Appl. Phys. Lett. **84** (2004) 879.

[41] H. Kamioka, H. Hiramatsu, K. Ueda, M. Hirano, T. Kamiya and H. Hosono, Opt. Lett. **29** (2004) 1659.

[42] H.B. Bartl and T. Scheller, Neues Jahrb. Mineral Monatsh **35** (1970) 547.

[43] R.W. Nurse, J.H. Welch and A. Majumdar. Trans. Br. Ceram. Soc. **64** (1965) 323.

[44] H. Hosono and Y. Abe, Inorg. Chem. **26** (1987) 1192.

[45] M. Lacerda, J.T.S. Irvine, F.P. Glasser and A.R. West, Nature **332** (1988) 525.

[46] H. Hosono, K. Yamazaki and Y. Abe, J. Am. Ceram. Soc. **70** (1987) 867.

[47] H. Hosono, N. Asada and Y. Abe, J. Appl. Phys. **67** (1990) 2840.

[48] H. Hosono, Int. J. Appl. Ceram. Tech. **5** (2004) 409.

[49] K. Hayashi, M. Hirano and H. Hosono, J. Mater. Res. **17** (2002) 1244.

[50] S. Watauchi, I. Tanaka, K. Hayashi, M. Hirano and H. Hosono, J. Cryst. Growth **237** (2002) 496.

[51] Y. Toda, M. Miyakawa, K. Hayashi, T. Kamiya, M. Hirano and H. Hosono, Thin Solid Films **445** (2003) 309.

[52] K. Hayashi, M. Hirano, S. Matsuishi and H. Hosono, J. Am. Chem. Soc. **124** (2002) 736.

[53] K. Hayashi, S. Matsuishi, N. Ueda, M. Hirano and H. Hosono, Chem. Mater. **15** (2003) 1851.

[54] S. Yang, J.N. Kondo, K. Hayashi, M. Hirano, K. Domen and H. Hosono, Chem. Mater. **16** (2003) 104.

[55] K. Hayashi, S. Matsuishi, M. Hirano and H. Hosono, J. Phys. Chem. B **108** (2004) 8920.

[56] K. Hayashi, N. Ueda, M. Hirano and H. Hosono, Solid State Ionics **179** (2004) 89.

[57] S. Matsuishi, K. Hayashi, M. Hirano, I. Tanaka and H. Hosono, J. Phys. Chem. B **108** (2004) 8920.

[58] S. Yang, J.N. Kondo, K. Hayashi, M. Hirano, K. Domen and H. Hosono, Appl. Catal. A **277** (2004) 234.

[59] Q. Li, K. Hayashi, M. Nishioka, H. Kashiwagi, M. Hirano, Y. Torimoto, H. Hosono and M. Sadakata: Appl. Phys. Lett. **80** (2002) 4259.

[60] K. Hayashi, M. Hirano, Q. Li, M. Nishioka, M. Sadakata, Y. Torimoto, S. Matsuishi and H. Hosono, Electorochem. Solid State Lett. **5** (2002) J13.

[61] Q. Li, K. Hayashi, M. Nishioka, H. Kashiwagi, M. Hirano, Y. Torimoto, H. Hosono and M. Sadakata, Jpn. J. Appl. Phys. **41** (2002) L530.

[62] Q. Li, H. Hosono, M. Hirano, K. Hayashi, M. Nishioka, H. Kashiwagi, Y. Torimoto and M. Sadakata, Surf. Sci. **527** (2003) 100.

[63] K. Hayashi, S. Matsuishi, T. Kamiya, M. Hirano and H. Hosono, Nature **419** (2002) 462.

[64] K. Hayashi, Y. Toda, T. Kamiya, M. Hirano, I. Tanaka, T. Yamamoto and H. Hosono, Appl. Phys. Lett. **86** (2005) 22109.

[65] P.V. Sushko, A.L. Shluger, K. Hayashi, M. Hirano and H. Hosono, Phys. Rev. Lett. **91** (2003) 126401.

[66] P.V. Sushko, A.L. Shluger, K. Hayashi, M. Hirano and H. Hosono, Thin Solid Films **445** (2003) 161.

[67] P.V. Sushko, A.L. Shluger, K. Hayashi, M. Hirano and H. Hosono, Appl. Phys. Lett. **86** (2005) 92101.

[68] S. Matsuishi, K. Hayashi, M. Hirano and H. Hosono, J. Am. Chem. Soc. **127** (2005) 12454.

[69] L. Dye, Inorg. Chem. **36** (1997) 3817.

[70] S. Matsuishi, Y. Toda, M. Miyakawa, K. Hayashi, M. Hirano, I. Tanaka and H. Hosono, Science **301** (2003) 626.

[71] S.-W. Kim, M. Miyakawa, K. Hayashi, M. Hirano and H. Hosono, J. Am. Chem. Soc. **127** (2005) 1370.

[72] M. Miyakawa, Y. Toda, K. Hayashi, T. Kamiya, M. Hirano, N. Matsunami and H. Hosono, J. Appl. Phys. **97** (2005) 23510.

[73] Y. Toda, S. Matsuishi, K. Hayashi, K. Ueda, T. Kamiya, M. Hirano and H. Hosono, Adv. Mater. **16** (2004) 685.

[74] Y. Toda, S-W Kim, K. Hayashi, M. Hirano, T. Kamiya. H. Hosono, T. Haraguchi and H. Yasuda, Appl. Phys. Lett. **87** (2005) 254103.

[75] M. Miyakawa, K. Hayashi, M. Hirano, Y. Toda, T. Kamiya and H. Hosono, Adv. Mater. **15** (2003) 1100.

[76] E.E.B Campbell, D. Ashkenasi and A. Rosenfeld, Mater. Sci. Forum. **301** (1999) 123.

[77] J. Kruger and W. Kautek, Appl. Surf. Sci. **96–98** (1996) 430.

[78] K.H. Davis, K. Miura, N. Sugimoto and K. Hirao, Opt. Lett. **21** (1996) 1729.

[79] D. von der Linde, K. Sokolowski-Tinten and J. Bialkowski, Appl. Surf. Sci. **109–111** (1997) 1.

[80] K. Hirao, T. Mitsu, J. Si and J. Qui, Active Glass for Photonic Devices Springer Series in Photonics. (2002) P47.

[81] B.C. Stuart, M.D. Feit, A.M. Rubenchik, B.W. Shore and M.D. Perry, Phys. Rev. Lett. **74** (1995) 2248.

[82] E.P. Ippen and C.V. Shank, Ultrashort Light Pulses, Ch.3, Ed. by S.L. Shapiro, New York: Springer-Verlag, 1997.

[83] H.M. Phillips and R.L. Sauerbrey, Optical Engineering **32** (1993) 2424.

[84] A.M. Ozkan, A.P. Malshe, T.A. Railkar, W.D. Brown, M.D. Shirk and P.A. Molian, Appl. Phys. Lett. **75** (1999) 3716.

[85] K. Kawamura, N. Sarukura, M. Hirano and H. Hosono, Appl. Phys. B **71** (2000) 119.

[86] K. Kawamura, N. Sarukura, M. Hirano and H. Hosono, Jpn. J. Appl. Phys. **39** (2000) L767.

[87] K. Kawamura, N. Sarukura, M. Hirano and H. Hosono, J. Appl. Phys. Lett. **78** (2001) 1038.

[88] H. Hosono, K. Kawamura, S. Matsuishi and M. Hirano, Nucl. Instrum. Methods in Phys. Res. B **191** (2002) 89.

[89] K. Kawamura, N. Motomitsu, M. Hirano and H. Hosono, J. Appl. Phys. **41** (2002) 4400.

[90] M. Hirano, K. Kawamura and H. Hosono, Applied Surface Science **197–198** (2002) 688.

[91] T. Kondo, S. Matsuo, S. Juodkazis and H. Misawa, Appl. Phys. Lett. **79** (2001) 725.

[92] J. Si, J. Qiu, J. Zhai, Y. Shan and K. Hirao, Appl. Phy. Lett. **80** (2002) 359.

[93] Y. Li, W. Watanabe, K. Yamada, T. Shinagawa, K. Itoh, J. Nishii and Y. Jiang, Appl. Phys. Lett. **80** (2002) 1508.

[94] K. Kawamura, N. Sarukura, M. Hirano and H. Hosono, Appl. Phys. Lett. **79** (2001) 1228.

[95] K. Kawamura, N. Ito, N. Sarukura, M. Hirano and H. Hosono, Rev. Sci. Instr. **73** (2002) 1711.

[96] M. Hirano, K. Kawamura and H. Hosono, Proc. of SPIE **5061** (2002) 89.

[97] K. Kawamura, M. Hirano, T. Kamiya and H. Hosono, Appl. Phys. Lett. **81** (2002) 1137.

[98] H. Kamioka, K. Kawamura, T. Miura, M. Hirano and H. Hosono, Proc. of SPIE **4760** (2002) 994.

[99] H. Kamioka, K. Kawamura, T. Miura, M. Hirano and H. Hosono, J. Nanoscience and Nanotechnology **2** (2002) 321.

[100] S. Backus, J. Peatross, Z. Zeek, A. Rundquist, G. Taft, M.M. Murnane and H.C. Kapteyn, Opt. Lett. **21** (1996) 665.

[101] C.W. Siders, N.C. Turner III, M.C. Downer, A. Babine, A. Stepanov and A.M. Sergeev, J. Opt. Soc. Am. B **12** (1996) 330.

[102] G. Marcus, A. Zigler and Z. Henis, J. Opt. Soc. Am. B **16** (1999) 792.

[103] S.K. Sharma, D.W. Matson, J.A. Philpotts and T.L. Roush, J. Non-Cryst. Solids **68** (1984) 99.

[104] F.N. Schwettman, D.J. Dexter and D.F. Cole, J. Electrochem. Soc. **120** (1973) 1566.

[105] J. Roberson. Adv. Phys. **35** (1986) 317.

[106] B.C. Stuart, M.D. Feit, S. Herman, A.M. Rubenchik, B.W. Shore and M.D. Perry, J. Opt. Soc. Am B **13** (1996) 459.

[107] M. Campbell, D.N. Sharp, M. Harrison, R.G. Denning and A.J. Turberfield, Nature **404** (2000) 53.

[108] K. Kawamura, M. Hirano, T. Kurobori, D. Takamizu, T. Kamiya and H. Hosono, Appl. Phys. Lett. **84** (2004) 311.

[109] T. Hurobori, T. Kitano, Y. Hirose, K. Kawamura, D. Takamizu, M. Hirano and H. Hosono, Radiation Measurements **38** (2004) 759.

[110] T. Kurobori, Y. Hirose, K. Kawamura, M. Hirano and Hosono, Physica Status Solidi (c) **2** (2005) 637.

[111] K. Kawamura, D. Takamizu, T. Kurobori, T. Kamiya, M. Hirano and H. Hosono, Nuc. Instr. and Met. in Phys. Res. B **218** (2004) 332.

CHAPTER 2

The Role of Lattice Defects in Oxides

Mitsuru Itoh

Abstract

Recent topics related to oxide materials are reviewed, focusing on the effect of defects in oxide materials mainly on perovskite-type oxides, except for strongly correlated systems. Spin crossover in oxides is discussed mainly with respect to Co^{3+} with d^6 configuration. Demazeau and Hagenmüller's pioneering work has pointed out the importance of the crystal field strength at the B ion site of perovskite. Competing colinear chemical bonds along directions perpendicular to the central transition metal ion breaks the local symmetry and consequently changes the crystal field strength. Following the discussion of previous results, recent studies on the spin state of Co^{3+} ions in $LaCoO_3$ are explained, based on our experiments. Spin crossover in $Pr_{0.5}Ca_{0.5}CoO_3$ is also explained. Next, the effect of lattice defects including compositional disorder is discussed for ferroelectric oxides. A discussion of the formation of dipolar entities due to point defects and compositional disorder in oxides is given, using $SrTiO_3$ as the example. In the final section, direct correlation of lattice defects with ionic conductivity is discussed for lithium ion conductors.

Keywords: **positional disorder, point defect, oxygen vacancy, perovskite, ferroelectricity, ferromagnetism, spin crossover.**

2.1 Introduction

Lattice defects in oxides, defined as lattice imperfections including point defects (cation vacancies, anion vacancies, interstitials, and compositional disorder), edge and screw dislocations, stacking faults, and so on, play an important role in the evolution of physical properties such as ferromagnetism, ferroelectricity, and ionic conductivity.

In the field of ferroelectrics and ionic conductors, compositional disorder can be a strong random source of ions because the dipolar field due to a dipolar entity cannot be screened by the conducting electrons. In these materials, the local ion arrangement as well as the lattice size can become a predominant factor for the evolution of high performance properties. As a starting point, we need to know the magnitude of the dipole interaction in magnetic and dielectric materials. Suppose that positive and negative unit charges, $e = 1.6 \times 10^{-19}$ C, are separated by 1×10^{-10} m, being the definition of dipole moment μ of 1 debye. Two dipoles with 1 debye, separated by 5 Å (5×10^{-10} m), have an energy $U = 3.7 \times 10^{-20}$ J, corresponding to 2.7×10^3 K. In the same situation, magnetic dipoles with $\mu = 1$ Bohr separated by 5 Å (5×10^{-10} m), have an energy $U \sim 1.0 \times 10^{-2}$ K. This value is small compared to the exchange interaction in magnetic materials. This means that dipolar interaction can act over a long range in the crystal. In this case, defects in insulating oxides can be a strong dipolar center, modifying the total properties of that crystal.

In this review, we focus on the effect of lattice defects of the properties of oxide materials. This review covers the following:

(1) Magnetic materials.

(2) Ferroelectric materials.

(3) Ion conducting materials.

(4) Concluding remarks.

(5) Clue to the design of new functional materials.

2.2 Magnetic Materials (Spin Crossover in Oxides)

Photo-induced spin-crossover is an important field for practical applications, especially for memory devices, although the transition temperature of a usual material showing a spin crossover is below 100 K [1]. Compared with complexes, spin crossover in oxides is not familiar. This is because the oxide lattice is too hard to allow the spin crossover of ions which usually accompanies a large volume change. Control of the spin state of transition metal ions is an art in solid state chemistry. Control of the spin state also implies control of the crystal field. In this research field, many systematic studies have been reported for perovskite-type oxides, mainly by Demazeau and Hagenmüller's group [2, 3].

Tanabe–Sugano diagrams [4] give the relative energy of vs. Dq/B of different spectroscopic terms corresponding to the electronic population d^n in O_h symmetry. In the case of lower symmetry, e.g., D_{4h}, a diagram giving the stability domains of different electronic configurations can be drawn [5], based on the following criteria: (a) the energy of the different terms varies with Dq; (b) the crystal field parameter Dq is directly proportional to $(d_{c-a})^{-5}$, d_{c-a} being the cation–anion distance; (c) the average of d_{c-a}, i.e., $\bar{d}_{c-a} = (2/3)d_{c-a(xy)} + (1/3)d_{c-a(z)}$ is constant; (d) $\overline{Dq}$ is equal to $(2/3)Dq_{xy} + (1/3)Dq_z$; (e) spin-orbit coupling is neglected; and (f) the distortion parameter θ of the MO_6 octahedron is defined as $\theta = d_{M-O(z)}/d_{M-O(xy)}$.

Elongation of the MO_6 coordination octahedron can induce such specific $3d$ configurations as high-spin $Fe^{4+}(t_{2g}^3 d_{z^2}^1)(^5A_{1g})$, low-spin $Co^{4+}(d_{yz}^2 d_{zx}^2 d_{x0}^1)(^2B_{2g})$,

Co^{3+} with intermediate electronic configuration $(d_{yz}^2 d_{zx}^2 d_{xy}^1 d_{z^2}^1 d_{x^2-y^2}^0)(^3B_{2g})$, low-spin $Ni^{3+}(t_{2g}^6 d_{z^2}^1)(^2A_{1g})$, or low spin $Cu^{3+}(t_{2g}^6 d_{z^2}^2 d_{x^2-y^2}^0)(^1A_{1g})$.

The energy gain in $\overline{Dq}/B$ units $\eta(\theta) = (E/B(\theta) - E/B(1))/(\overline{Dq}/B)$ resulting from the stabilizing effect of such an elongation appears in Fig. 2.1. To check this model, many of the oxides in Table 2.1 were synthesized and subjected to magnetic measurements.

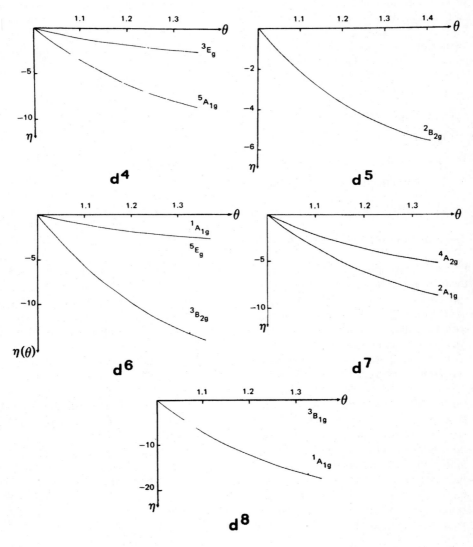

Fig. 2.1. Stabilizing effect of the D_{4h} elongation of terms of the d^4, d^5, d^6, d^7, and d^8 configuration [5].

Table 2.1.

Structure	Phase	Valence of Metal	Ref.
perovskite	La_2LiFeO_4	Fe^{5+}	6
K_2NiF_4	$Ca_{0.5}La_{1.5}Li_{0.5}Fe_{0.5}O_4$	Fe^{4+}	7
K_2NiF_4	$Sr_{0.5}La_{1.5}Li_{0.5}Fe_{0.5}O_4$	Fe^{4+}	7
K_2NiF_4	$Ba_{0.5}La_{1.5}Li_{0.5}Fe_{0.5}O_4$	Fe^{4+}	7
K_2NiF_4	$SrLaMg_{0.5}Fe_{0.5}O_4$	Fe^{4+}	8, 9
K_2NiF_4	$SrLaZn_{0.5}Fe_{0.5}O_4$	Fe^{4+}	8, 9
K_2NiF_4	$Ca_{0.5}La_{1.5}Li_{0.5}Fe_{0.5}O_4$	Fe^{4+}	8, 9
K_2NiF_4	$Sr_{0.5}La_{1.5}Li_{0.5}Fe_{0.5}O_4$	Fe^{4+}	8, 9
K_2NiF_4	$Ba_{0.5}La_{1.5}Li_{0.5}Fe_{0.5}O_4$	Fe^{4+}	8, 9
perovskite	$La_2(Li_{0.9}Fe_{0.5})O_{5.95}$	Fe^{4+}	10
K_2NiF_4	$SrLaFeO_4$	Fe^{3+}	8, 9
K_2NiF_4	$Sr_{0.5}La_{0.5}Li_{0.5}Co_{0.5}O_4$	Co^{4+}	7
K_2NiF_4	$La_2Li_{0.5}Co_{0.5}O_4$	Co^{3+}	7, 12
K_2NiF_4	$Ca_{0.5}La_{1.5}Mg_{0.5}Co_{0.5}O_4$	Co^{3+}	2, 11
K_2NiF_4	$Sr_{0.5}La_{1.5}Mg_{0.5}Co_{0.5}O_4$	Co^{3+}	2, 11
K_2NiF_4	$Ba_{0.5}La_{1.5}Mg_{0.5}Co_{0.5}O_4$	Co^{3+}	2, 11
K_2NiF_4	$SrLaAl_{0.5}Co_{0.5}O_4$	Co^{3+}	3
K_2NiF_4	$SrLaGa_{0.5}Co_{0.5}O_4$	Co^{3+}	3
K_2NiF_4	$Sr_{1.5}La_{0.5}Ti_{0.5}Co_{0.5}O_4$	Co^{3+}	3
K_2NiF_4	$Sr_{0.5}La_{1.5}Zn_{0.5}Co_{0.5}O_4$	Co^{3+}	3
K_2NiF_4	$La_2Li_{0.5}Ni_{0.5}O_4$	Ni^{3+}	7, 12, 14
K_2NiF_4	$Sr_{0.5}La_{1.5}Mg_{0.5}Ni_{0.5}O_4$	Ni^{3+}	13
K_2NiF_4	$Ca_{0.5}La_{1.5}Mg_{0.5}Ni_{0.5}O_4$	Ni^{3+}	14
K_2NiF_4	$Ba_{0.5}La_{1.5}Mg_{0.5}Ni_{0.5}O_4$	Ni^{3+}	14
perovskite	$LaNiO_3$	Ni^{3+}	15
K_2NiF_4	$La_2Li_{0.5}Cu_{0.5}O_4$	Cu^{3+}	7, 12
perovskite	$LaCuO_3$	Cu^{3+}	15

Most of the compounds are of the K_2NiF_4 structure, and some show a spin crossover from low to high spin with a second order character.

Figure 2.2 shows the ground state regions for d^4, d^5, d^6, d^7, and d^8 configurations vs. θ [4]. Using these figures, we can tell the spin state through the values of Dq and θ. The K_2NiF_4 structure is convenient for the design of an anisotropic environment for the transition metal ions along the c- and a-axis.

Figure 2.3 shows the competing bonds in the K_2NiF_4 structure by forming a 1:1 ordered arrangement in the BO_2 plane [13]. Such an ordered structure in the K_2NiF_4 structure enables one to realize an ideal two-dimensional structure with alternating competing B′–O and B″–O bonds in the c-plane. Figure 2.4 shows an example of Fe(V) ions [6]. Super-superexchange among Fe(V) is possible via the O2p π orbitals. We can tell the sign and magnitude of the magnetic interaction by the so-called Goodenough's rule.

In the K_2NiF_4 structure, the anisotropic environment around the transition metal ion is enhanced by the formation of a 1:1 ordered structure. In this case, competing A–O

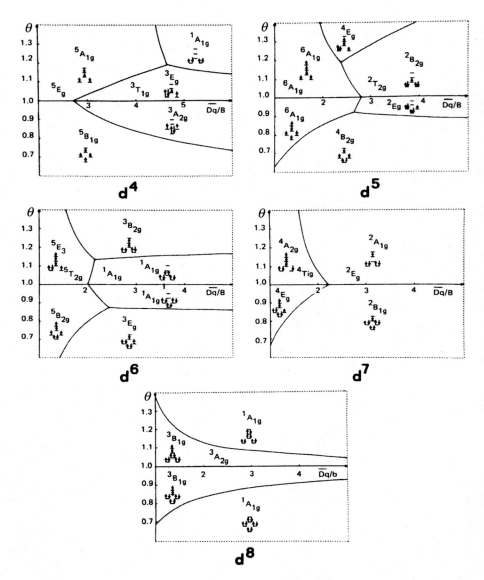

Fig. 2.2. Ground state regions for d^4, d^5, d^6, d^7 and d^8 configuration [5].

and B–O bonds and competing B′–O and B″–O bonds can create a strong anisotropic crystal field favorable to the stabilization of higher spin states. Work by Demazeau and Hagenmüller is a milestone of transition chemistry, and has succeeded in establishing quantitative control of the spin states of transition metal ions in insulating oxides.

LaCoO$_3$ is a well-known material which shows a spin state transition of Co^{3+} from low spin $^1A_1$ (t_{2g}^6) to high spin $^5T_2$ ($t_{2g}^4 e_g^2$) or an intermediate state ($t_{2g}^5 e_g^1$). Discussion of the

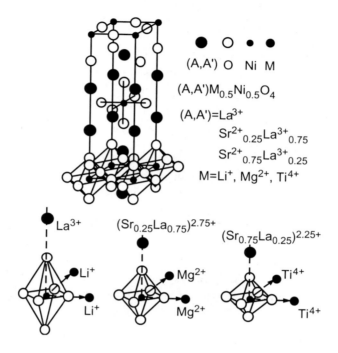

Fig. 2.3. Comparison of the distortions of the NiO$_6$ octahedron according to the nature of the chemical bonds in various Ni (III) oxides [13].

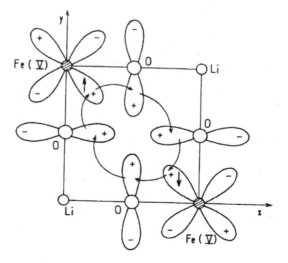

Fig. 2.4. Magnetic superexchange couplings between iron (V) t_{2g} orbitals involving two nearest neighboring oxygen atoms [6].

high spin or intermediate spin in the excited state has continued for about fifty years, but a convincing explanation has not been achieved yet.

Recently Kyomen et al. [16] have introduced a new concept of 'negative cooperative effect' for Co^{3+} in $LaCoO_3$. This is an idea that explains why a phase transition does not occur on heating. Figure 2.5 shows the magnetic susceptibility, heat capacity, net excitation energy, and fraction of Co ions in the high-spin-state.

The magnetic susceptibility and heat capacity due to the spin-state transition in $LaCoO_3$ was calculated by a molecular–field model in which the energy level diagram of the high-spin-state reported by Ropka and Radwanski [17] was assumed for the excited state, and the energy and entropy of mixing of high-spin Co ions and low-spin Co ions was introduced phenomenologically. The experimental data below 300 K were well reproduced by this model, which proposes that the high-spin excited state can be populated even if the energy of the high-spin state is much larger than that of the low-spin state, because the negatively large energy of mixing reduces the net excitation energy.

Figure 2.6 shows the proposed arrangement of excited Co ions in the low-spin matrix of $LaCoO_3$. Figure 2.6(a) shows that the high-spin Co ions are located as far apart as possible. Figure 2.6(b) shows how the intermediate-spin Co ions collect into domains.

The energy of mixing is negatively large, which reduces the total energy and thus the net excitation energy. The reduction of covalent energy of the Co–O bonds around the excited high-spin Co ion is proposed for the origin of the negative energy of mixing according to the Goodenough's suggestion. The negative energy of mixing indicates that repulsive interaction acts between the Co ions in the excited state and thus predicts that there is no domain or short-range order composed of high-spin Co ions such as those reported in spin-crossover complexes.

This result is important from the viewpoint of material design using functional oxides. In the case of Co^{3+}, low-spin Co^{3+} with t_{2g}^6 configuration has a stronger covalency with the surrounding oxygen ions. On the contrary, the high-spin or intermediate states have weaker covalency, reflected in the ionic size of Co^{3+} ions. There are a few oxides which show a spin crossover from low-spin to high-spin state. This is because the elastic strain induced by the appearance of a new phase during the transition is quite large. In oxide materials, the crossover is not known in stoichiometric systems. When an oxygen vacancy is introduced in such a way that an anisotropic structure is formed, spin crossover is allowed. This is the case in oxygen-deficient perovskite [18].

Recently, Tsubouchi et al. [19] have reported that stoichiometric $Pr_{0.5}Ca_{0.5}CoO_3$ shows a spin crossover of Co^{3+}, as shown in Fig. 2.7. In $Pr_{0.5}Ca_{0.5}CoO_3$, the valence of the cobalt ions at $T > 90$ K is considered to be 3.5 on average, so a 1:1 mixture of low-spin $Co^{4+}(t_{2g}^5)$ and intermediate spin $Co^{3+}(t_{2g}^5 e_g^1)$ is postulated which was confirmed by a configurational entropy calculation. Below 90 K, $Pr_{0.5}Ca_{0.5}CoO_3$ shows a spin-crossover from intermediate to low-spin-state (t_{2g}^6). When both Co^{3+} and Co^{4+} ions take a low-spin-state, $Pr_{0.5}Ca_{0.5}CoO_3$ shows a ferromagnetic transition below 5 K. As mentioned above, the spin state transition accompanies a large volume change in

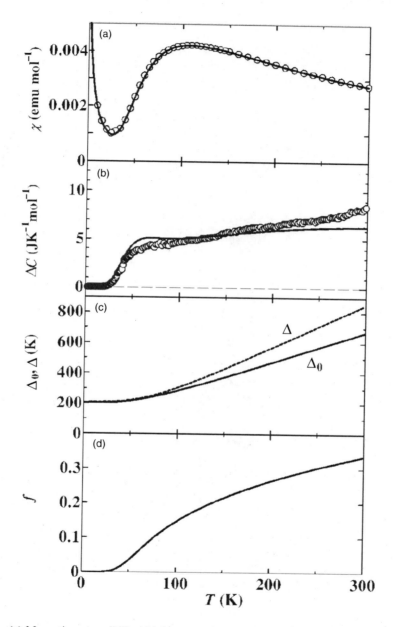

Fig. 2.5. (a) Magnetic susceptibility, (b) heat capacity, (c) net excitation energy, and (d) fraction of Co ions in the high-spin excited state: Circles and lines represent the experimental data and the calculated curves, respectively [16].

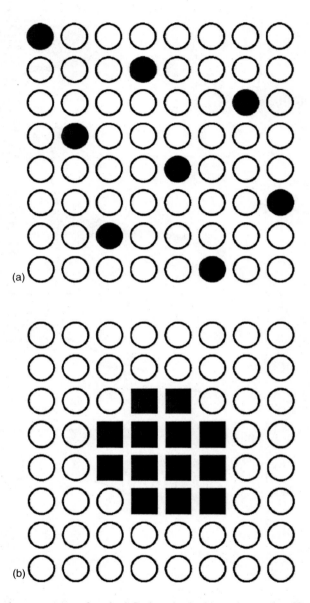

Fig. 2.6. Proposed arrangement of excited Co ions in the low-spin matrix of $LaCoO_3$. (a) High-spin Co ions are located as far apart as possible. (b) Intermediate-spin Co ions are corrected to form domains. Open circles, solid circles, and solid squares indicate low-spin, high-spin, and intermediate-spin Co ions, respectively [16].

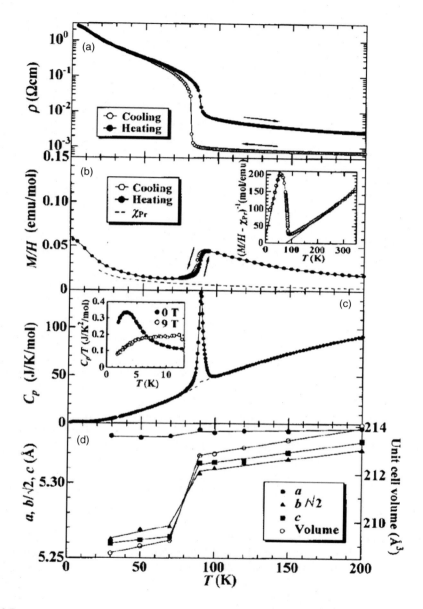

Fig. 2.7. (a) Resistivities, (b) DC magnetization, (c) heat capacities, and (d) lattice constants and unit cell volume of $Pr_{0.5}Ca_{0.5}CoO_3$. The inset of (b) shows the inverse of contribution to the magnetic susceptibility from Co atoms obtained by subtracting the Pr contributions [dashed line in (b)]. The solid lines in the inset indicate the results of fitting according to the Curie–Weiss law. The inset of (c) shows the C_p/T it vs. T plot in magnetic fields of 0 and 9 tesla [19].

the lattice. $Pr_{0.5}Ca_{0.5}CoO_3$ shows a 1.5% volume change at room temperature, which introduces cracks in the crystal. At the present time, only $Pr_{0.5}Ca_{0.5}CoO_3$ shows a spin state transition, of all the compounds in the $Ln_{0.5}A_{0.5}CoO_3$ (Ln = La, Pr, Nd, Sm; A = Ca, Sr, Ba) system. We should consider whether the Pr–O bond plays a role in the evolution of the spin state transition of Co^{3+}.

Further chemical design to optimize the chemical bond strength of the B–O bond via control of the A–O bond may enhance the transition temperature to above room temperature.

2.3 Ferroelectric Materials

In the field of ferroelectrics, the development of new material has almost ceased and studies are now focused on the fabrication of ferroelectric memories, the mechanism of the evolution of the ferroelectricity including first principle calculations, the piezoelectric response of ferroelectrics near the morphotropic phase boundary, and random field induced domain response in relaxor ferroelectrics. Dielectric impurities such as cation vacancies, oxygen vacancies, substitutional impurity, and compositional disorder can lead to the formation of dipolar entities. In this section, we will try to explain the role of dipolar impurities in the evolution of ferroelectricity in oxides.

2.3.1 Compositionally disordered perovskites [20]

Random lattice disorder produced by chemical substitution in ABO_3 perovskites can lead to the formation of dipolar impurities and defects that have a profound influence on the static and dynamic properties of these materials. In highly polarizable host lattices, dipolar entities form polar nanodomains whose size is determined by the dipolar correlation length, r_c, of the host and that exhibit dielectric relaxation in the applied field. In the very dilute limit, each domain behaves as a non-interacting dipolar entity with a single relaxation time. At a higher concentration of disorder, however, the domains can interact, leading to more complex relaxation behavior. A relaxor state or an ordered ferroelectric (FE) state appears for a sufficiently high concentration of overlapping domains.

After a brief discussion of the physics of random-site electric dipoles in dielectrics, this review explains the simplest cases, namely the relaxational properties of substitutional impurities in the quantum paraelectric $SrTiO_3$.

Figure 2.8 shows the possible origin of the dipolar entities produced by chemical substitution and oxygen vacancies in $KTaO_3$ perovskite [20]. Random lattice disorder produced by chemical substitution of ABO_3 perovskites can lead to the formation of dipolar impurities and defects gives a serious influence on the static and dynamic properties of the perovskites. Depending on the polarizability of the ABO_3 host lattice associated with the soft FE mode, dipolar entities polarize regions around them. The dipolar entities form polar nano/micro domains whose size is determined by the temperature-dependent correlation length, r_c, of the host. When the dipolar entities possess more than one equivalent

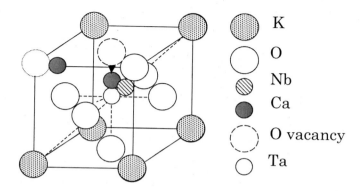

Fig. 2.8. The cubic perovskite lattice showing the location of some substituents and vacancies [20].

orientation, they may undergo dielectric relaxation in an applied AC field. In the very dilute limit (typically <0.1 at%), each polar domain behaves as a non interacting dipolar entity with a single relaxation time. At higher concentrations of dipolar entities, the polar domains can interact each other, in this case, more complex relaxation behavior should appear accompanying a distribution of relaxation times. Even with equivalent substitutions such as Na for K, Li for K, Nb for Ta, off-center dipoles can be accommodated in the lattice. This is also the case for the substitution of Ca for Sr in $SrTiO_3$.

In an ideal cubic lattice of Fig. 2.8, the disorder-induced dipolar entities tend to have several equivalent orientations. Where the interaction among dipoles can be neglected, the relaxation is expected to be Debye-like with a single dipolar relaxation time τ.

The total dielectric response of the system in an AC field is described by the complex dielectric function.

$$\varepsilon = \varepsilon' - i\varepsilon'' \quad (2.1)$$

Using the Debye approximation, the real part of ε' and imaginary part of ε'' are given by the expression

$$\varepsilon'(\omega) = \varepsilon'_\infty + \frac{\varepsilon'_s - \varepsilon'_\infty}{1 + \omega^2 \tau^2} \quad (2.2)$$

and

$$\varepsilon''(\omega) = (\varepsilon'_s - \varepsilon'_\infty) \frac{\omega \tau}{1 + \omega^2 \tau^2} \quad (2.3)$$

where ω is the angular frequency of the applied field, ε'_∞ is the high frequency dielectric constant, and ε_s is the static dielectric constant. ε' and ε'' are related by Kramers–Kronig dispersion relations. The dielectric loss is proportional to ε'' and exhibits a maximum for

$\omega\tau = 1$. The dissipation factor, tan δ, is given by

$$\tan \delta = \varepsilon''(\omega)/\varepsilon'(\omega) = \frac{\left(\varepsilon'_s - \varepsilon'_\infty\right)\omega\tau}{\varepsilon'_s + \varepsilon'_\infty \omega^2\tau^2} \tag{2.4}$$

For dielectrics with dilute dipolar impurities, $\varepsilon'_s \approx \varepsilon'_\infty$ the Eq. 2.4 reduces to

$$\tan \delta = \left(\frac{\Delta\varepsilon'}{\varepsilon'_s}\right) \frac{\omega\tau}{1 + \omega^2\tau^2} \tag{2.5}$$

Hopping among the thermally equivalent positions gives an Arrhenius law

$$\tau^{-1} = \tau_0^{-1} \exp\left(-E/kT\right) \tag{2.6}$$

where τ_0 is the reciprocal of the attempt frequency, ω_0, and E the activation energy.

Dipolar interactions are assumed for two extreme cases in an ordinary dielectric host and briefly polarizable host. This point has already been discussed by Samara [20]. Here only the conclusion is introduced. In a system of random dipoles in a weakly polarizable medium, anisotropic dipolar interaction is described by the Hamiltonian

$$H_{dd} = \frac{1}{\varepsilon_0} \sum_{ij} \frac{1}{r_{ij}^3} \left[d_i^* d_j^* - 3\left(n_{ij}d_i^*\right)\left(n_{ij}d_i^*\right)\right] \tag{2.7}$$

where $d_i^* = d_i + \left[1 + r\left(\varepsilon'_0 - 1\right)/3\right]$ is the effective impurity dipole moment of dipole i, r the local field correlation factor, d_i the permanent dipole moment of the impurity, r the local field correction factor, ε'_0 the static dielectric constant of the pure crystal, r_{ij} the separation between dipoles i and j. The random distribution of dipoles and the variable size of the dipole–dipole interaction energy causes the local fields at the impurities to have different orientations. Then, at low temperatures the dipoles freeze into a state with zero net polarization in the absence of a local field. This low temperature state is the dipolar glass state, which is analogous to the spin glass state. A most important result is that impurity dipoles in ordinary dielectrics cannot produce long-range ferroelectric order. In such a case the configurational average $\langle H_{dd} \rangle$ of Eq. 2.7 vanishes. Where the polarizability of the host is small, the fluctuations of the local field $\vec{E}_{loc}$ become greater than the average local field, $\langle \vec{E}_{loc} \rangle$. In ordinary dielectrics, the correlation length for dipolar interactions, r_c, is small, so that in the dilute limit $r_{ij} \gg r_c$. As the concentration n of dipoles increases, weak correlation among dipoles appears and produces a maximum in the temperature dependence of the dielectric constant. The peak temperature, T_m, is given by

$$kT_m \approx \frac{nd^{*2}}{\varepsilon'_0} \tag{2.8}$$

Dipolar glass behavior due to such a mechanism occurs in alkali halides with dipolar impurities, CN^- and OH^- in KCl and KBr and Li^+ in KCl. Li^+ in KCl can be an off-center dipole in the K site. With increasing concentration of dipolar impurities, $\varepsilon'(T)$ tends to show a peak and the D–E loop tends to show hysteresis.

In a system of random dipoles in a highly polarizable lattice, other important results are obtained. Within r_c, dipolar motion is correlated to form leading to the formation of polar nanodomains. In this case the dipolar entity is not isolated. In soft mode ferroelectrics, the correlation length is determined by the polarizability of the host which is inversely proportional to the soft mode frequency, ω_s, so that $\omega_s \to 0$ on decreasing temperature, r_c diverges.

In the case of very large dielectric constants, $\varepsilon'_0 \gg 1$, the effective dipole moment of the dipolar entity, becomes

$$d^* = \frac{r\varepsilon'_0 \vec{d}}{3} \tag{2.9}$$

through the consideration of the dipolar entities i and j [20].

The key result is that the non-zero configurational average for these polar phonon-mediated interactions makes it possible for polar nanoregions to interact cooperatively. The strength of this interaction, $J(r)$, varies exponentially with distance as

$$J(r) = J_0 \exp\left(\frac{-r}{r_c}\right) \tag{2.10}$$

The FE transition temperature is given by

$$kT_c = \left(\frac{4\pi}{3}\right) \frac{d^{*2} n}{\varepsilon'_0(T_c)} = \left(\frac{4\pi}{27}\right) r^2 d^2 n \varepsilon'_0(T_c) \tag{2.11}$$

where $\varepsilon'_0(T_c)$ is the value ε' at T_c. Vugmeister and Glinchuk [21] gave two limiting cases in terms of the quantity nr_c^3, where n is the impurity concentration.

<u>Case 1</u> $nr_c^3 < N^*$

where N^* is the characteristic quantity for a given material system. In this case, the dipole nanoregions form a dipolar glass-like or relaxor state at low temperatures.

<u>Case 2</u> $nr_c^3 > N^*$

In this case, FE phase transition appears accompanying the spontaneous polarization given by $\langle p \rangle = n \langle d^* \rangle$. The crossover between the two cases is given by the condition $nr_c^3 = N^*$. For an ordinary polarizable crystal $r_c \approx a$, but for a highly polarizable lattice $r_c \gg a$, and it is strongly T dependent. In this case

$$r_c(T) \propto \frac{1}{\omega_s(T)} \propto \sqrt{\varepsilon'(T)} \tag{2.12}$$

Examples

Figure 2.9 compares the dielectric properties of the typical ferroelectric triglycine sulphate (TGS) and the relaxor $Pb_{1/3}MgNb_{2/3}O_3$ (PMN) [22]. The ferroelectric TGS has macroscopic ferroelectric domains and shows a clear and distinct change in the

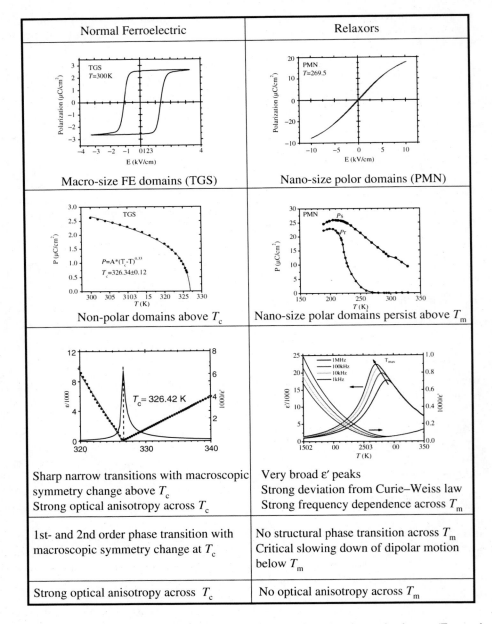

Fig. 2.9. Comparison between the properties of normal ferroelectrics and relaxors (Fu et al., unpublished [24]).

properties at T_c, which contrasts with the typical compositionally disordered relaxor ferroelectric PMN.

Figures 2.10 and 2.11 show the pressure dependence of the dielectric constants of the ferroelectric SrTi^{18}O$_3$ (STO18) [23] and relaxor Sr$_{0.993}$Ca$_{0.007}$TiO$_3$ (SCT) [24]. T_c and T_m shift to the lower temperature side with pressure and finally disappear. This result is consistent with the criteria given by Samara [20]. Figures 2.12 and 2.13 compare the

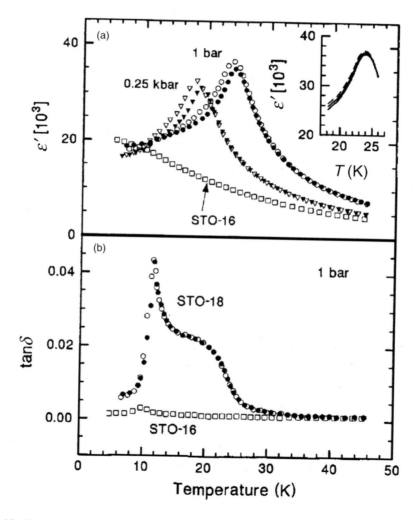

Fig. 2.10. Temperature dependence at 1 bar and 10^4 Hz of the dielectric constant ε' and dielectric loss tan δ for our STO-18 ($\geq 97\%$ ^{18}O) crystal compared with similar results for an unsubstituted STO-16 crystal. $\varepsilon'(T)$ results for STO-18 at 0.25 kbar are also shown to emphasize the observed thermal hysteresis in T_c and in the ferroelectric phase (solid symbols = heating; open symbols = cooling). The inset shows that the $\varepsilon'(T)$ response is essentially independent of frequency in the range 10^2–10^5 Hz [23].

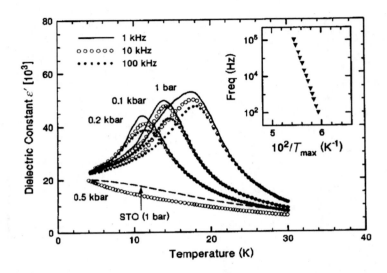

Fig. 2.11. The dielectric $\varepsilon'(T)$ response of SCT (0.007) at different pressures showing the frequency dispersion, which vanishes at 0.5 kbar. Also shown for comparison is the frequency-independent response of pure SrTiO$_3$ (STO) at 1 bar. The inset is an Arrhenius plot of the T dependence of the relaxational frequency (= inverse relaxation time) at 1 bar [24].

pressure dependence of the dielectric properties of the ferroelectric STO18 and relaxor SCT. Crossover from a relaxor state to a FE state in PMN can be demonstrated by applying electric field, which tunes the strength of the dipolar interaction in PMN across N*. Schematic phase diagram for paraelectric, FE, and dipolar states is given in (Fig. 4 in [20]) As shown in the insets of the figures, the slopes 'dT_c/dT' at 0 K are infinite and finite, respectively, for STO18 and SCT. This means that STO18 is a 'quantum' ferroelectric with a definite T_c above $S > S_c$, where S is an interaction parameter. On the other hand, SCT is a relaxor like PMN. Depending on the kinds of matrices, compositional disorder can induce apparently different phenomena, as seen in the above section. Physics on the compositionally disordered systems, including lead-based ferroelctrics or relaxor, are now in progress based on the new data of neutron diffraction [27] and optical measurement. Phase transition mechanism of STO18 is also discussed by many researchers in ferroelectric field. At the present time, the following is raised as new view points: (1) Ferroelectric optical soft mode shows a perfect softening at T_c. (2) Locally symmetry broken region (LSBR) appears at high temperatures far apart from T_c. (3) LSBR is estimated to form around the lattice defects. (4) LSBR itself may not directly turn into FE region. (5) Only the mass change in oxygen can cause the FE transition in STO. (6) Strength of the random field originating from the defects in STO is an order of a few tens of kelvin. Complete understanding for the phase transition mechanism of STO18 will give a hint for the material design in the FE field.

Lastly, we will show an example in which the polar cluster does not contribute to the dielectric properties. Figure 2.14 shows the frequency dependence of ε' and ε'' for hexagonal BaTiO$_3$ at room temperature [26]. This material is known as a ferroelectric below

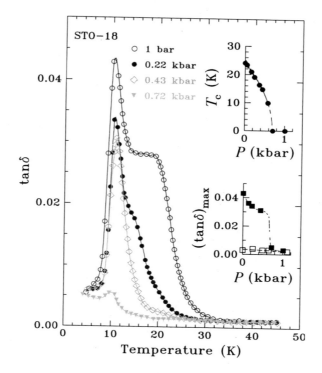

Fig. 2.12. Dielectric loss (tan δ) vs. temperature for STO-18 at 10 kHz under four pressures. Upper inset: FE T_c it vs. P; lower inset: tan δ maximum vs. P for STO-18 (solid squares) and STO-16 (open squares) [25].

70 K, and is paraelectric at room temperature. When oxygen vacancies are introduced into the crystal, it shows a colossal dielectric constant (Fig. 2.14). Further analysis reveals that both the external layer with a capacitance of 1.0 nF and the internal boundary layer with a capacitance of 1.4 nF are contributing to the total capacitance of this material. The capacitance of the bulk is only 1 pF. The permittivity, which exceeds 10^5, can be obtained by the capacitance of internal layer, which is considered to be composed of two-dimensional defects such as stacking faults, and normal single crystal. Recently, many studies have been reported on the colossal dielectric permittivity of paraelectric $CaCu_3Ti_4O_{12}$ [27]. We believe this is caused by the same mechanism as in hexagonal $BaTiO_3$. Defect control for hexagonal $BaTiO_3$ is easy and reproducible, making it a candidate for tunable dielectric.

2.4 Lithium Ion Conductivity in Oxides

Solid lithium conductors with conductivity as high as $10^{-1} \sim 10^{-2}$ S·cm^{-1}, (the same order as a liquid lithium electrolyte) are required for the development of reliable solid state lithium batteries with higher energy density. All solid state batteries have problems in the interfacial region between the electrode and electrolyte materials, but researchers

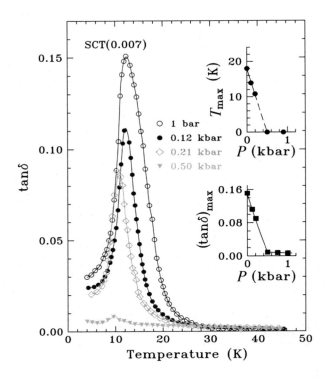

Fig. 2.13. Dielectric loss (tan δ) it vs. temperature for SCT(0.007) at 10 kHz under four pressures. Upper inset: FE T_c it vs. P; lower inset: tan δ maximum it vs. P [25].

in this field believe that the improved lithium conductivity of oxides may have some applications to industry other than batteries.

The report of A-site deficient perovskite titanate [28], (La, Li)TiO$_3$ (LLTO), 12 years ago, has played a role in stimulating the field of solid state chemistry. The ionic conductivity is discussed in more than 300 articles of LLTO. As expected from the existence of T_i^{4+}, LLTO does not have a resistance to reduction by lithium metal, which means that it cannot be used as a lithium ion conducting electrolyte for the lithium battery.

Despite this, perovskite-type oxides have many scientific merits; controllable structure, carrier concentration (defect concentration), and site percolation. The discovery of this material resulted in a comprehensive knowledge for the design of cation conductors. Since there are a number of review articles [29–32] on lithium ion conductors, this report reviews only recent studies on lithium ion conducting crystalline oxides from the view point of the dimensionality of the materials.

2.4.1 One and quasi-one dimensional lithium ion conductor

β-eucryptite [33–39]. The structure of β-eucryptite is a derivative of the high-quartz structure in the hexagonal system. In β-eucryptite, half of the SiO$_4$ teterahedra are

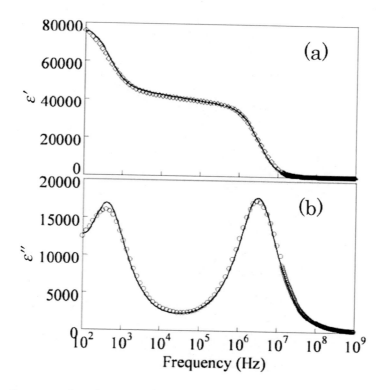

Fig. 2.14. Frequency dependences of ε'(a) and ε''(b) for oxygen-deficient hexagonal BaTiO$_3$ (J-D Yu et al., [26]).

replaced by AlO$_4$ tetrahedra. The framework is built up of layers of SiO$_4$ and AlO$_4$ tetrahedra stacked alternately along the c-axis. Figure 2.15 shows the structure projected on the (001) plane of the ideal β-eucryptite structure. In the adjacent channels, lithium ions occupy sites which are alternately coplanar with Si ions and Al ions as shown in Fig. 2.16. This ordering of lithium ions as well as the ordered arrangement of Si and Al ions results in a doubling of the lattice constant a_0 and c_0 compared to the value for high-quartz. Lithium ion conductivity along the c-axis of the lithium channel in β-eucryptite is higher by about three orders in magnitude than along the perpendicular direction. The lithium ion conductivity along the c-axis is $\sim 10^{-9}$ S·cm^{-1} at room temperature and $\sim 10^{-3}$ S·cm^{-1} at 573 K. When compositional modulation is introduced during crystal growth, the anisotropic ionic conductivity is decreased to $\sigma_c/\sigma_a \approx 10$. Studies on LiAlSiO$_4$–SiO$_2$ solid solutions Li$_{1-x}$Al$_{1-x}$Si$_{1+x}$O$_4$ with $x = 0, 0.07, 0.17$, and 0.22 have also been reported. With increasing SiO$_2$ content, the concentration of charge carriers decreases; however, the conductivity is independent of the chemical composition.

Lithium titanate Li$_2$Ti$_3$O$_7$ [40–42]. Li$_2$Ti$_3$O$_7$ has the ramsdellite α-MnO$_2$ structure, as shown in Fig. 2.17. Li$_2$Ti$_3$O$_7$ consists of disordered TiO$_6$ octahedra which link

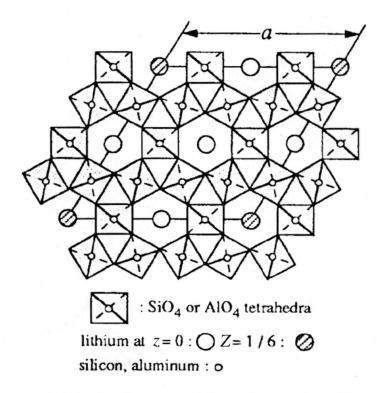

Fig. 2.15. The idealized structure of β-eucryptite; projection on (001).

with adjacent octahedra by sharing opposite edges to form double chains like ribbons. These ribbons share corners, forming rectangular channels along the c-axis. Since there are eight oxygens in the unit cell, the formula of $Li_2Ti_3O_7$ is written as $Li_{2.29}Ti_{3.43}O_8$.

Boyce and Mikkelson have measured the lithium ion conductivity of single crystals of $Li_2Ti_3O_7$, and report $\sigma_b/\sigma_a \approx 4$ and $\sigma_c/\sigma_b \approx 7$. This anisotropy is significant but is not sufficient to classify this channel-structured ramsdellite material as a one-dimensional conductor. The lithium ion conductivity along the c-axis is 7×10^{-6} S·cm^{-1} at room temperature and 1.8×10^{-2} S·cm^{-1} at 573 K. These authors try to explain the relationship between this anisotropy of the conductivity and the crystal structure. There are eight octahedral sites per unit cell, four in the channel and four on the ribbons of which 3.43 are occupied by Ti and 0.57 are vacant. This gives 0.57 non-channel sites of a total of four channel sites which are available for lithium conduction. When random occupation of these sites is assumed, there are 0.5 lithium ions per available octahedral site. So the probability that one channel site is occupied is 1/2 and the probability for a non-channel site is $1/2 \times 1/7 = 1/14$. Figure 2.18 shows a schematic depiction of the migration of ions along a- and b-directions. By a simple calculation, the conductivity ratio along different directions is obtained. For a jump in the a-direction, lithium migrates

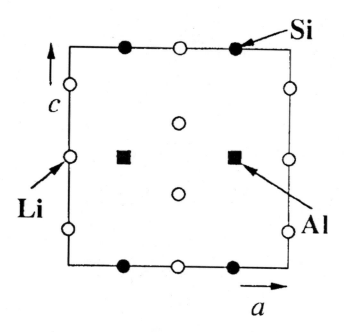

Fig. 2.16. Cation distribution in β-eucryptite (section of the structure along the AC plane).

a distance of approximately $a/2$ in both positive and negative directions. This motion yields

$$\sigma_a \sim \left(\frac{a}{2}\right)^2 \left(\frac{1}{14} + \frac{1}{14}\right) \qquad (2.13)$$

Along the b-direction, the lithium ion jumps a distance of $b/4$, encountering a channel site in one direction and a non-channel site in other direction. As a result,

$$\sigma_b \sim \left(\frac{b}{4}\right)^2 \left(\frac{1}{2} + \frac{1}{4}\right) \qquad (2.14)$$

Along the c-direction, all the channels are available for migration,

$$\sigma_c \sim c^2 \left(\frac{1}{2} + \frac{1}{2}\right) \qquad (2.15)$$

Consequently, the relations $\sigma_b/\sigma_a \approx 3.6$ and $\sigma_c/\sigma_b \approx 3$ are obtained.

The conduction is not one-dimensional due to the fact that one seventh of the lithium channel sites do not contain an immobile Ti ion. Thereby, a path for the mobile Li ions perpendicular to that channel is given.

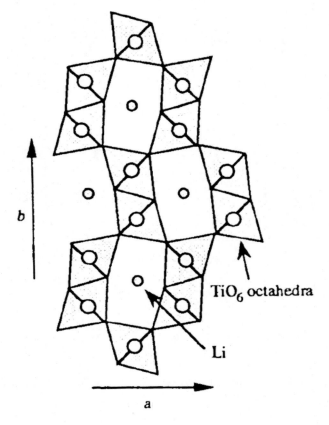

Fig. 2.17. The ramsdellite structure of $Li_2Ti_3O_7$.

2.4.2 Two-dimensional lithium ion conductor

Li-β-alumina [43–46]. Single crystal of Na–β–alumina with the ideal formula $Na_2O \cdot 11Al_2O_3$ exhibits high ionic conductivity of 3×10^{-2} S·cm^{-1} at room temperature. The unit cell of Na–β–alumina consists of two $Al_{11}O_{16}$ spinel-type blocks related to each other by a two-fold screw axis parallel to c-axis. Each block contains four layers of cubic closed packed oxygen ions. Aluminum atoms are sandwiched between each pair of oxygen layers in both octahedral and tetrahedral interstitial sites similar to Mg and Al ions in the structure of $MgAl_2O_4$ spinel. The adjacent spinel-type blocks are held apart by A-O-Al spacer units. Between the adjacent blocks, loosely packed conduction planes pass through the spacer oxygen atom. There is one sodium atom in each plane and consequently two sodium atoms in the unit cell. Na–β–alumina has a planar structure, and as a result, Na ions are expected to move in these two dimensional planes. Li–β–alumina can be obtained by ion exchange using molten $LiNO_3$ or LiCl. The conductivities and activation energy for a single crystal are 2.7×10^{-3} S·cm^{-1} and 0.24 eV, respectively. Li–β–alumina has a lower lithium ion conductivity and higher activation energy than Na–β–alumina. Since the lithium ion is smaller than the sodium

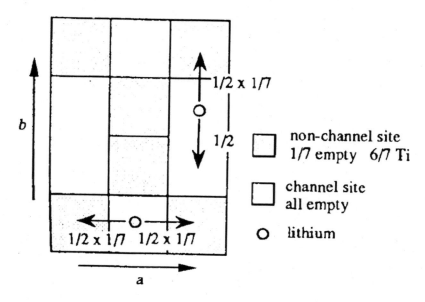

Fig. 2.18. A schematic depiction of the migration of ions along a- and b-direction in $Li_2Ti_3O_7$.

ion it cannot fit near the mid-plane position equidistant from the closed packed oxygen planes (spinel blocks) as the sodium ion does and so it occupies sites in the walls of the conduction plane as shown in Fig. 2.19. This model is supported by the fact that the ionic conductivity increases by applying pressure which enables the lithium ion to sit less tightly in the positional potential well and move towards the mid-plane position.

2.4.3 Three-dimensional lithium ion conductor

Li_4SiO_4 and related compounds [47–52]. The compounds Li_4SiO_4 and Li_4GeO_4 have moderately high lithium conductivities, $\sim 10^{-4}$ S·cm^{-1} at 573–673 K, and are suitable for carrier doping. These compounds consist of isolated tetrahedral SiO_4 or GeO_4 anion group with eight lithium ions distributed over 18 available cation sites in the stoichiometric composition. Substitutions have led to an improvement in the ionic conductivity of several orders in magnitude.

$Li_{14}ZnGe_4O_{16}$, which has been designated LISICON, shows $\sigma = 10^{-1}$ S·cm^{-1} at 573 K. The highest room temperature conductivity of the Li_4SiO_4-related compounds occurs in $Li_{3.5}V_{0.5}Ge_{0.5}O_4$ for which $\sigma = 7 \times 10^{-5}$ S·cm^{-1}.

Lithium ion conductors based on NASICON-type structure [53–65]. Sodium (Na) Super Ion Conductor (NASICON) with the chemical formula $Na_{1+x}Zr_{2-x}P_{3x}O_{12}$ is known to be a fast ion conductor. The structure consists of a three-dimensional skeletal network of PO_4 tetrahedra sharing with corners of ZrO_6 octahedral sites in the interstitial space.

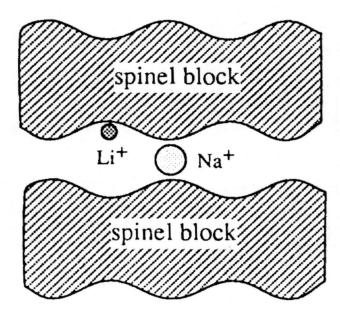

Fig. 2.19. Schematic representation comparing ionic sizes relative to the size of the channel in β-alumina.

Each octahedron is connected to six PO$_4$ tetrahedra. Each tetrahedron is linked to four ZrO$_6$ octahedra as shown in Fig. 2.20. Many studies on lithium ion conductors with this NASICON-type structures have been reported. Since the NASICON-type structure allows for the substitution by various ions, the ionic conductivity can be changed over a wide range.

In NASICON-type lithium ion conductor LiM$_2$(PO$_4$)$_3$ (M: tetravalent ion ; Ti, Zr, Hf, and Ge), the Li(1) (6b) sites are completely filled by lithium ions while the Li(2) (18e) sites are empty. By partially substituting M'$^{3+}$ (Al^{3+}, In^{3+}, Ga^{3+}, Sc^{3+}, Cr^{3+}) or M''$^+$ (Nb^{5+} and Ta^{5+}) for M^{4+} ions to form Li$_{1+x}$M'$_x$M$_{2-x}$(PO$_4$)$_3$ or Li$_{1-x}$M''$_x$M$_{2-x}$(PO$_4$)$_3$, lithium ions partially occupy the Li(2) sites or randomly occupy the Li(1) and Li(2) sites in order to maintain electrical neutrality. Since the lithium ions in Li(1) sites can migrate to the other sites only via Li(2) sites, the occupation ratio in each site strongly influences the conductivity. At the same time the substitution of another cation for M^{4+} may change the size of the conduction channel (bottleneck). Researchers explain that changes in ionic conductivity are attributed to changes in the lithium content, the size of ion conducting channel as mentioned above, the difference of the electronegativity of the constituent ions, and the porosity of the samples. The data for room temperature lithium ion conductivity of a NASICON type lithium ion conductor are summarized in Table 2.2.

Perovskite-type Oxides. The perovskite structure, which has the formula ABX$_3$, consists of a framework of BX$_6$ octahedra linked by their corners, with a large A-cation occupying a cavity of the same size as the X ions as shown in Fig. 2.21. When viewing only the arrangement of the A-ions, they form a simple cubic lattice.

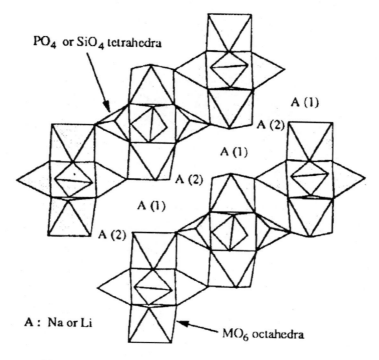

Fig. 2.20. A schematic figure of NASICON-type structure.

Table 2.2. Room temperature ion conductivity of a NASICON-type lithium ion conductor.

Compound	σ at room temperature (S·cm^{-1})	Ref.
$Li_{0.85}Ta_{0.15}Zr_{1.85}(PO_4)_3$	4.7×10^{-6}	[55]
$Li_{0.8}Nb_{0.2}Zr_{1.8}(PO_4)_3$	4.6×10^{-7}	[55]
$Li_{1.4}In_{0.4}Ti_{1.6}(PO_4)_3$	1.9×10^{-4}	[56–59]
$Li_{1.6}Ga_{0.6}Ti_{1.4}(PO_4)_3$	1.3×10^{-5}	[57]
$Li_{1.6}Mg_{0.3}Ti_{1.4}(PO_4)_3$	2.8×10^{-6}	[57]
$Li_{1.0}Ti_{0.5}Zr_{1.5}(PO_4)_3$	6.2×10^{-8}	[58]
$Li_{1.3}Sc_{0.3}Ti_{1.7}(PO_4)_3$	2.9×10^{-5}	[61]
$Li_{1.8}Cr_{0.8}Ti_{1.2}(PO_4)_3$	2.4×10^{-4}	[61]
$Li_{1.5}Al_{0.5}Ge_{1.5}(PO_4)_3$	3.5×10^{-5}	[62]
$Li_{1.3}Al_{0.3}Ti_{1.7}(PO_4)_3$	7.0×10^{-4}	[64]
$Li_{0.1}Nb_{0.9}Zr_{1.1}(PO_4)_3$	1.0×10^{-5}	[65]

Few studies of lithium ion conductivity in perovskite-type oxides have been reported, to our knowledge. Latie et al. investigated the ionic conductivity of $Li_xLn_{1/3}Nb_{1-x}Ti_xO_3$ (Ln = La, Na and $x \leq 0.1$) with a perovskite-related structure. In addition, it has been reported that lanthanum lithium titanates with perovskite structure shows bulk ionic conductivity as high as 1×10^{-3} S·cm^{-1} at room temperature. This high conductivity is

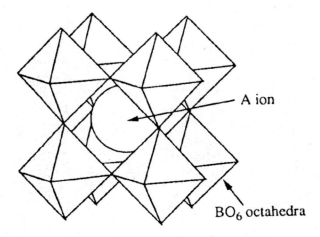

Fig. 2.21. A schematic of Perovskite Structure.

considered to originate from the presence of a vacancy in the A-site with many equivalent sites for lithium ion to occupy and to migrate through, via the vacant A-site in the perovskite structure. The conduction channel is thought to be three-dimensionally linked. The smallest cross-sectional areas of an interstitial passageway, named 'bottlenecks' are located in the space between the two adjacent A-sites, which are surrounded by four oxygens (in a cubic lattice with space group Pm3m, corresponding to 3c site) [Fig. 2.22]. Lithium ion conductivity in perovskite oxides has been discussed as follows:

a) Influence of lithium ion vacancy concentration on lithium ion conductivity

In the $(La_{2/3-x}Li_{3x}[\ ]_{1/3-2x})TiO_3$ system, the lithium conductivity changes with changes in the lanthanum content i.e. lithium and vacancy contents as shown in Fig. 2.23. This change can be explained as follows. The ionic conductivity σ is given by the equation

$$\sigma = |e|n\mu \tag{2.16}$$

where e is the charge, n is the concentration and μ is the mobility of the mobile species. According to the electrical neutrality condition, the n-values for Li$^+$ and vacancies in the A-sites in $(La_{2/3-x}Li_{3x}[\ ]_{1/3-2x})TiO_3$ are $3x$ and $1/3-2x$, respectively. Since it is expected that lithium ions are bound within the A-sites and are mobile only through the vacant A-sites, their mobility is proportional to the concentration of vacancies, $1/3-2x$. Therefore the ionic conductivity is proportional to the factor

$$\sigma = |e|n\mu \infty (3x)\left(\frac{1}{3} - 2x\right) = \frac{2}{-3}\left(3x - \frac{1}{4}\right)^2 + \frac{1}{24} \tag{2.17}$$

The Role of Lattice Defects in Oxides

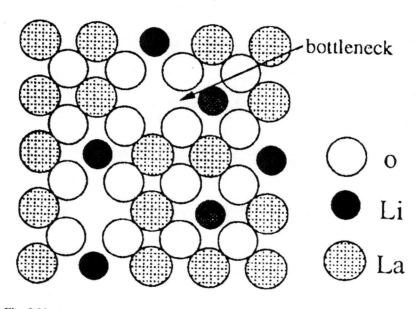

Fig. 2.22. A schematic structure in the cross section of lanthanum lithium titanate.

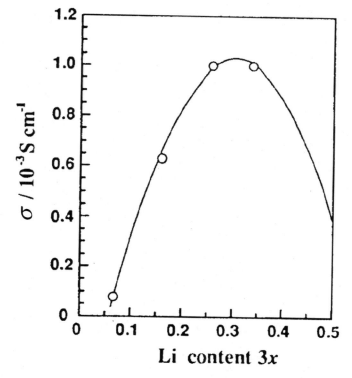

Fig. 2.23. Ionic conductivity at 300 K as a function of lithium content for $(La_{2/3-x}Li_{3x}[\ ]_{1/3-2x})TiO_3$.

This equation implies that ionic conductivity is a parabolic function of the Li fraction $3x$ with a maximum at $3x = 0.25$. In fact, the experimental data can be fitted to the parabolic function.

According to percolation theory, the critical value for percolation in a simple cubic lattice is 0.312. Therefore, it is expected that when the total concentration of lithium ions and vacancies is greater than this, the conduction channel is linked infinitely, resulting in the appearance of ionic conductivity. Since the total concentration of lithium ions and vacancies is larger than 1/3 in $(La_{2/3-x}Li_{3x}[\ ]_{1/3-2x})TiO_3$, the conduction channel is thought to be linked. Therefore, high ionic conductivity appears. Since $(La_{1/2}Na_{1/2})TiO_3$ shows no ionic conductivity, sodium ions cannot migrate in the lattice. Therefore lanthanum and sodium ion is regarded as an obstacle in $(La_{1/2}Li_{1/2})_{1-x}(La_{1/2}Na_{1/2})_x TiO_3$ and the ionic conductivity decreases with increasing x, decreasing especially rapidly between $x = 0.25$ and 0.4 as shown in Fig. 2.24. At 300 K the ionic conductivity in $x = 0.4$ is smaller than 1×10^{-7} S·cm^{-1}. When $x = 0.40$, the total concentration of lithium ion and vacancies is about 0.3, being smaller than the critical value. The rapid decrease in the conductivity is explained by site percolation.

b) Influence of bottleneck size on lithium-ion conductivity

The substitution of other lanthanide ions with smaller ionic radii (Ln = Pr, Nd, and Sm) for La in $La_{1/2}Li_{1/2}TiO_3$ ($Ln_{1/2}Li_{1/2}TiO_3$) decreases the ionic conductivity and increases the activation energy, as shown in Fig. 2.25 [69], while the substitution of Sr with a larger ionic radius for La in $La_{1/2}Li_{1/2}TiO_3$ ($[(La_{1/2}Li_{1/2})_{1-x}Sr_x]TiO_3$) increases the ionic conductivity [70]. These results indicated that the A-site space contraction reduces the bottleneck size and consequently disturbs lithium ion migration via A-site vacancy, while the A-site space dilatation increases the bottleneck size and consequently promotes ion migration. Lanthanum lithium titanate with 5 mol% Sr substitution shows the highest room temperature lithium ion conductivity $\sigma = 1.5 \times 10^{-3}$ S·cm^{-1} of the perovskite oxides. The application of external hydrostatic pressure induces a decrease in the bottleneck size, resulting in a decrease in the conductivity [71].

c) Observation of the glass transition by calorimetry[72]

The heat capacities and electrical moduli of $La_{0.51}Li_{0.35}TiO_3$ were measured below 300 K. A glass transition observed at 102 K is due to the freezing-in of the positional disorder of lithium ions. A close relation was suggested to exist in general between high ionic conductivity and the positional disorder of mobile ions.

Since the perovskite structure is very simple and can tolerate different valence states in the A- and B-sites [73–78], there are many possibilities for developing and studying other lithium conductors.

Lithium ion conductivities in various lithium ion conductors. Figure 2.26 shows the temperature dependence of the conductivity for various oxides lithium ion conductors. As the value of the conductivity at room temperature is higher, the activation energy for migration becomes smaller. The conductivity seems to converge to a saturated value, ~ 1 S·cm^{-1}, at high temperature for each compound. This value is substantial for oxide lithium conductors and needs to be clarified in the future from a chemical viewpoint.

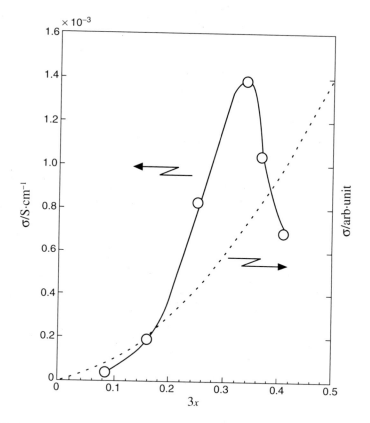

Fig. 2.24. Variation of the ionic conductivity at 300 K it vs. the lithium concentration $3x$ in quenched $La_{2/3-x}Li_{3x}[\ ]_{1/3-2x}TiO_3$. Dotted line represents the ionic conductivity expected from the site percolation.

Figure 2.27 shows the scattering amplitude distribution in the vicinity of the (002) plane ($0.35 < z < 0.65$) of $La_{0.62}Li_{0.16}TiO_{3.0}$ at 77 K and room temperature [83]. This result gives a slightly different view of the diffusion path of lithium ions in oxides from that given above. This matter is now under discussion.

We conclude here that our first report on LLTO had a large impact on the field of solid state ionics. This is due to the fact that the local structure, carrier concentration, and dimensionality of perovskite can readily be modified for the purposed experiment.

2.5 Concluding Remarks

We have briefly reviewed recent studies on spin state transition in oxides, mainly K_2NiF_4 and perovskite-type oxides in the *first* Section. The isolated high-spin or intermediate

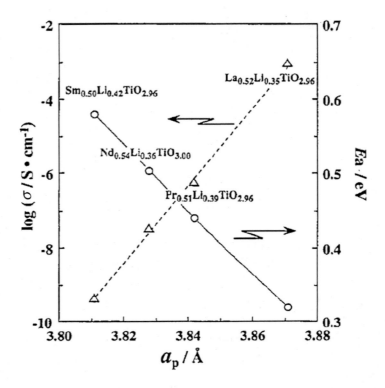

Fig. 2.25. Plots of conductivity at 300 K and activation energy for lithium ion conduction vs. perovskite parameter a_p for $Ln_{1/2}Li_{1/2}TiO_3$ (Ln = La, Pr, Nd, and Sm).

entity itself in the low-spin matrix is a kind of defect because the chemical bonding character is completely different from that of the matrix, although the atomic species are the same. Due to a negative cooperation effect among higher spin atoms through the chemical bond with oxygen, the higher spin region does not develop to become a domain, which would result in a phase transition. The isolated region possessing different chemical bonding character from the matrix can give an insight into the design of ferroelectrics and magnetics. The introduction of such ideas may assist the design of quite new materials.

In the *second* Section, we discuss the strength of the dipolar entity produced by lattice defects including compositional disorder. This has not been noticed by researchers in other fields because dipolar interaction in insulators is a fairly long-range effect. When we do experiments involving atomic substitutions in oxides, which is a normal procedure in ceramics technology, we must consider the dipoles produced by the off-centered atoms whose sizes are different from the matrix, and also the dipoles produced by compositional disorder when atoms with different valence are substituted. This is the reason why the physics of ferroelectric-related materials becomes difficult.

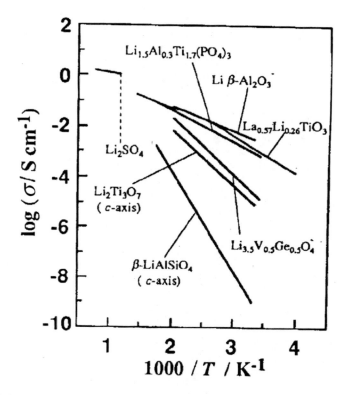

Fig. 2.26. Temperature dependence of the conductivity of various lithium ion conductor.

In the *last* Section, we discuss lithium ion conducting oxides. The concept of the ionic conductor itself overlaps with that of defect chemistry because the migration of atoms accompanies the formation of off-center dipoles or defects. Except for a few researchers in this field, most people do not take account of the change in the chemical bond during hopping, especially in a short time scale, e.g. femtoseconds. The introduction of the idea of a local field in ferroelectrics is also indispensable for the design of new ionic conductors.

2.6 Clue to the Design of New Functional Oxide Materials

This review has focused on the role of the minor species with different chemical characters compared to that of the matrix phase in oxides.

Spin crossover in oxides itself is quite well known phenomena in materials but the quantitative study has not been carried out for oxides because application of the crystal field theory to the quantitative control of the spin state of transition metal ions in oxides had been considered to be difficult. To my opinion, the work by Demazeau and Hagenmuller group has proved that the control is easy through the chemical bonding

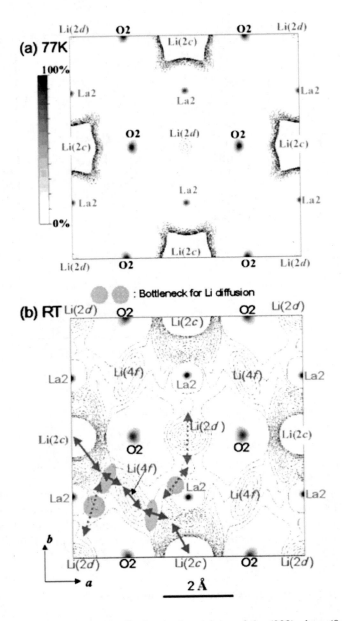

Fig. 2.27. Scattering amplitude distribution in the vicinity of the (002) plane ($0.35 < z < 0.65$) of $La_{0.62}Li_{0.16}TiO_{3.0}$ at (a) 77 K and (b) room temperature. Contour in the range from -1.00–0.05 fm/Å^2 (0.05 fm/Å^2 per step). (c) Atomic arrangement on the (002) plane where the ionic radii after Shannon are used to draw the atomspheres. Li atoms are placed at the 2c, 2d, and 4f sites with interatomic distances between the Li and oxygen atoms. The arrow indicates the diffusion path. The hatched region denotes a possible bottleneck for Li cation diffusion. The dashed arrows are the other possible diffusion paths with a bottleneck indicated by the orange hatched region. The dashed line with arrows is the diffusion path from the literature [83].

aspects. This is quite easy to apply to the complex oxides by considering the competing chemical bonding in them. The bi-stability of the competing two electronic states of transition metal ions containing one transition metal ion can give a controllability to the material. Spin crossover usually accompanies a quite large change in volume around the minor spin species. The negative correlation among the higher spin Co^{3+}-centered clusters in $LaCoO_3$ has proven that elastic and chemical energies, including the kinetic energy of electrons, are compatible in this material. This hints that a trivial extrinsic fluctuation results in a first order phase transition into a new state. In a certain case, ordered arrangement among nanoclusters (precipitates) is known. This can be applied to the other situation like hetero interfaces in oxides. The controlability of the size and distribution of the hetero phase with nano size in the matrix phase can be applicable to the design of the optical devices, optical magnetic devices, molecular magnet, and also of the ferro(piezo)electric materials.

The role of the defects in ferroelectric materials is manifold. When the defects are isolated with each other, they can be a dipolar entity. Such a dipolar center can create polar nanoclusters in ferroelectrics at $T > T_c$ and in paraelectrics. The point defects in ferroelectrics can also become pinning centers for the domain. When the defects aggregate, they form hetero interface with the normal matrix with few defects. When electron carrier is introduced into such a state, apparent large dielectric response is observed. Comparison of the dielectric properties for the defect-free and defect-abundant ferroelectric materials will be indispensable for the full understanding of the true nature of them. For example, we can expect completely different behavior between a perfect single crystal of $SrTiO_3$ and that with $10^7 - 10^9$ cm^{-2} of the dislocation density, which is typical value for Verneuil-grown single crystal.

From the comprehensive studies for lithium conducting oxides, especially for perovskites, we could first draw out the role of the independent parameters of (1) carrier density, (2) site percolation, (3) bottle neck size, and (4) the degree of the covalency between metal and oxygen. These results are also effective for the design and evaluation for ion conductors other than lithium. We also concluded that lithium ion conductivity is highest in perovskite-type oxides, with the order of 10^{-3} S·cm^{-1} at room temperature. This means that all solid state lithium battery with high energy density, comparable to that of the lithium battery using liquid electrolyte, composed of lithium anode, oxide solid electrolyte, and oxide cathode, can not be obtained. The strategy for finding a new material with higher lithium ion conductivity should be directed to other material group, such as sulfides, selenides, or tellurides as a target.

References

[1] S. Decurtins, P. Gütlich, C.P. Köhler, H. Spiering and A. Hauser, Chem. Phys. Lett. **105** (1984) 1.

[2] G. Demazeau, M. Pouchard, Z. Li-Ming and P. Hagenmuller, z. anorg. allg. Chem. **555** (1987) 64.

[3] G. Demazeau, S.H. Byeon, P. Hagenmuller and J.H. Choy, z. anorg. allg. Chem. **610** (1992) 91.

[4] Y. Tanabe and S. Sugano, J. Phys. Soc. Jpn. **9** (1954) 753.

[5] B. Buffat, G. Demazeau, M. Pouchard and P. Hagenmuller, Proc. Indian Acad. Sci. (Chem. Sci.) **93** (1984) 313. G. Demazeau, P.H. Courbin, G. Le Flem, M. Pouchard, P. Hagenmuller, J.L. Souberyroux, I.G. Main and G.A. Robins, Nouveau Journal de Chimie **3** (1979) 171.

[6] G. Demazeau, B. Buffat, F. Menil, L. Fournes, M. Pouchard, J.-M. Dance, P. Fabritchnyi and P. Hagenmuller, Mat. Res. Bull. **16** (1981) 1465.

[7] G. Demazeau, M. Pouchard, M. Thomas, J.-F. Colombet, J.-C. Grenier, L. Fournes, J.-L. Soubeyroux and P. Hagenmuller, Mat. Res. Bull. **15** (1980) 451.

[8] G. Demazeau, Z. Li-Ming, L. Fournes, M. Pouchard and P. Hagenmuller, J. Solid State Chem. **72** (1988) 31.

[9] G. Demazeau, N. Chevreau, L. Fournes, J.-L. Soubeyroux, Y. Takeda, M. Thomas and M. Pouchard, Revue de Chimie minerale **20** (1983) 155.

[10] J.-H. Choy, G. Demazeau and S.H. Byeon, Solid State Commun. **80** (1991) 683.

[11] S.H. Byeon, G. Demazeau, L. Fournes, J.-M. Dance and J.-H. Choy, Solid State Commun. **80** (1991) 457.

[12] G. Villeneuve, T. Rojo, G. Demazeau and P. Hagenmuller, Mat. Res. Bull. **23** (1988) 1787.

[13] Z.-L. Ming, G. Demazeau, M. Pouchard, J.-M. Dance and P. Hagenmuller, J. Solid State Chem. **78** (1989) 46.

[14] S.-H. Byeon, G. Demazeau, J.M. Dance and J.-H. Choy, Eur. J. Solid State Inorg. Chem. **28** (1991) 643.

[15] J.B. Goodenough, N.F. Mott, M. Pouchard, G. Demazeau and P. Hagenmuller, Mat. Res. Bull. **8** (1973) 647.

[16] T. Kyomen, Y. Asaka and M. Itoh, Phys. Rev. B **67** (2003) 144424. T. Kyomen, Y. Asaka and M. Itoh, Phys. Rev. B **71** (2005) 024418.

[17] Z. Ropka and R.J. Radwanski, Physica B **312–313** (2002) 777. R.J. Radwanski and Z. Ropka, *ibid*, **281–282** (2000) 507.

[18] C. Martin, A. Maignan, D. Pelloquin, N. Nguyen and B. Raveau, Appl. Phys. Lett. **71** (1997) 1421. A. Maignan, C. Martin, D. Pelloquin, N. Nguyen and B. Raveau, J. Solid State Chem. **142** (1999) 247.

[19] S. Tsubouchi, T. Kyomen, M. Itoh, P. Ganguly, M. Oguni, Y. Shimojo, Y. Morii and Y. Ishii, Phys. Rev. B **66** (2002) 052418.

[20] G.A. Samara, J. Phys.: Condens. Matter **15** (2003) R367.

[21] B.E. Vugmeister, and M.D. Glinchuk, Rev. Modern Phys. **62** (1990) 993.

[22] D. Fu, Private communication.

[23] E.L. Venturini, G.A. Samara, M. Itoh and R. Wang, Phys. Rev. B **69** (2004), 184105.

[24] E.L. Venturini, G.A. Samara and W. Kleemann, Phys. Rev. B **67** (2003) 214102.

[25] E.L. Venturini, W. Kleemann, G.A. Samara and M. Itoh, unpublished.

[26] J. Yu, T. Iskikawa, Y. Arai, S. Yoda, M. Itoh and Y. Saita, Appl. Phys. Lett. **87** (2005) 252904.

[27] M.A. Subramanian, D. Dong, N. Duan, B.A. Reisner and A.W. Sleight, J. Solid State Chem. **151** (2000) 323.

[28] Y. Inaguma, C. Liquan, M. Itoh, T. Nakamura, T. Uchida, H. Ikuta and M. Wakihara. Solid State Commun. **86** (1993) 689.

[29] S. Stramare, V. Thangadurai and W. Weppner, Chem. Mat. **15** (2003) 3974.

[30] Y. Inaguma and M. Itoh, Rep. Res. Lab. Erg. Mater. **20** (1995) 71.

[31] H. Yamamura and Y. Iwahara, Eds., Crystal Lattice Defects and the Application in Materials Development, IPC, Tokyo 2002 (in Japanese).

[32] A.W. West, Solid State Chemistry and its Applications, John Wiley & Sons, 1984.

[33] H.G.F. Winkler, Acta Cryst. **1** (1948) 27.

[34] V. Tscherry, H. Schulz and F. Laves, Z. Krist. **135** (1972) 161.

[35] V. Tscherry, H. Schulz and F. Laves, *ibid.* **135** (1972) 175.

[36] H. Schulz and V. Tscherry, Acta Cryst. B **28** (1972) 2174.

[37] H. Bohm, Phys. Stat. Sol. A **30** (1975) 531.

[38] U.V. Alpen, H. Schulz, G.H. Talat and H. Bohm, Solid State Commun. **23** (1977) 911.

[39] W. Nagel and H. Bohm, Solid State Commun. **42** (1982) 625.

[40] B. Morosin and J.C. Mikkelsen, Jr., Acta Cryst. B **35** (1979) 798.

[41] I. Abrahams, P.G. Bruce, W.I.F. David and A.R. West, J. Solid State Chem. **78** (1989) 170.

[42] J.B. Boyce and J.C. Mikkelsen, Jr., Solid State Commun. **31** (1979) 741.

[43] Y.F. Yu Yao and J.T. Kummer, I. Inorg. Nucl. Chem. **29** (1967) 2453.

[44] G.C. Farrington, B.S. Dunn and J.L. Briant, Solid State Ionics **3/4** (1981) 405.

[45] G.A. Samara, Solid State Physics **38** (1984) 1.

[46] R.H. Radzilowski and J.T. Kummer, J. Electrochem. **118** (1971) 714.

[47] H.Y-P. Hong, Mat. Res. Bull. **13** (1978) 117.

[48] A.R. West, Z. Krist. **141** (1975) 422.

[49] A.R. West and F.P. Glasser, J. Solid State Chem. **4** (1972) 20.

[50] J. Kuwano and A.R. West, Mat. Res. Bull. **15** (1980) 1661.

[51] Y.-W. Hu, I.D. Raistrick and R.A. Huggins, J. Electrochem. **124** (1977) 1240.

[52] A.R. West, J. Appl. Electrochem. **3** (1973) 327.

[53] J.B. Goodenough, H.Y.-P. Hong and J.A. Kafalas, Mat. Res. Bull. **11** (1976) 203.

[54] H.Y.-P. Hong, Mat. Res. Bull. **11** (1976) 173.

[55] B.E. Taylor, A.D. English and T. Berzins, Mat. Res. Bull. **12** (1977) 171.

[56] L. Shi-Chun and L. Zu-Xiang, Solid State Ionics **9/10** (1983) 835.

[57] Z.X. Lin, H.J. Yu, S.C. Li and S.B. Tian, Solid State Ionics **18/19** (1986) 549.

[58] M.A. Subramanian, R. Subramanian and A. Clearfield, Solid State Ionics **18/19** (1986) 562.

[59] S. Hamdoune, D. Tranqui and E.J.L. Schouler, Solid State Ionics **18/19** (1986) 587.

[60] D. Petit, Ph. Colomban, G. Collin and J.P. Boilot, Mat. Res. Bull. **21** (1986) 365.

[61] Z.X. Lin, H.J. Yu, S.C. Li and S.B. Tian, Solid State Ionics **31** (1988) 91.

[62] S.C. Li, J.Y. Cai and Z.X. Lin, Solid State Ionics **28–30** (1988) 1265.

[63] M. de L. Chavez, P. Quintana and A.R. West, Mat. Res. Bull. **21** (1986) 1411.

[64] H. Aono, E. Sugimoto, Y. Sadaoka, N. Imanaka and G. Adachi, J. Electorchem. Soc. **136** (1989) 590.

[65] B.V.R. Chowdari, K. Radhakrishanan, K.A. Thomas and G.V. Subba Rao, Mat. Res. Bull. **24** (1989) 221.

[66] L. Latie, G. Villeneuve, D. Conte and G.L. Flem, J. Solid State Chem. **51** (1984) 293.

[67] A.G. Belous, G.N. Novitskaya, S.V. Polyanetskaya and Yu. I. Gornikov, Izv. Akad. Nauk SSSR, Neorg. Mater. **23** (1987) 470.

[68] Y. Inaguma, L. Chen, M. Itoh, T. Nakamura, T. Uchida, M. Ikuta and M. Wakihara, Solid State Commun. **86** (1993) 689.

[69] Y. Inaguma, L. Chen, M. Ioth and T. Nakamura, Solid State Ionics **70/71** (1994) 196.

[70] M. Itoh, Y. Inuguma, W.H. Jung, L. Chen and T. Nakamura, Solid State Ionics **70/71** (1994) 203.

[71] Y. Inaguma, J. Yu, Y. Shan, M. Itoh and T. Nakamura, J. Electrochem, **42** (1995) L8.

[72] M. Oguni, Y. Inaguma, M. Itoh and T. Nakamura, Solid State Commun. **91** (1994) 627.

[73] Y. Inaguma, Y. Matsui, Y.-J. Shan, M. Itoh and T. Nakamura, Solid State Ionics **79** (1995) 91.

[74] Y. Inaguma and M. Itoh, Solid State Ionics **86–88** (1996) 257.

[75] T. Katsumata, Y. Matsui, Y. Inaguma and M. Itoh, Solid State Ionics **86–88** (1996) 165.

[76] Y. Inaguma, Y. Matsui, J.-D. Yu, Y.-J. Shan, T. Nakamura and M. Itoh, J. Phys. Chem. Solids **58** (1997) 843.

[77] Y. Inaguma, J.-D. Yu, T. Katsumata and M. Itoh, J. Ceram. Soc. Jpn. **105** (1997) 548.

[78] T. Katsumata, Y. Inaguma, M. Itoh and K. Kawamura, J. Ceram. Soc. Jpn. **107** (1999) 615.

[79] Y. Inaguma, T. Katsumata and M. Itoh, Electrochem. **68** (2000) 534.

[80] S. Kunugi, T. Kyomen, Y. Inaguma and M. Itoh, Electrochem. Solid-State Lett. **5** (2002) A131.

[81] Y. Inaguma, T. Katsumata, M. Itoh and Y. Morii, J. Solid State Chem. **166** (2002) 67.

[82] T. Katsumata, Y. Inaguma, M. Itoh and K. Kawamura, Chem. Mater. **14** (2002) 3930.

[83] M. Yashima, M. Itoh, Y. Inaguma and Y. Morii, J. Am. Chem. Soc. **127** (2005) 3491.

CHAPTER 3

Size Effect of Ferroelectric and High Permittivity Thin Films

Hiroshi Funakubo

Abstract

Ferroelectric and dielectric property degradation with decreasing film thickness (the 'size effect') is the most critical issue in scaling these properties down to the nanometer-size range. In particular, property degradation of polycrystalline film is of greatest concern to devices such as ferroelectric random access memories (FeRAM) and film capacitors for decreasing the operating voltage and increasing the capacitance, respectively. In this section, previous data and recent progress on the 'size effect' are introduced and our recent results on polycrystalline films are discussed.

Keywords: **size effect, ferroelectric property, dielectric property, simple perovskite, bismuth layer structured dielectrics.**

3.1 Introduction

3.1.1 Background

Capacitors play an important role in many key devices such as gate dielectrics in field effect transistors (FET), dynamic random access memories (DRAM) and so on [1]. To progress these devices, thin films of materials with high dielectric constant are required to achieve the necessary large capacitance. For example, thin films of materials of dielectric constant higher than SiO_2 and thicknesses less than 10 nm have been widely investigated for the gate materials, for example, HfO_2 and ZrO_2 [2]. These materials have also been investigated as capacitance materials for DRAM devices [3]. However, thin film research into materials with dielectric constants higher than 200 has found little commercial application in films of this thickness. This is due to the degradation of the dielectric constant with decreasing film thickness, termed as the 'size effect' of the dielectric constant [4]. A related ferroelectric phenomenon is also observed in ferroelectric materials [5], similar to the degradation of ferromagnetism with downsizing. It should be noted that ferroelectric thin films show the largest dielectric constants of all dielectric materials. Downsizing of the ferroelectric thin film is also very important for ferroelectric random access memories 'FeRAM'. FeRAM is the only commercialized memory integrated on Si using the memory effect of material itself rather than the device structure. The requirement for downsizing (decreasing the film thickness) is also essential for FeRAM applications to enable the operating voltage to be decreased to 1.5 V.

3.1.2 Research on the 'size effect' in dielectric and ferroelectric properties

Epitaxial film research for modeling. The 'size effect' in ferroelectric properties has been widely investigated not only in thin films but also in particles [6]. Before the 1990s, the research was concentrated on particles, but this situation has been dramatically changed by the emergence of a novel type of FeRAM [7]. The degradation of ferroelectricity with film thickness dramatically decreases last decay [5]. In particular, the research on epitaxial films clearly shows large extrinsic contributions to the 'size effect' which are limited to the very thin film region.

One of the most typical results was reported by Fong *et al.* [8], who made the *in-situ* X-ray observations on epitaxial $PbTiO_3$ thin films using Synchrotron radiation, ($PbTiO_3$ is the host material of $Pb(Zr,Ti)O_3$, the most widely used ferroelectric material for FeRAM applications). These workers showed that the ferroelectric phase is stable down to 1.2-nm thickness, corresponding to 3 unit cells of $PbTiO_3$. Based on the fact that the crystal structure of the top layer is realigned by the surface reconstruction, the underling two unit cells were considered to constitute the intrinsic 'size effect'.

Another recent study on the 'size effect' of ferroelectricity was carried out by Kim *et al.*, on $BaTiO_3$ thin film, another typical ferroelectric material [9]. They clearly showed that the experimental data for (100)-oriented epitaxial $BaTiO_3$ films grown on $(100)SrRuO_3//(100)SrTiO_3$ substrates took almost the same degradation values as expected from *ab-initio* calcurations of the same system, taking account of the depolarization field effect [10].

The most impressive research related to this topic, but not on thin films, was reported by Saad *et al.* [11,12], who showed the electrical property of $BaTiO_3$ single crystal down to 75 nm was exactly the same as for the bulk. This shows that the most of widely recognized 'size effect' is not the intrinsic characteristic of the $BaTiO_3$, but an extrinsic effect, such as crystal imperfection of the film, interface degradation, etc.

Polycrystalline research for actual applications. The degradation of ferroelectricity with decreasing film thickness is sometimes accompanied by an increase in the coercive field (E_c) and a decrease in the dielectric constant. These can be modeled by the series connection of a ferroelectric layer with high dielectric constant and a non-ferroelectric layer with a low dielectric constant at the interface with the electrode. This low dielectric constant layer is generally called the 'dead layer' and the origin of this dead layer has long been discussed [13].

In spite of the result that the 'size effect' is observed only in the limited thickness region, data observed for polycrystalline films including those with single-axis orientation are quite different from those of epitaxial films. For example, C. Parker et al. [14] showed the complexity of the 'size effect' of $BaTiO_3$ film, not only as a degradation of the dielectric constant at room temperature, but also as a complex change in the temperature dependence of the dielectric constant with decreasing film thickness.

Much of the literature clearly shows that the realistic solution of the 'size effect problem' lies not only in the development of materials with 'size-effect-free' characteristics, but also in a 'less-size-effect' approach, by taking account of the dielectric material in combination with the ferroelectric and electrode materials. This is because the 'size effect' is related to the interface property between the film and the electrode.

In the following section, a capacitor structure showing 'less-size-effect' characteristics is demonstrated, and a novel candidate material with 'size-effect-free' characteristics is introduced.

3.2 Size Effect of Ferroelectricity in PZT Thin Films

3.2.1 Background

$Pb(Zr,Ti)O_3$ is known to have a large remanent polarization (P_r) and thin films 100–200-nm-thick are already used in commercial FeRAM applications [15]. The key issue of the present FeRAM is the lower operation voltage achieved by downsizing the film thickness. However, the degradation of the ferroelectricity with decreasing film thickness is the main drawback to this approach.

3.2.2 Formation of PZT capacitors operating at 1.5 V

To achieve low voltage operation, we have proposed the following:

(i) development of a preparation method for PZT films with good ferroelectric properties;

(ii) choice of a lower electrode providing a good interface with the PZT.

We have proposed a pulsed metal organic chemical vapor deposition (MOCVD) process for this purpose [16,17]. This involves the pulsed introduction of a mixture of source gases into the reaction chamber, which is very effective for making films with good ferroelectricity and low-leakage current density.

Further, we have proposed an Ir lower electrode instead of conventional Pt to lower the operating voltage [18,19]. Figure 3.1 shows the changes in the XRD patterns with film thickness for films prepared at 540°C on $(111)Ir/TiO_2/SiO_2/(100)Si$ and $(111)Pt/TiO_2/SiO_2/(100)Si$ substrates. The (001) and (100) orientations were clearly observed for the films deposited on $(111)Pt//TiO_2/SiO_2/(100)Si$ substrates, in good

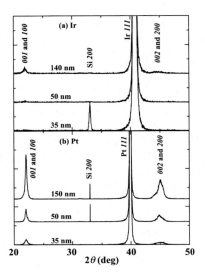

Fig. 3.1. Change in the XRD pattern for films prepared at 540°C on (111) Ir/TiO$_2$/SiO$_2$/(100)Si and (111)Pt/TiO$_2$/SiO$_2$/(100)Si substrates.

agreement with previous reports of rhombohedral PZT with a Zr/(Zr + Ti) ratio of 0.68 [20]. On the other hand, peaks related to PZT were not observed in films deposited on (111)Ir/TiO$_2$/SiO$_2$/(100)Si substrates except for small PZT *001* and PZT *100* peaks.

Figure 3.2 shows the XRD reciprocal space mappings for 50-nm-thick PZT films deposited on (111)Pt/TiO$_2$/SiO$_2$/(100)Si and (111)Ir/TiO$_2$/SiO$_2$/(100)Si substrates. (100) and (001) out-of-plane orientation was clearly observed for the film on the (111)Pt/TiO$_2$/SiO$_2$/(100)Si substrate (Fig. 3.2(a)). On the other hand, the PZT phase also showed (100) and (001) orientations by XRD-RSM for the film on the (111)Ir//TiO$_2$/SiO$_2$/(100)Si substrate (Fig. 3.1(b)), but this was tilted about 15° from the direction normal to the surface. In addition, peak broadening along the psi axis was observed, suggesting that the film was weakly oriented. By comparing the integrated intensity of the PZT phase along the psi axis [21], its peak intensity was found to be almost the same, suggesting similar crystallinity.

Figure 3.3 compares the insulating characteristics of the films shown in Fig. 3.2. The leakage current densities of films deposited on (111)Ir/TiO$_2$/SiO$_2$/(100)Si and (111)Pt/TiO$_2$/SiO$_2$/(100)Si substrates were 1.0×10^{-6} and 1.7×10^{-4} A/cm^2, respectively at an applied electric field of +150 kV/cm. This suggests that the insulation characteristics of the films deposited on (111)Ir/TiO$_2$/SiO$_2$/(100)Si substrates were better than those on (111)Pt/TiO$_2$/SiO$_2$/(100)Si substrates.

The surface roughness of these films was 2.4 nm mean square roughness (R_{ms}) for the (111)Ir/TiO$_2$/SiO$_2$/(100)Si substrates and 3.1 nm for the (111)Pt/TiO$_2$/SiO$_2$/(100)Si substrates (Fig. 3.4). The better surface flatness of the film on the (111)Ir/TiO$_2$/SiO$_2$/(100)Si substrate than on the (111)Pt/TiO$_2$/SiO$_2$/(100)Si substrate produced better insulation

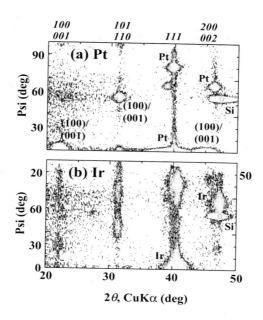

Fig. 3.2. XRD reciprocal space mappings for 50-nm-thick PZT films deposited on (a) (111)Pt/TiO$_2$/SiO$_2$/(100)Si and (b) (111)Ir/TiO$_2$/SiO$_2$/(100)Si substrates.

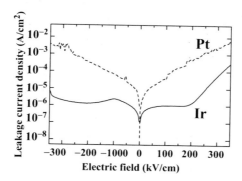

Fig. 3.3. Leakage current density as a function of applied electric field for the films shown in Fig. 3.2.

characteristics, because the leakage characteristics of MOCVD-PZT films are strongly dependent on the surface roughness of the films [16].

Figure 3.5(a) shows the relative dielectric constant (ε_r) as a function of film thickness, indicating that PZT films deposited on (111)Ir/TiO$_2$/SiO$_2$/(100)Si substrates show a larger ε_r value than those on (111)Pt/TiO$_2$/SiO$_2$/(100)Si substrates. The ε_r value of

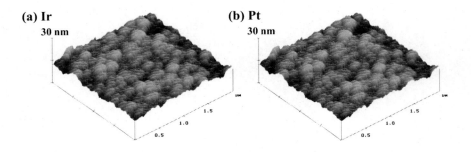

Fig. 3.4. SPM images of 50-nm-thick films deposited on (a) (111)Ir/TiO$_2$/SiO$_2$/(100)Si and (b) (111)Pt/TiO$_2$/SiO$_2$/(100)Si substrates.

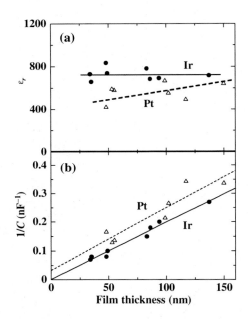

Fig. 3.5. (a) Relative dielectric constant (ε_r) and (b) inverse of capacitance as a function of the film thickness for PZT films deposited on (111)Ir/TiO$_2$/SiO$_2$/(100)Si and (111)Pt/TiO$_2$/SiO$_2$/(100)Si substrates.

the 35-nm-thick film was plotted for the (111)Ir/TiO$_2$/SiO$_2$/(100)Si substrate but not for (111)Pt/TiO$_2$/SiO$_2$/(100)Si because of the higher leakage current density of the 35-nm film deposited on the (111)Pt/TiO$_2$/SiO$_2$/(100)Si substrate.

On the other hand, as shown in Fig. 3.5(b), the plot of 1/C passes through zero for the films on (111)Ir/TiO$_2$/SiO$_2$/(100)Si substrates, but not for films on (111)Pt/TiO$_2$/SiO$_2$/(100)Si substrates, although the slopes of the straight lines are virtually identical. This shows

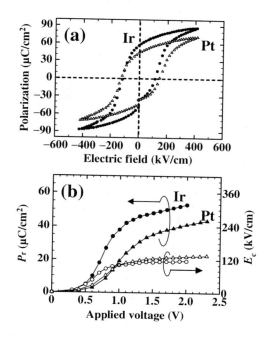

Fig. 3.6. (a) Polarization–electric field (P–E) hysteresis loops measured at 20 Hz and (b) saturation properties of P_r and E_c for films prepared on (111)Ir/TiO$_2$/SiO$_2$/(100)Si substrates by pulsed-MOCVD.

that the 'dead layer' contribution is small in films on (111)Ir/TiO$_2$/SiO$_2$/(100)Si substrates. These data show that the combination of the pulsed-MOCVD process and (111)Ir/TiO$_2$/SiO$_2$/(100)Si substrates produces films with good ferroelectric properties.

Figure 3.6 shows the polarization–electric field (P–E) hysteresis loops together with the saturation characteristics of the remanent polarization (P_r) and the coercive field (E_c) for 50-nm-thick films. Good saturation of P_r and E_c was established in both films, the larger P_r value of 47 μC/cm^2 being achieved at 1.2 V for the films on (111)Ir/TiO$_2$/SiO$_2$/(100)Si substrate, while the E_c value of 125 kV/cm at 1.2 V was almost the same in each case.

Figure 3.7 shows the saturation properties of the P_r and E_c values for films of various thickness deposited on both substrates. The P_r value saturated at low voltage for the films deposited on (111)Ir/TiO$_2$/SiO$_2$/(100)Si substrates, suggesting that the (111)Ir/TiO$_2$/SiO$_2$/(100)Si substrate is useful for low-voltage operation. Moreover, the P_r value was almost unchanged with film thickness for the films on (111)Ir/TiO$_2$/SiO$_2$/(100)Si substrates, but this parameter slightly decreased for Pt-top substrates when the film thickness decreased. On the other hand, the E_c value gradually increased with decreasing film thickness in both films.

Figure 3.8 shows the change of P_r and E_c values for the films deposited on both substrates. The P_r values are larger and the E_c values are smaller for the films on

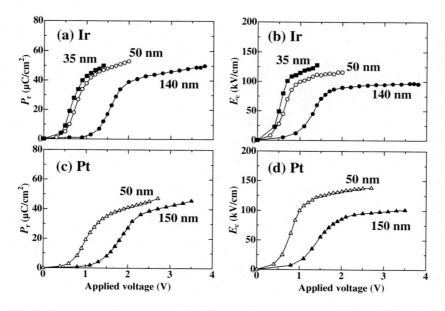

Fig. 3.7. Saturation properties of (a),(c) P_r and (b),(d) E_c as a function of the applied voltage for the films deposited on (a),(b) (111)Ir/TiO$_2$/SiO$_2$/Si and (c),(d) (111)Pt/TiO$_2$/SiO$_2$/Si substrates.

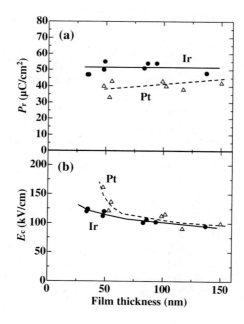

Fig. 3.8. (a) P_r and (b) E_c of the films deposited on (111)Ir/TiO$_2$/SiO$_2$/(100)Si and (111)Pt/TiO$_2$/SiO$_2$/(100)Si substrates.

(111)Ir/TiO$_2$/SiO$_2$/(100)Si substrates than for the films on (111)Pt/TiO$_2$/SiO$_2$/(100)Si substrates regardless of the film thickness. Moreover, the rate of decrease of P_r and the increase of E_c with decreasing film thickness for the films deposited on (111)Ir/TiO$_2$/SiO$_2$/(100)Si substrates is smaller than for the (111)Pt/TiO$_2$/SiO$_2$/(100)Si substrates, especially in the thin-film region below 50 nm. This is thought to be due to the smaller interdiffusion between the PZT film and the Ir bottom electrode due to the good diffusion barrier characteristics of Ir.

3.3 Size-Effect-Free Characteristics of Bismuth Layer Structured Dielectrics

3.3.1 Introduction to bismuth layer-structured ferroelectric dielectrics (BLD)

Bismuth layer-structured dielectrics (BLD) were first reported in 1949 [22]. Their most distinctive structural feature is the intergrowth of the perovskite-like slab (pseudoperovskite-layer) and bismuth-oxide layers along the c-axis as shown in Fig. 3.9. The chemical formula is usually described as $(Bi_2O_2)^{2+}(A_{m-1}B_mO_{3m+1})^{2-}$, where A represents the mono-, di-, or trivalent ions (Na$^+$, Sr^{2+}, Ba^{2+}, Pb^{2+}, Ca^{2+}, Bi^{3+}, lanthanide ions etc.), B represents the tetra-, penta-, or hexavalent ions (Ti^{4+}, V^{5+}, Ta^{5+}, Nb^{5+}, Mo^{6+}, and W^{6+} etc.) and m represents the number of BO$_6$ octahedra in the pseudoperovskite layer. Figure 3.10 shows the crystal structures of the representative compounds Bi$_2$VO$_{5.5}$ ($m = 1$), SrBi$_2$Ta$_2$O$_9$ ($m = 2$) and Bi$_4$Ti$_3$O$_{12}$ ($m = 3$). Almost all BLDs have orthorhombic symmetry at room temperature (that of Bi$_4$Ti$_3$O$_{12}$

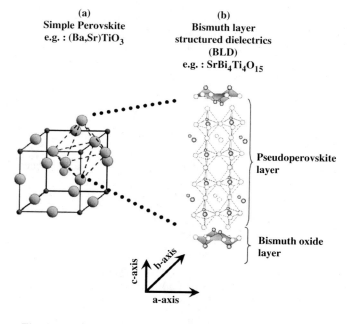

Fig. 3.9. Crystal structures of simple perovskite and BLD.

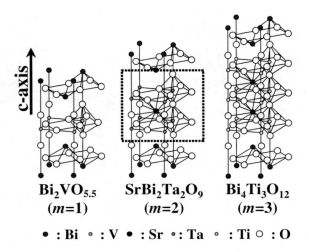

Fig. 3.10. Schematic illustration of $Bi_2VO_{5.5}$ ($m = 1$), $SrBi_2Ta_2O_9$ ($m = 2$) and $Bi_4Ti_3O_{12}$ ($m = 3$) crystal structures, showing a half-unit cell along the c-axis.

is strictly monoclinic). More than 50 materials with different m-numbers ranging from 1 to 5, or combinations of these have been reported (Table 3.1), and are classed as ferroelectrics [23].

Research on thin-film BLDs increased dramatically after the discovery of the fatigue-free characteristics of the ferroelectric behavior in $SrBi_2Ta_2O_9$ films after multiple switching cycles, because fatigue was the most severe problem in the PZT films originally used in FeRAM applications [24]. After $SrBi_2Ta_2O_9$, novel materials based on $Bi_4Ti_3O_{12}$, including $(Bi, La)_4Ti_3O_{12}$ [25], $(Bi, Nd)_4Ti_3O_{12}$ [26,27] and $(Bi, Nd)_4(Ti, V)_3O_{12}$ [28] were developed to overcome the problems of low ferroelectricity and high processing temperature of $SrBi_2Ta_2O_9$. This led to the design concept of ferroelectrics called the 'site-engineered concept' [29,30].

3.3.2 Orientation dependence of the electrical properties of BLD

Because of the lack of single crystal data, the orientation dependence of epitaxial BLD films has been systematically investigated by many groups including our group [31–34]. By this process, the novel 'size-effect-free' concept was developed.

Epitaxial films of $Bi_2VO_{5.5}$, $SrBi_2Ta_2O_9$, and $Bi_4Ti_3O_{12}$ were systematically grown on (100), (110), and (111)$SrTiO_3$ substrates covered by epitaxially-grown $SrRuO_3$ bottom electrode layers with the same orientation as the $SrTiO_3[SrRuO_3//SrTiO_3]$. The layers were grown by MOCVD to have different m-numbers, by adjusting the deposition conditions [34]. Figure 3.11 shows an example of the XRD reciprocal space mappings of $Bi_4Ti_3O_{12}$ films grown on $SrRuO_3//SrTiO_3$ substrates with different orientations. (001)-, (118)-, and (104)-oriented films were epitaxially grown on (100), (110), and (111)$SrRuO_3//SrTiO_3$ substrates.

Table 3.1. Family of BLDs family layer perovskite.

m	1	2	3	4	5
Materials	$Bi_2VO_{5.5}$	$SrBi_2Ta_2O_9$	$Bi_4Ti_3O_{12}$	$SrBi_4Ti_4O_{15}$	$Sr_2Bi_4Ti_5O_{18}$
	Bi_2MoO_6	$CaBi_2Ta_2O_9$	$(Bi_{4-x}Ln_x)Ti_3O_{12}$	$BaBi_4Ti_4O_{15}$	$Ba_2Bi_4Ti_5O_{18}$
	Bi_2WO_6	$BaBi_2Ta_2O_9$	$Bi_{4-x/3}Ti_{3-x}V_xO_{12}$	$PbBi_4Ti_4O_{15}$	$Pb_2Bi_4Ti_5O_{18}$
		$PbBi_2Ta_2O_9$	$Bi_{4-2x/3}Ti_{3-x}W_xO_{12}$	$CaBi_4Ti_4O_{15}$	$Bi_6Ti_3Fe_2O_{18}$
		$SrBi_2Nb_2O_9$	$(Bi,Ln)_4(Ti,V)_3O_{12}$	$Bi_5Ti_3FeO_{15}$	
		$CaBi_2Nb_2O_9$	$Bi_2Sr_2Nb_2MnO_{12-x}$	$Na_{0.5}Bi_{3.5}Ti_4O_{15}$	
		$BaBi_2Nb_2O_9$	$(Bi_{4-x}Sr_x)(Ti_{3-x}Nb_x)O_{12}$	$K_{0.5}Bi_{3.5}Ti_4O_{15}$	
		$PbBi_2Nb_2O_9$	$(Bi_{4-x}Ba_x)(Ti_{3-x}Nb_x)O_{12}$		
		Bi_3TiNbO_9	$(Bi_{4-x}Pb_x)(Ti_{3-x}Nb_x)O_{12}$		
		Bi_3TiTaO_9			

m	1, 2	2, 3	3, 4
Materials	$Bi_2WO_6-Bi_3Ti_{1.5}W_{0.5}O_9$	$Bi_3Ti_{1.5}W_{0.5}O_9-Bi_4Ti_3O_{12}$	$Bi_4Ti_3O_{12}-SrBi_4Ti_4O_{15}$
	$Bi_2WO_6-Bi_3TiNbO_9$	$Bi_3TiNbO_9-Bi_4Ti_3O_{12}$	$Bi_4Ti_3O_{12}-BaBi_4Ti_4O_{15}$
	$Bi_2WO_6-Bi_2(Bi_{0.5}Na_{0.5})NbO_9$	$Bi_3TiTaO_9-Bi_4Ti_3O_{12}$	$Bi_4Ti_3O_{12}-PbBi_4Ti_4O_{15}$
	$Bi_2WO_6-Bi_3TaTiO_9$	$SrBi_2Nb_2O_9-Bi_4Ti_3O_{12}$	$Bi_4Ti_3O_{12}-(Na_{0.5}Bi_{0.5})Bi_4Ti_4O_{15}$
		$BaBi_2Nb_2O_9-Bi_4Ti_3O_{12}$	

Ln: Lanthanide.

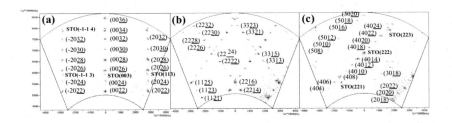

Fig. 3.11. XRD reciprocal space mappings of $Bi_4Ti_3O_{12}$ films deposited on (a)(100), (b)(110), and (c)(111) $SrRuO_3$//$SrTiO_3$ substrates.

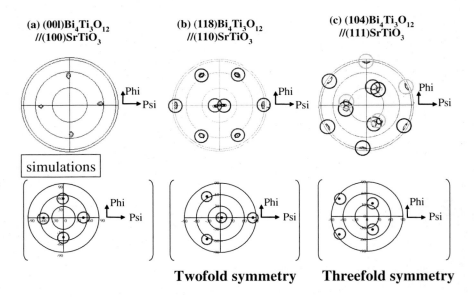

Fig. 3.12. XRD pole figure plots of the films shown in Fig. 3.11 together with the simulation results assuming the orientations shown in Fig. 3.11.

In addition, analysis of the XRD pole figure plots and comparison with the simulation shown in Fig. 3.12 for $Bi_4Ti_3O_{12}$ shows that the films grown on (100), (110), and (111) $SrRuO_3$//$SrTiO_3$ substrates have the 1, 2, and 3 in-plane variants shown schematically in Fig. 3.13.

Table 3.2 summarizes the orientation of the epitaxial films. An equivalent epitaxial relationship between the film and the substrate was obtained, irrespective of m-number.

Figure 3.14 shows the TEM images of (001)- and (116)-oriented $SrBi_2Ta_2O_9$ films grown directly on (100)- and (110) $SrTiO_3$ substrates without $SrRuO_3$ layers. Figs. 3.14(a) and (b) agree well with the schematic drawing of Fig. 3.13(a) and (b) [35,36].

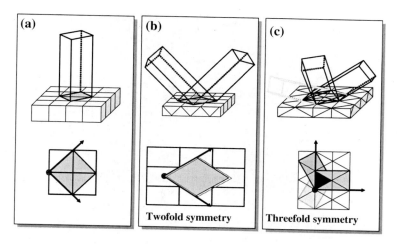

Fig. 3.13. Schematic drawing of the variants of the BLD films grown on (a)(100), (b)(110), and (c)(111) SrRuO$_3$//SrTiO$_3$ substrates.

Table 3.2. Crystal growth relationship between epitaxial BLD films and SrTiO$_3$ substrates.

m	Material	Orientation		
		(100)SrTiO$_3$	(110)SrTiO$_3$	(111)SrTiO$_3$
3	Bi$_4$Ti$_3$O$_{12}$	(001)	(118)	(104)
2	SrBi$_2$Ta$_2$O$_9$	(001)	(116)	(103)
1	Bi$_2$VO$_{5.5}$	(001)	(114)	(102)
Tilting angle of c-axis (deg.)		0	45	55

Figure 3.15 shows the AFM images of the (001)-, (118)-, and (104)-oriented Bi$_4$Ti$_3$O$_{12}$ films deposited on SrRuO$_3$//SrTiO$_3$ substrates. In the case of the (001)-oriented film, a plate like morphology was observed as shown in Fig. 3.15(a) and the root mean square of the surface roughness (R_a) was 0.547 nm. Fig. 3.15(d) shows the cross section of the film shown in Fig. 3.15(a). The gap of the height of each plate-like region is about 1.6 nm, corresponding to half of the c-axis lattice parameter, indicating that this film has a flat surface. On the other hand, columnar grains growing in one direction and triangular grains were observed for the (118)- and (104)-oriented films, respectively (Fig. 3.15(b,c)). The R_a values were 3.4 and 55.5 nm for the (118)- and (104)-oriented films, respectively. The roughness of the Bi$_4$Ti$_3$O$_{12}$ film surfaces increased in the following order: (001)-, (118)-, and (104)-orientation; this orientation dependency was also observed for Bi$_2$VO$_{5.5}$ and SrBi$_2$Ta$_2$O$_9$ films. Thus, this orientation dependency of the surface roughness originates in the growth mode of the BLD films; the growth rate along the a–b plane is faster than along the c-axis.

Figure 3.16 shows the ε_r values of films with the c-axis perpendicular to the substrates and one tilted at 45°. Films tilted at 55°, such as (102)-oriented Bi$_2$VO$_{5.5}$, (103)-oriented

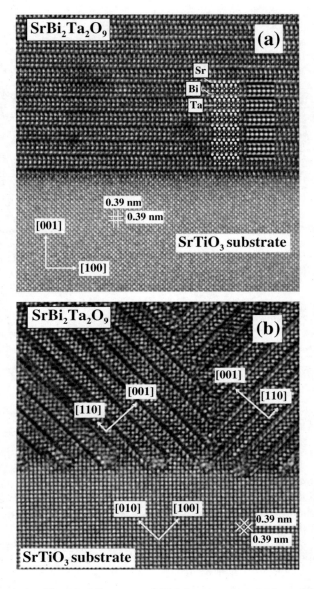

Fig. 3.14. Transmission Electron Microscope (TEM) images of (a) (001)- and (b) (116)-oriented $SrBi_2Ta_2O_9$ films grown on (100) and (110)$SrTiO_3$ substrates, respectively.

$SrBi_2Ta_2O_9$, and (104)-oriented $Bi_4Ti_3O_{12}$ were not measured due to their large leakage currents. As the m-number increased, the ε_r value increased irrespective of the film orientation. Moreover, 45°-tilted films showed higher ε_r values than c-axis-oriented films. This is a common feature of BLD materials and indicates that the c-axis has a smaller ε_r value than the a- and b-axes, in good agreement with the data reported for single crystals [37].

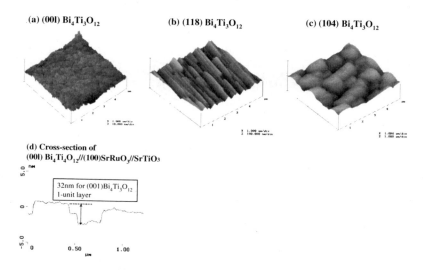

Fig. 3.15. AFM images of (a) (00l)-, (b) (118)-, and (c) (104)-oriented $Bi_4Ti_3O_{12}$ films and (d) the cross section of (001)-oriented $Bi_4Ti_3O_{12}$ films.

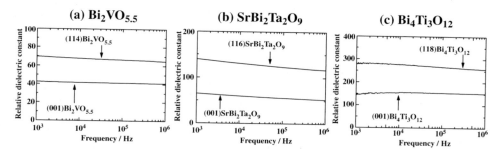

Fig. 3.16. Relative dielectric constant (ε_r) of (a) $Bi_2VO_{5.5}$, (b) $SrBi_2TaO_9$, and (c) $Bi_4Ti_3O_{12}$ films grown on (100) and (110)$SrRuO_3$//(100)$SrTiO_3$ substrates.

Figure 3.17 shows the leakage current density for the films shown in Fig. 3.16. The c-axis films tilted at 45° show larger leakage current densities than c-axis-oriented films. Fouskova et al. reported that the leakage current of a BIT single crystal along the a-axis was 30 times larger than along the c-axis, in good agreement with the present study [38]. This is also supported by the comparison of the (100)/(010)- and (001)-oriented epitaxial $SrBi_2Ta_2O_9$ and $Bi_4Ti_3O_{12}$ films shown in Fig. 3.18 [39]. This can be explained by the high resistivity of the $(Bi_2O_2)^{2+}$ layer, an important characteristic, especially for making thin films.

P–E hysteresis loops of the films of Fig 3.17 are shown in Fig. 3.19. The ferroelectricity of the (00l)- and (114)-oriented $Bi_2VO_{5.5}$ film was confirmed by a scanning

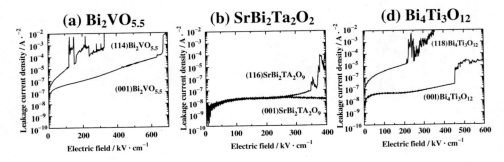

Fig. 3.17. Leakage current density as a function of electric field for (a) $Bi_2VO_{5.5}$, (b) $SrBi_2TaO_9$, and, (c) $Bi_4Ti_3O_{12}$ films grown on (100) and (110)$SrRuO_3$//(100)$SrTiO_3$ substrates.

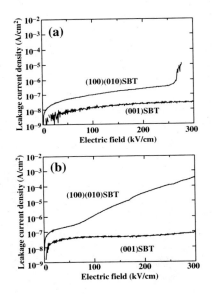

Fig. 3.18. Leakage current density as a function of electric field for (100)/(001) and (001)-oriented (a) $SrBi_2TaO_9$ and (b) $Bi_4Ti_3O_{12}$ films [39].

nonlinear dielectric microscope; its remanent polarization was less than 1 $\mu C/cm^2$, in good agreement with the reported value for $Bi_2VO_{5.5}$ single crystal [40]. A large ferroelectric anisotropy was confirmed for $SrBi_2TaO_9$ and $Bi_4Ti_3O_{12}$ films as shown in Fig. 3.19(b),(c),(e), and (f).

Furthermore, as predicted from the crystal structure shown in Fig. 3.20, $Bi_4Ti_3O_{12}$ ($m = 3$, odd) showed ferroelectricity but not $SrBi_2TaO_9$ ($m = 2$, even) along the c-axis. The estimated spontaneous polarization (P_s) values along the a- and c-axes were 0 and

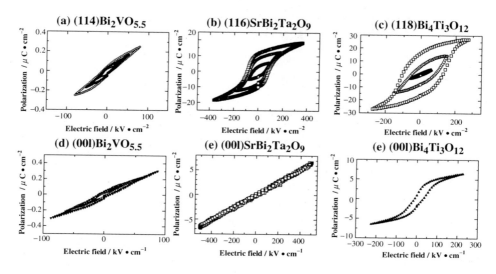

Fig. 3.19. Polarization–electric field (P–E) relationships for (a) (14)-oriented $Bi_2VO_{5.5}$ film, (b) (116)-oriented $SrBi_2TaO_9$ film, (c) (118)-oriented $Bi_4Ti_3O_{12}$ film, (d) (001)-oriented $Bi_2VO_{5.5}$ film, (e) (001)-oriented $SrBi_2TaO_9$ film, and (f) (001)-oriented $Bi_4Ti_3O_{12}$ film.

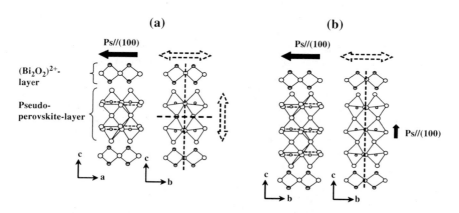

Fig. 3.20. Schematic drawing of bismuth layer structured dielectrics having (a) even m number, and (b) odd m number.

22 µC/cm² for the SBT film and, 4.0 and 48.4 µC/cm² for the $Bi_4Ti_3O_{12}$ film. Thus, the P_s value of the BLD film is in approximate agreement with that of a single crystal and the calculation based on the crystal structure [37,41], suggesting that a small contribution of strain remains in the film. In fact, the lattice constants of the epitaxial $SrBi_2TaO_9$ film were found to be almost the same as reported for $SrBi_2TaO_9$ powder [42]. The films, although epitaxial, are apparently not coherently strained to match the lattice spacing of the $SrTiO_3$ substrates. Therefore, this strain-free character of BLD films is suitable for

116   Nanomaterials: From Research to Applications

device applications because these characteristics are known to be very sensitive to the strain in the case of simple perovskites [43].

3.3.3 Epitaxial (001)-oriented SBT film

In the previous section, we established the unique characteristics of BLD films with c-axis orientation. These can be summarized as follows

(a) *Good leakage characteristics*
This is a common characteristic of BLD (Figs. 3.17 and 3.18),

(b) *Good surface flatness*
This is observed for single crystals, so it is taken to be a common characteristic of BLD. This is useful for making thin films with low-leakage characteristics because it strongly depends on the leakage characteristics as shown in PZT thin films [16],

(c) *Good electric field independence of the dielectric constant* (for even values of m).
As shown in Fig. 3.19(e), the P–E hysteresis is a straight line, suggesting a stable dielectric constant with electric field for BLD films of even m number.

The above characteristics (a) and (b) are considered to originate in the stacked structure of bismuth oxide and the pseudoperovskite layers (Fig. 3.21). On the other hand, characteristic (c) arises from the c-axis of BLD being nonferroelectric. This c-axis characteristic of BLD is unique because the electrical property of the lack of a polar axis in ferroelectric materials has never been investigated in thin films since it is very difficult to grow this direction selectively in normal perovskites. In the case of BLD, we have already demonstrated that the c-axis oriented film can be grown even on widely used incoherent substrates such as (111)Pt/TiO$_2$/SiO$_2$/(100)Si [44]. These data suggest that a thin film capacitor with good surface flatness and a dielectric constant independent of the electric field can be deposited. Next, we checked the film-thickness dependence of the dielectric constant of BLD films with even m numbers. SrBi$_2$Ta$_2$O$_9$ was selected for this demonstration because it has been widely investigated not only in thin-film form, but also in single crystals due to its practical application to the FeRAM technology [45].

Figure 3.22 shows the XRD patterns of SrBi$_2$Ta$_2$O$_9$ films of various thicknesses prepared on (100)SrRuO$_3$//(100)SrTiO$_3$ substrates. Only the 00l diffraction peaks of SrBi$_2$Ta$_2$O$_9$ were observed, in addition to the peaks from the substrate. This figure shows that c-axis-oriented SrBi$_2$Ta$_2$O$_9$ films were grown on these substrates. In addition, the XRD pole figures show that these films were grown epitaxially.

Figure 3.23 shows the film-thickness dependence of the rocking curve, the full width at half maximum (FWHM) values of 00$\underline{18}$ SrBi$_2$Ta$_2$O$_9$. The FWHM values are independent of the film thickness and are less than 0.13°. This result indicates that the perfection of the crystal orientation is very similar, irrespective of the film thickness.

Figure 3.24 shows the cross-sectional TEM image of the c-axis-oriented SrBi$_2$Ta$_2$O$_9$ film grown on a (100)SrRuO$_3$//(100)SrTiO$_3$ substrate. The stacking structure of the (Bi$_2$O$_2$)$^{2+}$

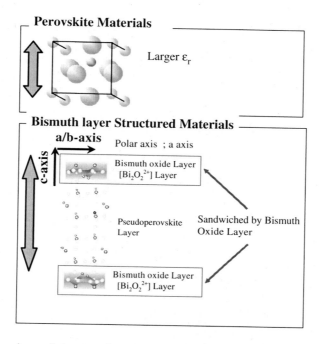

Fig. 3.21. Comparison of the crystal structures of simple perovskite and the bismuth layer-structured dielectrics.

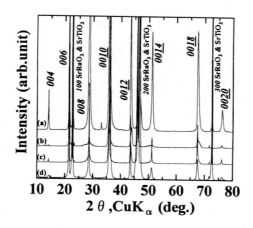

Fig. 3.22. XRD patterns of the (a) 170-nm-, (b) 100-nm-, (c) 65-nm-, and (d) 20-nm-thick SrBi$_2$Ta$_2$O$_9$ films grown on (100)SrRuO$_3$//(100)SrTiO$_3$ substrates.

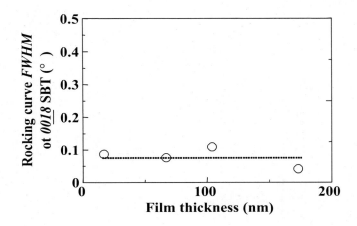

Fig. 3.23. Film thickness dependence of the rocking curve FWHM of the $00\underline{18}$ $SrBi_2Ta_2O_9$ thin films.

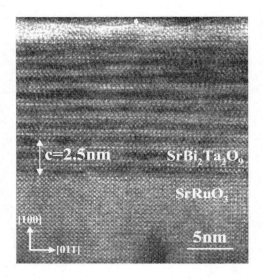

Fig. 3.24. Cross-sectional TEM image of the c-axis-oriented $SrBi_2Ta_2O_9$ film grown on a $(100)SrRuO_3//(100)SrTiO_3$ substrate.

and $(SrTa_2O_7)^{2-}$ layers is clearly observed. Moreover, the interfacial layer between the $SrBi_2Ta_2O_9$ film and $SrRuO_3$ was not detected.

Figure 3.25 shows relative dielectric constant (ε_r) and the reciprocal of the capacitance ($1/C$) as a function of the film thickness for c-axis-oriented epitaxial $SrBi_2Ta_2O_9$ films. The relative dielectric constant retained a value of about 55 even when the film thickness

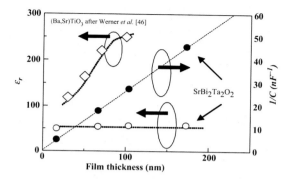

Fig. 3.25. Relative dielectric constant (ε_r) and the reciprocal of the capacitance ($1/C$) as a function of film thickness of c-axis-oriented epitaxial $SrBi_2Ta_2O_9$ films. Data for $(Ba,Sr)TiO_3$ were taken from Werner et al. [46].

was decreased to 20 nm. On the other hand, $(Ba,Sr)TiO_3$ films reported by Werner et al. [46], which are also shown in Fig. 3.25, strongly depend on the film thickness. Moreover, the film thickness dependence of $1/C$ shows a linear relationship passing through zero and the relative dielectric constant calculated from the slope of this line is 55, suggesting that the so-called 'dead-layer' is barely present. The relative dielectric constant of $SrBi_2Ta_2O_9$ thin films is about 55, but the relative dielectric constants of c-axis-oriented BLD samples are known to increase to about 200 with increasing m number [47] (Fig. 3.16).

Figure 3.26(a) shows capacitance and applied electric field (C–E) characteristics of c-axis-oriented epitaxial $SrBi_2Ta_2O_9$ thin films of various thicknesses. The change of capacitance with applied electric field was found to be small and independent of the film

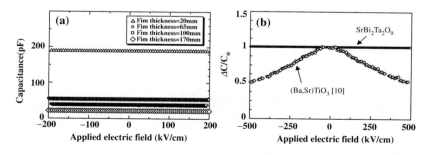

Fig. 3.26. (a) Capacitance and applied electric field (C–E) characteristics of c-axis-oriented epitaxial $SrBi_2Ta_2O_9$ thin films, and (b) $\Delta C/C_0$–E characteristics of $(Ba,Sr)TiO_3$ film vs. 65-nm-thick $SrBi_2Ta_2O_9$ film. Data for $(Ba,Sr)TiO_3$ were taken from Shin et al. [48]. C_0 represents the capacitance at 0 kV/cm.

thickness. Fig. 3.26(b) compares the $C-E$ characteristic of 80-nm-thick (Ba, Sr)TiO$_3$ film reported by Shin *et al.* [48] and a 65-nm-thick SrBi$_2$Ta$_2$O$_9$ thin film. This figure shows that the changes of capacitance of the (Ba,Sr)TiO$_3$ and SrBi$_2$Ta$_2$O$_9$ plots are 51% and 0.017%, respectively at applied electric fields in the range -500 to 500 kV/cm. These results indicate that the capacitance of a SrBi$_2$Ta$_2$O$_9$ film is more stable towards the applied electric field than for a (Ba,Sr)TiO$_3$ film.

Figure 3.27(a) shows the leakage current density of c-axis-oriented SrBi$_2$Ta$_2$O$_9$ films as a function of the applied electric field. The leakage current characteristics with applied field of SrBi$_2$Ta$_2$O$_9$ films 20–170-nm-thick are very similar, with leakage current densities of the order of 10^{-8} A/cm^2. Figure 3.27(b) shows the leakage current densities at 150 kV/cm as a function of the film thickness of SrBi$_2$Ta$_2$O$_9$ film. The leakage current densities of the SrBi$_2$Ta$_2$O$_9$ thin films are almost constant, and less than 4.0×10^{-8} A/cm^2.

Figure 3.28 shows AFM images of 20-nm-, 65-nm-, and 100-nm-thick c-axis-oriented SrBi$_2$Ta$_2$O$_9$ thin films. In epitaxially grown Bi$_4$Ti$_3$O$_{12}$ films, the leakage current density is strongly affected by the surface roughness of the film [31]. The average surface

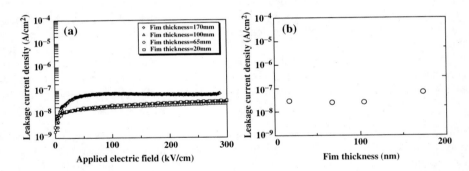

Fig. 3.27. (a) Leakage current density as a function of applied electric field for SrBi$_2$Ta$_2$O$_9$ thin films, and (b) the leakage current density at 150 kV/cm for the Pt/SrBi$_2$Ta$_2$O$_9$/SrRuO$_3$/(100)SrTiO$_3$ structure as a function of SrBi$_2$Ta$_2$O$_9$-film thickness.

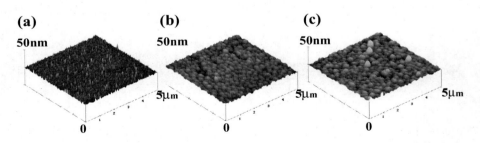

Fig. 3.28. AFM images of (a) 20-nm-, (b) 65-nm-, and (c) 100-nm-thick c-axis-oriented SrBi$_2$Ta$_2$O$_9$ thin films grown on (100)SrRuO$_3$//(100)SrTiO$_3$ substrates.

roughness (R_a) of c-axis-oriented 20-nm-, 65-nm-, and 100-nm-thick $SrBi_2Ta_2O_9$ thin films estimated from Fig. 3.28 was 1.2, 1.5, and 1.5 nm, respectively. This indicates that these $SrBi_2Ta_2O_9$ thin films have good surface flatness irrespective of film thickness and this tendency is the same as found for the leakage current densities (Fig. 3.27), because of the preferential a–b plane growth due to the crystal structural anisotropy [49].

This 'size-effect-free' character is very large because the pseudoperovskite layers which mainly contribute to the total dielectric constant, are separated by bismuth oxide layers. The thickness of these separated pseudoperovskite layers is known to produce the desired decrease in the dielectric constant by comparison with a slab of infinite thickness. This shows that the present 'size-effect-free' characteristics along the c-axis of BLD is a characteristic of the stacking structure of the pseudoperovskite and bismuth oxide layers. The detailed mechanism of these 'size-effect-free' characteristics is under investigation.

3.3.4 Expansion to single-axis oriented films

From a practical point of view, an increase in the dielectric constant above 200 is essential to extend the applications. In addition, the same electrical properties must be realized not only in epitaxial films but also in polycrystalline materials, at least in single-axis oriented films. This case corresponds to c-axis orientation. This is because epitaxial films are not suitable for use with widely-used substrates such as (100)Si, metal foil and the polycrystalline ceramic substrates. One way of increasing the dielectric constant is to increase the m number [47], so we selected $SrBi_4Ti_4O_{15}$ with $m = 4$ as a target material.

Figure 3.29 shows the XRD θ-2θ patterns of 30-nm-thick $SrBi_4Tl_4O_{15}$ films deposited by MOCVD on (a) (111)$Pt/TiO_2/SiO_2/$(100)Si substrate covered by an $LaNiO_3$ layer, i.e., (100)$_c LaNiO_3/$(111)$Pt/TiO_2/SiO_2/$(100)Si, and (b) (111)$Pt/TiO_2/SiO_2/$(100)Si substrates

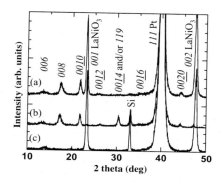

Fig. 3.29. XRD patterns of 30-nm-thick $SrBi_4Ti_4O_{15}$ films deposited on (a) (100)$_c LaNiO_3/$(111)$Pt/TiO_2/SiO_2/$(100)Si and (b) (111)$Pt/TiO_2/SiO_2/$(100)Si substrates, and XRD patterns of (c) (100)$_c LaNiO_3/$(111)$Pt/TiO_2/SiO_2/$(100)Si substrate before $SrBi_4Ti_4O_{15}$ deposition.

together with (c) the XRD θ-2θ patterns of a $(100)_c$LaNiO$_3$/(111)Pt/TiO$_2$/SiO$_2$/(100)Si substrate. Conductive LaNiO$_3$ films with a lattice parameter of 0.3861 nm [50] and the perovskite structure were deposited by RF magnetron sputtering to serve as an oxide buffer layer on (111)Pt. These demonstrated self-$(100)_c$-orientation, as shown in Fig. 3.29(c). Only the 00l peaks of SrBi$_4$Ti$_4$O$_{15}$ were observed, in addition to those of the substrate, for a film prepared on this $(100)_c$LaNiO$_3$/(111)Pt/TiO$_2$/SiO$_2$/(100)Si substrate (Fig. 3.29(a)). On the other hand, a SrBi$_4$Ti$_4$O$_{15}$ film deposited directly on a (111)Pt/TiO$_2$/SiO$_2$/(100)Si substrate showed peak intensities almost as strong as the 00l peaks of the SrBi$_4$Ti$_4$O$_{15}$ phase, as shown in Fig. 3.29(b). However, the relative intensity of the diffraction peak at about $2\theta = 30°$ corresponding to 00$\underline{14}$ SrBi$_4$Ti$_4$O$_{15}$ is much higher than that of the standard SrBi$_4$Ti$_4$O$_{15}$ powder, suggesting the coexistence of (119)-oriented SrBi$_4$Tl$_4$O$_{15}$ or the presence of impurity phases as well as the c-axis-oriented phase.

To identify the constituent phase and the orientation of the deposited films, an X-ray reciprocal space mapping method was used [51]. Figure 3.30(a)–(c) shows the X-ray reciprocal space maps of the SrBi$_4$Ti$_4$O$_{15}$ films and the $(100)_c$LaNiO$_3$/(111)Pt/TiO$_2$/SiO$_2$/(100)Si substrate shown in Fig. 3.29. Figure 3.30(d)–(f) shows the integrated patterns of the psi (ψ) scans in the 2θ range from 29 to 32°. The horizontal and vertical axes in these maps correspond to the 2θ angle in a conventional θ-2θ scan and the tilt angle (ψ) perpendicular to the scattering plane, respectively.

All of the diffraction spots were identified as single-phase SrBi$_4$Ti$_4$O$_{15}$, and no impurity phase was detected in Fig. 3.30(a) and (b). Several diffraction spots in Fig. 3.30(a) and (b) indicate that these SrBi$_4$Ti$_4$O$_{15}$ films are highly oriented, based on the comparison with the results shown in Fig. 3.30(c). The spots in Fig. 3.30(a) correspond to a purely (001)-oriented SrBi$_4$Ti$_4$O$_{15}$ phase. This was also ascertained for Fig. 3.30(d), in which a large peak was observed at $\psi = 0$ and 47.3°, and both diffraction peaks originate from 0014 SrBi$_4$Ti$_4$O$_{15}$ with a surface-normal (001) orientation. However, several additional diffraction spots were observed at about $2\theta = 30°$ (Fig. 3.30(b)), indicating the coexistence of non-c-axis-oriented SrBi$_4$Ti$_4$O$_{15}$ phases. Detailed analysis showed that the SrBi$_4$Ti$_4$O$_{15}$ film deposited directly on the (111)Pt/TiO$_2$/SiO$_2$/(100)Si substrate has orientations not only to (001) but also to (119), (111), and (101) perpendicular to the surface normal direction. This difference in the constituent orientations of the SrBi$_4$Ti$_4$O$_{15}$ phases is considered to originate in the lattice matching between the SrBi$_4$Ti$_4$O$_{15}$ films and the $(100)_c$-oriented LaNiO$_3$ and (111)Pt layers. The lattice mismatch between (001) SrBi$_4$Ti$_4$O$_{15}$ and (100)LaNiO$_3$ is estimated to be as little as 0.05% when the orientation relationship is (001)$\langle 100 \rangle$SrBi$_4$Ti$_4$O$_{15}$//$(100)_c\langle 110 \rangle_c$LaNiO$_3$. This low lattice mismatch is considered to be the origin of the perfect c-axis-orientation of SrBi$_4$Ti$_4$O$_{15}$ films on $(100)_c$LaNiO$_3$ bottom electrodes. Based on this analogy, the (100)-orientation of the Pt layer should also constitute a suitable bottom electrode for obtaining c-axis-oriented SrBi$_4$Ti$_4$O$_{15}$ films.

However, (111)-oriented Pt was used instead of (100)Pt in Fig. 3.30(b) and (d) because Pt is self-oriented to (111). The results shown in Figs. 3.29 and 3.30 clearly indicate that the $(100)_c$-self-oriented LaNiO$_3$ buffer layer is essential for the preparation of perfectly c-axis-oriented fiber-structured SrBi$_4$Ti$_4$O$_{15}$ films on (111)Pt-coated Si substrates.

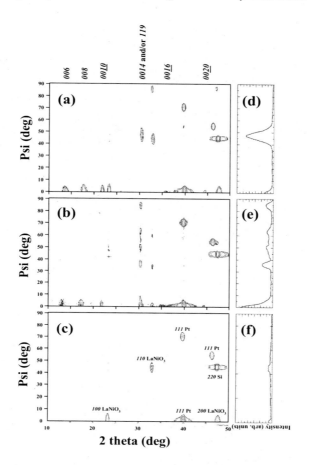

Fig. 3.30. X-ray reciprocal space maps [(a)–(c)] and integrated patterns of ψ scans in the 2θ range from 29 to 32° [(d)–(f)] for the $SrBi_4Ti_4O_{15}$ films and $(100)_c LaNiO_3/(111)Pt/TiO_2/SiO_2/(100)Si$ substrate shown in Fig. 3.29 (a)–(c).

Figure 3.31 shows AFM images with a different scan range for the same 30-nm-thick $SrBi_4Ti_4O_{15}$ films shown in Fig. 3.29(a) and (b), deposited on $(100)_c LaNiO_3/(111)Pt/TiO_2/SiO_2/(100)Si$ and $(111)Pt/TiO_2/SiO_2/(100)Si$ substrates. The $SrBi_4Ti_4O_{15}$ film deposited on a $(100)_c LaNiO_3/(111)Pt/TiO_2/SiO_2/(100)Si$ substrate consists of uniform-sized square grains, rotated with respect to each other (Fig. 3.31(a) and (b)), indicating a single-c-axis-oriented fiber-structure. In addition, this film has a smooth surface, with an average roughness, R_a, of 4.28 nm. However, the $SrBi_4Ti_4O_{15}$ film deposited directly on a $(111)Pt/TiO_2/SiO_2/(100)Si$ substrate contained grains of various sizes and irregular shapes (Fig. 3.31(c) and (d)). The surface was very rough, with an R_a value of 32.36 nm.

Figure 3.32 shows a cross-sectional TEM image of a 60-nm-thick $SrBi_4Ti_4O_{15}$ film deposited on a $(100)_c LaNiO_3/(111)Pt/TiO_2/SiO_2/(100)Si$ substrate. A stacked structure

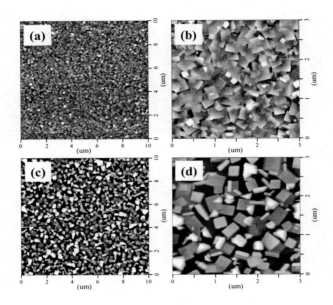

Fig. 3.31. AFM images of the same 30-nm-thick $SrBi_4Ti_4O_{15}$ films as shown in Fig. 3.1 deposited on (a),(b) $(100)_c LaNiO_3/(111)Pt/TiO_2/SiO_2/(100)Si$ and (c),(d) $(111)Pt/TiO_2/SiO_2/(100)Si$ substrates.

Fig. 3.32. Cross-sectional TEM image of 60-nm-thick $SrBi_4Ti_4O_{15}$ film deposited on a $(100)_c LaNiO_3/(111)Pt/TiO_2/SiO_2/(100)Si$ substrate.

without any noticeable interfacial layer between the $SrBi_4Ti_4O_{15}$ and $LaNiO_3$ is clearly observed. The average grain size of the $SrBi_4Ti_4O_{15}$ film is larger than that of the $LaNiO_3$ (about 300 nm) and is in good agreement with the grain size estimated from the AMF image shown in Fig. 3.31(b). This indicates that crystal growth along the a–b plane is more dominant than along the c-axis.

Figure 3.33 shows the capacitance and the dielectric loss as a function of the bias voltage for a 30-nm-thick $SrBi_4Ti_4O_{15}$ film deposited on a $(100)_c LaNiO_3/(111)Pt/TiO_2/SiO_2/Si$ substrate. The change in capacitance between 0 and ± 330 kV/cm is as little as -0.44%.

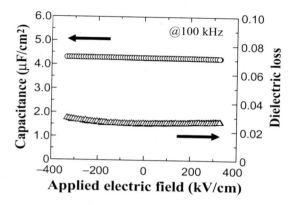

Fig. 3.33. Capacitance and dielectric loss as a function of the applied electric field for a 30-nm-thick $SrBi_4Ti_4O_{15}$ film deposited on a $(100)_cLaNiO_3/(111)Pt/TiO_2/SiO_2/(100)Si$ substrate.

The stability of this capacitance to the applied electric field is a common feature of c-axis-oriented BLD films and the epitaxial $SrBi_2Ta_2O_9$ films shown in Fig. 3.26(a), and is quite different from that of the perovskite-structured oxide. Capacitance degradation with increasing applied electric field has been reported in thin films of $(Ba,Sr)TiO_3$ and $SrTiO_3$ [46,52–54] to be about 36% over the electric field range shown in Fig. 3.33 for $(Ba,Sr)TiO_3$ films [54]. The dielectric loss is also independent of the applied electric field (Fig. 3.33). These stable capacitance and dielectric loss characteristics of c-axis-oriented BLD films are favorable for thin film capacitor applications.

Figure 3.34(a) shows the relative dielectric constant and the dielectric loss at 0 kV/cm as a function of the thickness of $SrBi_4Ti_4O_{15}$ films deposited on $(100)_cLaNiO_3/(111)Pt/TiO_2/SiO_2/(100)Si$ substrates. The relative dielectric constant of these films maintained a constant value of about 140 for film thicknesses down to 30 nm. This indicates the expected no-degradation characteristics with decreasing film thickness. Moreover, the film thickness dependency of the reciprocal of capacitance shows a linear relationship passing through the zero point (Fig. 3.34(b)). This suggests the absence of an interfacial layer with a low dielectric constant between the $SrBi_4Ti_4O_{15}$ film and the $LaNiO_3$ bottom electrode. As a result, c-axis-oriented fiber-structured $SrBi_4Ti_4O_{15}$ films showed the 'size-effect-free' characteristics of epitaxially-grown $SrBi_2Ta_2O_9$ films.

3.3.5 Improvement of Interface property: Electrode modification

In the previous section, single-axis oriented films with degradation-free characteristics of the dielectric constant were obtained by using (100)-self oriented $LaNiO_3$ buffer layers. However, the dielectric constant of $SrBi_4Ti_4O_{15}$ with $m = 4$ is smaller than expected, being almost the same as $Bi_4Ti_3O_{12}$ with $m = 3$, even for large values of m. This is considered to be due to the low crystallinity of the $LaNiO_3$ bottom electrode. High-temperature-deposition and post-heat-treatment results in both the formation of the La-rich phase La_2NiO_4, and the development of significant surface roughness [55].

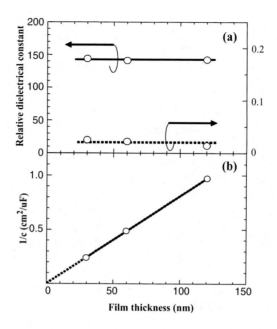

Fig. 3.34. (a) Relative dielectric constant and dielectric loss, and (b) reciprocal of capacitance ($1/C$) as a function of film thickness for $SrBi_4Ti_4O_{15}$ films deposited on $(100)_c LaNiO_3/(111)Pt/TiO_2/SiO_2/(100)Si$ substrates.

It is thus difficult to improve the crystal quality of the $LaNiO_3$ films. However, improvement of the crystallinity of the bottom electrode is critical to obtain dielectric films of higher crystal quality.

To solve this problem, we selected a 100_c-oriented $SrRuO_3$ film as an additional buffer layer on the $LaNiO_3$ because it can be prepared at high temperature and is also a conducting oxide with the perovskite structure.

In this section, the effect is reported of a $SrRuO_3/LaNiO_3$ doubly-stacked bottom electrode with an additional $SrRuO_3$ layer on the characteristics of deposited c-axis-oriented $CaBi_4Ti_4O_{15}$ films, representing the BLD family with $m = 4$. RF magnetron sputtering was employed for the preparation of the $CaBi_4Ti_4O_{15}$ films rather than chemical vapor deposition (CVD), since the former is preferred for industrial applications.

Figure 3.35 shows (a) the rocking curve FWHM of $SrRuO_3$ 200_c and $LaNiO_3$ 200_c, and (b) the average roughness as a function of heat-treatment temperature of the $LaNiO_3/Pt$ substrates. 100-nm-thick $LaNiO_3$ and 30-nm-thick $SrRuO_3$ was deposited at 360 and 700°C, respectively. As shown in Fig. 3.35(a), the rocking curve FWHM of both $SrRuO_3$ and $LaNiO_3$ films decreases with increasing heat-treatment temperature of the $LaNiO_3/Pt$ substrates even though the deposition temperature of $SrRuO_3$ was held constant at 700°C. This result clearly indicates that the crystal quality of $SrRuO_3$ films is affected by that

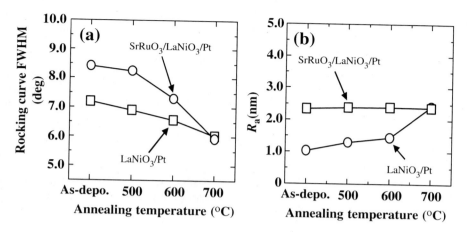

Fig. 3.35. (a) Rocking curve FWHM of SrRuO$_3$ 200_c and LaNiO$_3$ 200_c, and (b) R_a as a function of heat-treatment temperature of LaNiO$_3$/Pt substrates.

of the LaNiO$_3$ films. As previously reported, the surface roughness of the LaNiO$_3$ films increases with increasing heat-treatment temperature as shown in Fig. 3.35(b) [55]. However, the surface roughness of the SrRuO$_3$ films deposited on LaNiO$_3$ remained almost constant irrespective of the of LaNiO$_3$ films (Fig. 3.35(b)).

Figure 3.36 shows the rocking curves of SrRuO$_3$ 200_c for the same SrRuO$_3$/LaNiO$_3$/Pt substrate, where the SrRuO$_3$ was deposited after heat-treatment of the LaNiO$_3$/Pt at 700°C and the LaNiO$_3$ 200_c was deposited on the LaNiO$_3$/Pt substrate after heat-treated at 700°C for 30 min in air. Each spectrum was normalized to the theoretical intensity of the appropriate material. As shown in Fig. 3.36(a), the FWHM of the rocking curves of SrRuO$_3$ and LaNiO$_3$ is almost the same, indicating a similar degree of fluctuation of the crystal orientation in the SrRuO$_3$ and LaNiO$_3$ films. However, the integrated peak

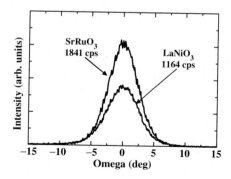

Fig. 3.36. Rocking curves of SrRuO$_3$ 200_c of the same SrRuO$_3$/LaNiO$_3$/Pt substrate where SrRuO$_3$ was deposited after heat treatment at 700°C and the LaNiO$_3$ 200_c was deposited on the LaNiO$_3$/Pt substrate after heat-treatment at 700°C for 30 min in air.

intensities of the rocking curves are different (Fig. 3.36), indicating that the SrRuO$_3$ film is of higher crystallinity than the LaNiO$_3$ film, mainly due to the high deposition temperature of the SrRuO$_3$ films.

Figure 3.37 shows the XRD patterns of (a) SrRuO$_3$/LaNiO$_3$/Pt and (b) LaNiO$_3$/Pt substrates after heat-treatment at 800°C for 30 min under O$_2$ atmosphere. As previously reported [55], LaNiO$_3$ films are unstable to heat-treatment under a high oxygen partial pressure, pO_2, readily forming NiO (Fig. 3.37(b)). However, the formation of NiO is suppressed by the SrRuO$_3$ capping layer as shown in Fig. 3.37(a).

Figure 3.38 shows the XRD patterns and pole figures of 58-nm-thick CaBi$_4$Ti$_4$O$_{15}$ films deposited at 600°C on (a) SrRuO$_3$/LaNiO$_3$/Pt and (b) LaNiO$_3$/Pt substrates. The fixed 2θ angle of the XRD pole figure measurement corresponds to 119 CaBi$_4$Ti$_4$O$_{15}$. Only the

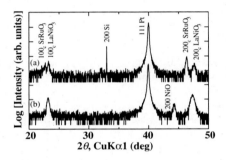

Fig. 3.37. XRD patterns of (a) SrRuO$_3$/LaNiO$_3$/Pt and (b) LaNiO$_3$/Pt substrates after heat-treatment at 800°C for 30 min under O$_2$ atmosphere.

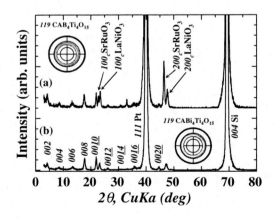

Fig. 3.38. XRD patterns and pole figures of 58-nm-thick CaBi$_4$Ti$_4$O$_{15}$ films deposited on (a) SrRuO$_3$/LaNiO$_3$/Pt and (b) LaNiO$_3$/Pt substrates.

$CaBi_4Ti_4O_{15}$ *00l* diffraction peaks were observed in addition to the peaks originating from the substrates for both films, suggesting perfect *c*-axis-orientation. In-plane random orientation with out-of-plane *c*-axis-orientation perpendicular to the substrate surface (a single-axis-oriented structure) was confirmed for both films by the XRD pole figure measurement (Fig. 3.38).

Figure 3.39 shows the rocking curve of the *008* diffraction peaks of the same $CaBi_4Ti_4O_{15}$ films as shown in Fig. 3.38 deposited on $SrRuO_3/LaNiO_3$/Pt and $LaNiO_3$/Pt substrates. The FWHM of the rocking curves was almost the same for the two films. However, the integrated intensity of the rocking curve was different, indicating higher crystallinity of the $CaBi_4Ti_4O_{15}$ films deposited on $SrRuO_3/LaNiO_3$/Pt substrates. This agreed with the observation of the bottom electrode (Fig. 3.36). Hence, the higher crystallinity of the $CaBi_4Ti_4O_{15}$ films on $SrRuO_3/LaNiO_3$/Pt substrates compared with those on $LaNiO_3$/Pt substrates is due to the improvement of crystallinity of the bottom electrodes by insertion of the additional $SrRuO_3$ buffer layer on the $LaNiO_3$.

Figure 3.40 shows the thickness dependence of R_a of *c*-axis-oriented $CaBi_4Ti_4O_{15}$ films deposited on $SrRuO_3/LaNiO_3$/Pt and $LaNiO_3$/Pt substrates. Both films showed similar R_a values as those of the substrates, irrespective of the film thickness. This result is due to the faster growth rate of the $CaBi_4Ti_4O_{15}$ films along the *a–b* plane than in the out-of-plane *c*-axis-direction.

Figure 3.41 shows the ε_r- and tan δ-frequency characteristics of *c*-axis-oriented fiber-structured $LaNiO_3$/Pt substrates. The change in ε_r over the frequency range 10^3–10^6 Hz was 2.86% for $CaBi_4Ti_4O_{15}$ films deposited on $SrRuO_3/LaNiO_3$/Pt and 0.68% for $LaNiO_3$/Pt substrates. In addition, the ε_r of a $CaBi_4Ti_4O_{15}$ film deposited on a $SrRuO_3/LaNiO_3$/Pt substrate is greater than on a $LaNiO_3$/Pt substrate.

Figure 3.42 shows the film thickness dependence of the ε_r value of *c*-axis-oriented $CaBi_4Ti_4O_{15}$ films deposited on $SrRuO_3/LaNiO_3$/Pt and $LaNiO_3$/Pt substrates. The ε_r value is essentially unchanged with decreasing film thickness down to 30 nm in the case of both films. Moreover, the film thickness dependence of the 1/C plots is linear, passing

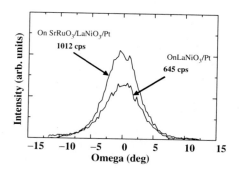

Fig. 3.39. Rocking curve of *008* diffraction peaks of the $CaBi_4Ti_4O_{15}$ films shown in Fig. 3.38. deposited on $SrRuO_3/LaNiO_3$/Pt and $LaNiO_3$/Pt substrates.

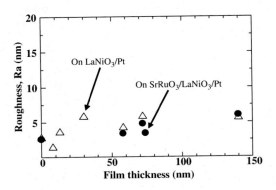

Fig. 3.40. Thickness dependence of the average roughness, R_a, of c-axis-oriented fiber-structured $CaBi_4Ti_4O_{15}$ films deposited on $SrRuO_3/LaNiO_3/Pt$ and $LaNiO_3/Pt$ substrates.

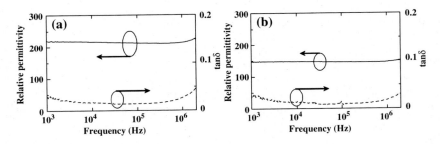

Fig. 3.41. ε_r- and $\tan\delta$-frequency characteristics of c-axis-oriented fiber-structured $CaBi_4Ti_4O_{15}$ films deposited on (a) $SrRuO_3/LaNiO_3/Pt$ and (b) $LaNiO_3/Pt$ substrates.

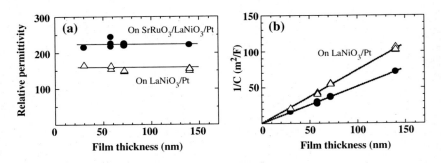

Fig. 3.42. Film thickness dependence of (a) ε_r and (b) reciprocal of the capacitance of c-axis-oriented fiber-structured $CaBi_4Ti_4O_{15}$ films deposited on $SrRuO_3/LaNiO_3/P$ and $LaNiO_3/Pt$ substrates.

through the zero point. This suggests that an interfacial layer with low permittivity (the so-called 'dead layer') is scarcely present in these films. The ε_r values calculated from the slopes of the straight lines of Figure 3.42(b) were 150 and 220 for the films on SrRuO$_3$/LaNiO$_3$/Pt and LaNiO$_3$/Pt substrates respectively, in good agreement with the ε_r values shown in Fig. 3.42(a). This suggests that the ε_r values of the CaBi$_4$Ti$_4$O$_{15}$ films themselves are different on SrRuO$_3$/LaNiO$_3$/Pt and LaNiO$_3$/Pt substrates, due to the improvement of crystallinity of the CaBi$_4$Ti$_4$O$_{15}$ films by the presence of the SrRuO$_3$ film (Fig. 3.39).

3.4 Summary and Future Aspects

Novel dielectric thin films with almost 'size-effect-free' characteristics have been developed, and the 'size effect' in the most widely used PZT films has been diminished by improvements to the interfaces between the dielectric and the electrode. A knowledge of the origin of these characteristics is very important for the development of new and novel materials. Superlattice structures have been investigated for the purpose of optimizing both the dielectric properties and the ferroelectric and optical properties [56]. BLD thin films have a natural superlattice structure with the advantages of greater thermal stability and stability in mass-production.

In addition, methods for modifying the properties of BLD films have already been proposed as 'site engineering concepts' in the case of the ferroelectric properties. We have already demonstrated the design of the dielectric constant and temperature dependence of a *TCC* (temperature coefficient of capacitor) [57]. Therefore, BLD is a promising material for 'designable capacitors' with dielectric constants larger than 150 and no degradation of the dielectric constant when the thickness is scaled down. These materials open the door to nanometer scale applications of dielectric materials with high ε_r values, such as MOS–FET devices, etc.

The present chapter also shows that the size affect of the dielectric constant depends strongly on both the dielectric material itself and the electrodes. Therefore, it is clear that the interface between the dielectric film and the electrode is a key issue in the 'size-effect-free' characteristics. The development of both electrode and dielectric material is another important part of future research.

Acknowledgment

I gratefully acknowledge the important contributions of the students in my laboratory.

References

[1] R. Waser, Ed. Nanoelectronics and Information Technology, Willey-VCH,

[2] X. Wu, D. Landheer, G.I. Sproule, T. Quance, M.J. Graham and G.A. Botton, J. Vac. Sci. Technol. **A20** (2002) 1141.

[3] K. Pollard and R. Puddephatt, Chem. mater. **11** (1999) 1069.

[4] D. Damjanovic, Rep. Prog. Phys. **61** (1998) 1267.

[5] H. Kohlstedt, N.A. Pertsev and R. Waser, Mater. Res. Symp. Proc. **688** (2002) C.6.5.1.

[6] K. Ishikawa, T. Nomura, N. Okada and K. Takada, Jpn. J. Appl. Phys. **35** (1996) 5196.

[7] F. Scott and C. Araujo, Science **246** (1989) 1400.

[8] D. Fong, B. Stephenson, S. Streiffer, J. Eastman, O. Auciello, P. Fuoss and C. Thompson, Science **304** (2004) 1650.

[9] Y. Kim, D. Kim, Y. Chang, T. Noh, J. Kong, K. Char, Y. Park, S. Bu, J. Yoon and J. Chung, Appl. Phys. Lett. **86** (2005) 102907.

[10] J. Junquera and P. Ghosez, Nature **422** (2003) 506.

[11] M. Saad, P. Baxter, R. Bowman, J. Greeg, F. Morrison and J. Scott, J. Phus. Cond. Matter **16** (2004) L451.

[12] M. Saad, R. Bowman and J. Gregg, Appl. Phys. Lett. **84** (2004) 1159.

[13] L.J. Sinnamon, R.M. Bowman and J.M. Gregg, Appl. Phys. Lett. **78** (2001) 1724.

[14] C. Parker, J. Maria and A. Kingon, Appl. Phys. Lett. **81** (2002) 340.

[15] O. Auciello, J. Scott and R. Ramesh, Phys. Today **51** (1998) 22.

[16] K. Nagashima, M. Aratani and H. Funakubo, Jpn. J. Appl. Phys. **39** (2000) L996.

[17] H. Funakubo, K. Nagashima, M. Aratani, K. Tokita, T. Oikawa, T. Ozeki, G. Asano and K. Saito, Proc. Mater. Res. Soc. **688** (2002) (C1.1)3.

[18] T. Oikawa, H. Morioka, A. Nagai and H. Funakubo, Appl. Phys. Lett. **85** (2004) 1754.

[19] T. Oikawa, H. Funakubo, H. Morioka and K. Saito, Integ. Ferro. **59** (2003) 1421.

[20] K. Tokita, M. Aratani, T. Ozeki and H. Funakubo, Key Eng. Mater. **216** (2002) 83.

[21] K. Saito, T. Oikawa, T. Kurosawa, T. Akai and H. Funakubo, Jpn. J. Appl. Phys. **41** (2002) 6730.

[22] B. Aurivillius, Ark. Kemi **1** (1949) 463.

[23] T. Watanabe and H. Funakubo, J. Appl. Phys., *Submitted*.

[24] C. Araujo, J. Cuchiaro, L. McMillan, M. Scott and J. Scott, Nature **374** (1995) 627.

[25] B.H. Park, B.S. Kang, S.D. Bu, T.W. Noh, L. Lee and W. Joe, Nature **401** (1999) 682.

[26] T. Kojima, T. Sakai, T. Watanabe, H. Funakubo, K. Saito and M. Osada, Appl. Phys. Lett. **80** (2002) 2746.

[27] T. Kojima, T. Watanabe, H. Funakubo, K. Saito, M. Osada and M. Kakihana, J. Appl. Phys. **93** (2003) 1707.

[28] H. Uchida, H. Yoshikawa, I. Okada, H. Matsuda, T. Iijima, T. Watanabe, T. Kojima and H. Funakubo, Appl. Phys. Lett. **81** (2002) 2229.

[29] T. Watanabe, T. Kojima, T. Sakai, H. Funakubo, M. Osada, Y. Noguchi and M. Miyayama, J. Appl. Phys. **92** (2002) 1518.

[30] H. Funakubo, T. Watanabe, T. Kojima, T. Sakai, Y. Noguchi, M. Miyayama, M. Osada, M. Kakihana and K. Saito, J. Crystal Growth **248** (2003) 180.

[31] T. Watanabe, K. Saito and H. Funakubo, J. Mater. Res. **16** (2001) 303.

[32] K. Ishikawa and H. Funakubo, Appl. Phys. Lett. **75** (1999) 1970.

[33] K. Ishikawa, K. Saito, T. Suzuki, Y. Nishi, M. Fujimoto and H. Funakubo, J. Appl. Phys. **87** (2000) 8018.

[34] T. Watanabe, T. Sakai, A. Saiki, K. Saito, T. Chikyo and H. Funakubo, Mater. Res. Soc. Proc. **655** (2001) CC1.9.1.

[35] T. Suzuki, Y. Nishi, M. Fujimoto, K. Ishikawa and H. Funakubo, Jpn. J. Appl. Phys. **38** (1999) L1261.

[36] T. Suzuki, Y. Nishi, M. Fujimoto, K. Ishikawa and H. Funakubo, Jpn. J. Appl. Phys. **38** (1999) L1265.

[37] S. Cummins and L. Cross, J. Appl. Phys. **39** (1968) 2268.

[38] A. Fouskova and L.E. Cross, J. Appl. Phys. **41** (1970) 2834.

[39] T. Watanabe, T. Sakai, H. Funakubo, K. Saito, M. Osada, M. Yoshimoto, A. Sasaki, J. Liu and M. Kakihana, Jpn. J. Appl. Phys. **41** (2002) L1478.

[40] K. Shantha and K. Varma, Mater. Res. Bull. **32** (1997) 1581.

[41] Y. Shimakawa, Y. Kubo, Y. Nakagawa, T. Kamiyama, H. Asano and F. Izumi, Appl. Phys. Lett. **74** (1999) 1904.

[42] K. Saito, K. Ishikawa, I. Yamaji, T. Akai and H. Funakubo, Integ. Ferro. **33** (2001) 59–69.

[43] J. Haeni, P. Irvin, R. Uecker, P. Reiche, Y. Li, S. Choundhury, W. Tian, M. Tian, M. Hawley, B. Craigo, A. Tagantsev, X. Pan, S. Streiffer, L. Chen, S. Klrchoefer, J. Levy and D. Schlom, Science **430** (2004) 758.

[44] T. Watanabe and H. Funakubo, Jpn. J. Appl. Phys. **39** (2000) 5211.

[45] T. Kojima, Y. Sakashita, T. Watanabe, K. Kato and H. Funakubo, Mater. Res. Soc. Symp. **748** (2003) U15.2.1.

[46] M. Werner, I. Banerjee, P. McIntyre, N. Tani and M. Tanimura, Appl. Phys. Lett. **77** (2000) 1209.

[47] H. Tabata, M. Hamada and T. Kawai, Mater, Res. Soc. Symp. Proc **401** (1996) 73.

[48] J.C. Shin, C.S. Hwang and H.J. Kim, Appl. Phys. Lett. **76** (2000) 1609.

[49] R. Iijima, Appl. Phys. Lett. **79** (2001) 2240.

[50] JCPDS 33-0710, H. Wustenberg, Hahn, Inst. fur Kristallogr., Technische Hochschule, Aachen, Germany, ICDD Grant-in-Aid, 1981.

[51] K. Saito, M. Mitsuya, N. Nukaga, I. Yamaji, T. Akai and H. Funakubo, Jpn. J. Appl. Phys. **39** (2000) 5489.

[52] S. Komatsu, K. Abe and N. Fukushima, Jpn. J. Appl. Phys. **37** (1998) 5651.

[53] S. Yamamichi, T. Sakuma, K. Takemura and Y. Miyasaka, Jpn. J. Appl. Phys. **30** (1991) 2193.

[54] P. Padmini, T.R. Taylor, M.J. Lefevre, A.S. Nagra, R.A. York and J.S. Speck, Appl. Phys. Lett. **75** (1999) 3186.

[55] K. Takahashi, M. Suzuki, T. Oikawa, H. Chen and H. Funakubo, Mater. Res. Soc. Symp. Proc. **784** (2004) G1.9.1.

[56] K. Takahashi, M. Suzuki, T. Kojima, T. Watanabe, K. Kato, O. Sakata and H. Funakubo, **12** (2006) 136.

[57] H. Lee, H. Christen, M. Chisholm, C. Rouleau and D. Lowndes, Nature **433** (2005) 395.

PART II

State-of-the-Art Polymers

CHAPTER 4

Photonic Devices using Liquid Crystal Nanostructures

Hideo Takezoe

Abstract

Photonic devices using periodic liquid crystal (LC) nanostructures are reviewed. Following a general introduction to the photonic effect and its application to LCs, two LC photonic devices, lasers and optical diodes, are introduced. These devices consist of cholesteric LC multilayers, sometimes with an intermediate nematic LC layer. Light waves are controlled by selective reflection of cholesteric LCs and phase retardation by the anisotropic nematic LC layer. Tunability is one of the attractive features of these LC devices; this includes electrotunability of lasing wavelength in laser devices and electrotunability of transmittance in optical diode devices.

Keywords: **photonic effect, liquid crystal, cholesteric liquid crystal, dispersion, lasing, DFB, defect mode, optical diode, tunability.**

4.1 Photonic Effect

4.1.1 Introduction

An electron in free space has an energy ε proportional to the square of a wave number k; the dispersion between ε and k is parabolic. It is well known in solid state physics that the dispersion for an electron existing in periodic structures such as crystals opens energy band gaps (EBG) at the edges of the Brillouin zones, where the Bragg condition is satisfied. Electrons cannot exist and electron waves cannot propagate within the gaps. The situation is similar for electromagnetic waves, although the dispersion relation between the angular frequency ω and k without a periodic structure is not quadratic but linear; $\omega = ck$, where c is the light velocity.

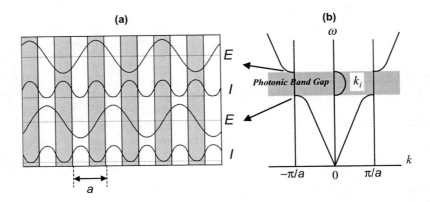

Fig. 4.1. (a) Standing waves localized in high and low dielectric layers when the Bragg condition is satisfied. (b) Dispersion relation for light propagating in a 1D periodic medium with a pitch a.

Consider a one-dimensional system consisting of alternating layers with high and low dielectric constants ε_H and ε_L with a periodicity a. The periodicity leads to two standing waves localized in the high and low dielectric layers (Fig. 4.1(a)) when the Bragg condition is satisfied; $n\lambda = a(\varepsilon_H^{1/2} + \varepsilon_L^{1/2})$. Since the two standing waves have different energies from that of light propagating in free space, bandgaps are introduced at the edges of the Brillouin zones, $k = n\pi/a$, as shown in Fig. 4.1(b). This gap is sometimes called the photonic band gap (PBG). Within the gap, k becomes complex; $k = k_r + ik_{im}$. Consider a monochromatic plane wave propagating along the x direction, $E(\omega) = E_0 \exp(ikx)$. The field can be written $E(\omega) = E_0 \exp(ik_r x)\exp(-k_{im} x)$ within the PBG, so that the electromagnetic wave is a damping wave with an amplitude change $E_0 \exp(-k_{im} x)$ with x. If this situation is satisfied three-dimensionally, i.e., the band gap appears for all $\mathbf{k}$, the electromagnetic field of light cannot be allowed to exist in this frequency range. This is the precise definition of the PBG [1–3]. This perfect PBG, i.e., energy gaps existing for all $\mathbf{k}$, is realized only in restricted crystals such as those with a deformed fcc structure [4] and diamond structure [5].

Thus the PBG is a photon analogue of the electron EBG. They are quite similar, but the PBG is more complicated and thus sometimes more attractive. Its complexity arises from the polarization of light and the strong frequency dependence of the refractive index in the vicinity of the PBG. One of the interesting photonic effects is a superprism effect [6], which originates from anomalous dispersion and group velocity in the vicinity of the PBG; i.e., the refraction angle for the incidence of light into photonic crystals seriously depends on the wavelength, and very large dispersion appears for light beams with wavelengths close to each other.

A crystal possessing a PBG is called a photonic crystal (PC), and phenomena observed in PCs are called photonic effects. So far, a variety of photonic structures with 1D, 2D, and 3D orders shown in Fig. 4.2 have been proposed and fabricated. As can easily be imagined, however, the fabrication particularly of 2D and 3D structures is not easy and requires sophisticated fabrication technologies. By contrast, as will be discussed

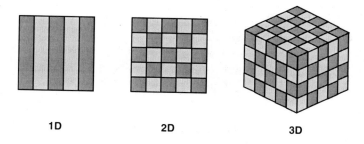

Fig. 4.2. Schematic illustrations of 1D, 2D, and 3D photonic structures.

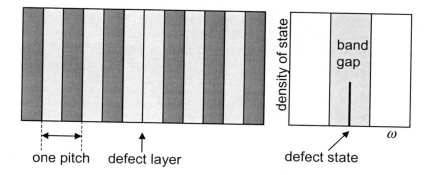

Fig. 4.3. An example of a defect structure and the defect state.

below, liquid crystals (LCs) often constitute ideal photonic structures which are formed spontaneously. Moreover, photonic structures of LCs are tunable by a variety of external stimuli such as temperature, electric field, and light irradiation, which makes LC photonic devices attractive.

Before proceeding to LC photonic structures, the concept of defect structure is described. As defect centers in perfect crystals induce electronic defect levels, physical imperfections introduced in periodic PC structures can cause spectrally narrow resonant modes (defect modes) inside the PBG, as shown in Fig. 4.3. The defect structure can localize light at the defect state, whereas the PBG can damp light propagation and confine light within the PBG. This property enables us to make waveguides [7] in which light propagates along the defect line or surfaces. In the application of photonic structures to lasers discussed in Section 4.2, the defect mode is important, since the localization of light at a defect state can suppress spontaneous emission and enhance stimulated emission, bringing about thresholdless lasing [8]. Some trials using LCs will be introduced later.

4.1.2 Photonic liquid crystals

The science and engineering of liquid crystals (LCs) has been extensively developed partly because of their practical applications to LC displays. LCs represent intermediate states between crystals and isotropic liquids and thus have intermediate physical properties, i.e., the fluidity of a liquid and the anisotropy of a crystal [9,10]. Because of these properties, further applications are expected. The use of LCs as photonic crystals is one such promising application. Anisotropy can provide periodic structures, and fluidity makes tunability easy. Furthermore, some LCs spontaneously develop photonic structures. The periodic structures are helical, caused by the introduction of chirality. These helical structures can be tuned from less than 100 nm [11] to infinity. For use as photonic devices, the helical pitch must be in the visible range. In this sense, the nanostructures for the application of LCs as photonic devices are in the range of visible wavelengths.

Figure 4.4 shows a variety of helical structures observable in LCs; (a) cholesteric LC (CLC) phase, (b) chiral smectic C (SmC*) phase, (c) chiral SmC_A* phase, (d) twist grain boundary (TGB) phase, and (e) blue phase (BP). In the CLC phase, molecules locally align parallel to a particular average direction (director; **n**) and form a macroscopic helical structure with the helical axis perpendicular to **n**. CLCs are realized in pure chiral molecular systems and also in nematic LCs doped with chiral dopants. The pitch is dependent on the temperature and the amount of dopant. Using the dielectric anisotropy of materials, the pitch can be expanded by applying an electric field. The SmC* and

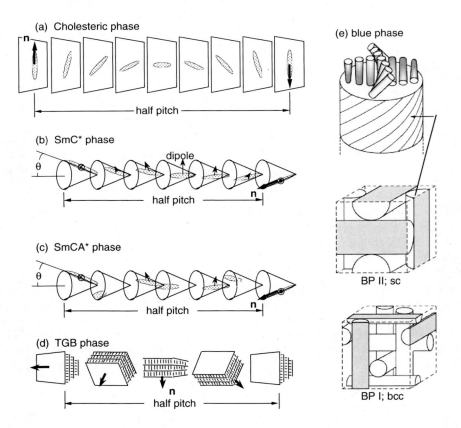

Fig. 4.4. Helical structures in liquid crystals (LCs).

SmC$_A$* phases have uniform smectic layers in which molecules are tilted from the layer normal locally uniformly (SmC*) and anticlinically (SmC$_A$*) from layer to layer. This tilt plane precesses along the layer normal to form the helical structure. These phases can exhibit ferroelectric [12] and antiferroelectric [13,14] electrooptic response, so that fast modulation of periodic structures should be possible. In the TGB phase [15], the layers twist discontinuously with an array of screw dislocations at the plane making the discontinuous layer twist. Only the BP [16] has 3D structures. Helices along two perpendicular directions form a rod and sets of these rods are stabilized by orientational defects. Usually BP emerges in a very narrow temperature range of about 1°C between the isotropic and CLC phases. Recently, however, Kikuchi has succeeded in stabilizing the phase over several tens of degrees using photopolymerizable LCs [17].

4.1.3 Selective reflection in cholesteric liquid crystals

The best-known and important photonic effect in LCs is selective reflection; circularly polarized light is totally reflected when it is incident along the helical axis of a CLC and the wavelength λ and helical handedness are the same as those of the CLC [9,10]. The reflection band is located at a wavelength $\lambda_R = np$, where p is the helical pitch and n the average refractive index, $\{(n_o^2 + n_e^2)/2\}^{1/2}$ with n_o and n_e for ordinary and extraordinary light, respectively. The band width $\Delta\lambda$ is proportional to the optical anisotropy $\Delta n = n_e - n_o$; $\Delta\lambda = \lambda_R \Delta n/n$.

Figure 4.5(a) shows experimental selective reflection spectra for oblique incidence of light on a right-handed (R-) CLC [18]. One may notice several characteristic features:

(1) The selective reflection band shifts to a shorter wavelength with increasing incidence angle due to the Bragg's law.

(2) An additional reflection band emerges both for right- and left-handed circularly polarized light (R-CPL and L-CPL) in the middle of the selective reflection band, where only R-CPL is reflected. This is called the total reflection band, in contrast to the selective reflection band.

(3) In addition to fine interference fringes, additional large modulated structures such as swell and beat are observed. The oblique incidence gives a second-order reflection consisting of three bands with characteristic polarization characteristics, as shown in Fig. 4.5(b) [19]. In normal incidence of light, the spatial variation of the refractive index change is perfectly sinusoidal, so that no higher reflection bands appear, and neither does the total reflection band.

Light propagation along the helical axis of CLC can rigorously be described and analytical solutions have been given [9,10]. Here we only briefly describe the essence of the derivation. Let us now consider the propagation of an electromagnetic wave of frequency ω along the helical axis, z, where the dielectric constant ε varies with the helix wavenumber q_0 along z preserving the local dielectric anisotropy $\Delta\varepsilon = \varepsilon_e - \varepsilon_0$.

$$\hat{\varepsilon}(z) = \left(\frac{\varepsilon_e + \varepsilon_0}{2}\right)\begin{pmatrix} 1 & 0 \\ 0 & 1 \end{pmatrix} + \frac{\Delta\varepsilon}{2}\begin{pmatrix} \cos 2q_0 z & \sin 2q_0 z \\ \sin 2q_0 z & -\cos 2q_0 z \end{pmatrix} \quad (4.1)$$

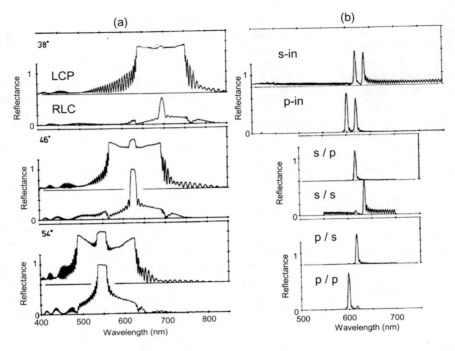

Fig. 4.5. Reflection spectra of the (a) first and (b) second orders under oblique incidence.

To derive the explicit solution it is convenient to introduce the circular base of electric field.

$$E^{\pm} = E_x + iE_y \tag{4.2}$$

with non-zero field components

$$E_x = \text{Re}\left\{E_x(z) \cdot e^{-i\omega t}\right\}, \tag{4.3a}$$

$$E_y = \text{Re}\left\{E_y(z) \cdot e^{-i\omega t}\right\} \tag{4.3b}$$

Then the Maxwell equation reduces to

$$-\frac{d^2 E^+}{dz^2} = k_0^2 E^+ + k_1 \exp(2iq_0 z) E^-, \tag{4.4a}$$

$$-\frac{d^2 E^-}{dz^2} = k_1 \exp(-2iq_0 z) E^+ + k_0^2 E^- \tag{4.4b}$$

where

$$k_0^2 = \left(\frac{\omega}{c}\right)^2 \left(\frac{\varepsilon_e + \varepsilon_0}{2}\right), \tag{4.5a}$$

$$k_1^2 = \left(\frac{\omega}{c}\right)^2 \left(\frac{\Delta\varepsilon}{2}\right) \tag{4.5b}$$

By solving Eq. (4.4) we can find the form of the modes

$$E^+ = a \exp\{i(k+q_0)z\}, \tag{4.6a}$$

$$E^- = b \exp\{i(k-q_0)z\} \tag{4.6b}$$

with

$$\left(-k_0^2 + k^2 + q_0^2\right)^2 - 4q_0^2 k^2 - k_1^4 = 0 \tag{4.7}$$

For fixed ω, k_0 and k_1, Eq. (4.7) gives four possible values of k (real or complex). These are two distinct branches, which we call (+) and (−). At $k = 0$, this gives

$$\omega_+(0) = \frac{cq_0}{n_0} \quad \text{and} \quad \omega_-(0) = \frac{cq_0}{n_e} \tag{4.8}$$

The eigenmodes are generally elliptically polarized with the ellipticity,

$$\rho = \frac{-a+b}{a+b} = \frac{-2kq_0}{\pm s^2 - k_1^2} \tag{4.9}$$

$$s^2 = +\sqrt{k_1^4 + 4q_0^2 k^2} \tag{4.10}$$

At high- and low-energy bandgap edges, the eigenmode is linearly polarized perpendicular and parallel to the local director and is guided by the helix. Elsewhere, ρ is almost ± 1, i.e., nearly circularly polarized (see Fig. 4.6).

In this way, selective reflection takes place for normal incidence and is analytically described. We now consider what happens when light with wavelength λ within the PBG is emitted from inside the CLC. Since the wavenumber k has an imaginary component, the emitted light suffers damping. Actually, as shown in Fig. 4.7, a dip is clearly observed in the emission spectrum from dyes embedded in CLC, but this dip is not observed in the spectrum of dyes in chloroform [20]. The dip position shifts to shorter wavelength with increasing temperature, indicating a temperature dependence of the helical pitch. This experiment suggests that light emitted from inside a CLC is affected by the PBG and may result in lasing, if the dyes are excited using intense light.

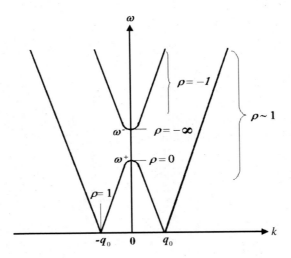

Fig. 4.6. Dispersion relation in CLC. Ellipticity ρ of two elliptically polarized eigenmodes is also shown.

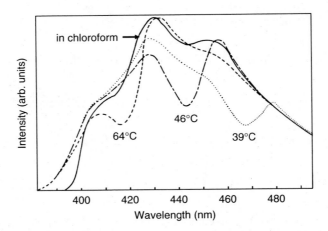

Fig. 4.7. Emission spectra of a dye embedded in CLC at various temperatures. A spectrum of the same dye in chloroform is also shown.

4.1.4 Lasing from LC nanostructures

Principle of the LC dye laser. Nomally organic dyes have an emission band with a wide bandwidth over various vibrational and rotational states, so that dye lasers are used as wavelength tunable lasers. Dye molecules have a strong absorption band due to a singlet–singlet transition and normally an emission occurs from the vibrational ground state within the excited singlet state. During the relaxation process among the vibronic levels, population inversion occurs and stimulated emission should occur. Such a process is available for pulsed excitation, during which energy transfer to triplet states would not occur. Hence for the continuous lasing of dye lasers, quenching of the triplet states is required.

Let us consider the conditions for efficient lasing. In an isotropic medium, the rate R of photon emission from an excited molecule is described by Fermi's Golden Rule:

$$R_{iso} \sim M_{iso} |E \cdot d|^2, \qquad (4.11)$$

where M_{iso} is the density of state (DOS), d the transition dipole moment, and E the electric field. In isotropic media, the DOS is independent of the polarization and radiation directions.

In anisotropic media, emission depends on the orientation of the transition dipole moment d with respect to the polarization of light E. When emission occurs from excited CLC molecules, light propagates as one of the two eigenmodes E_1 and E_2. Then the emission rate for an eigenmode $E_i(R_i)$ is described as

$$R_i \sim M_i |E_i \cdot d|^2, \qquad (4.12)$$

where M_i is the DOS associated with the eigenmode E_i. Fluorescent molecules embedded in CLCs have some degree of nematic order, resulting in an anisotropic orientation distribution of the transition dipole moment. Hence for a large value of R_i, d should be parallel to the polarization of the eigenmode E_i. Another factor giving a large R_i value is the DOS M, which is defined as

$$M = \left| \frac{d}{d\omega} \operatorname{Re}(k) \right|. \qquad (4.13)$$

Figure 4.8 shows a simulated transmittance spectrum and the corresponding DOS. The DOS (M) shows maxima at the PBG edges where the group velocity approaches zero. We will see how we can achieve a large M and consequently large R_i values in the following sections by considering practical model structures.

Distributed feedback (DFB) CLC laser. When light propagates in periodic media with the same periodicity as the light wavelength, the light suffers reflection due to the PBG. Hence, if CLC is doped with dyes, emitted light within the PBG is confined and amplified in the CLC, resulting finally in lasing. This type of cavity without the use of mirrors is called a distributed feedback (DFB) cavity. Lasers using DFB cavities are called DFB lasers. The DFB cavity is widely used in semiconductor lasers, in which active materials are deposited on substrates with periodic refractive index changes, as shown in Fig. 4.9. The guided wave senses the periodicity and reflection takes place.

Dowling *et al.* [21] predicted that DFB lasing will occur at the edge of the PBG for 1D periodic structures with sufficiently large refractive index modulation. They demonstrated that the photon group velocity approaches zero near the band edge of a one-dimensional

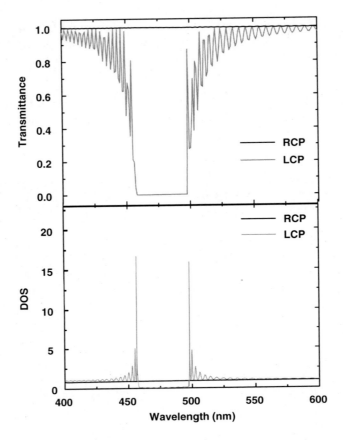

Fig. 4.8. Simulated transmittance spectrum and DOS for R-CP and L-CP.

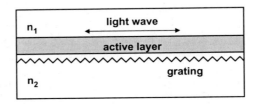

Fig. 4.9. Structure of the DFB cavity of a semiconductor laser.

PBG structure. This effect implies an exceedingly long optical path length in this structure, and the photon dwell time for incident waves at the band edge is significantly increased.

CLCs act as spontaneously formed DFB cavities, though the optical properties and cell geometries are different from semiconductor lasers. Thus, in CLCs the refractive index change is due to the helical structure of the dielectric ellipsoid and is thus sinusoidal, whereas in semiconductor lasers it is discrete, $\ldots n_H/n_L/n_H/n_L \ldots$.

M in Eq. (4.13) is the absolute inverse slope of the dispersion relation or reciprocal form of the group velocity. Since the emission rate R is proportional to the DOS, the emission rate becomes maximum when the group velocity approaches zero, which occurs at the edges of the PBG (Fig. 4.8). Thus, a low-threshold and mirrorless CLC laser is realized at the edges of the PBG, where the DOS gives rise to a maximum.

Defect mode lasing. As mentioned in 4.1.1, defect mode lasing is very important to realize low-threshold lasing [8]. Many types of defect mode have been studied in 1D [22–24], 2D [25], and 3D [26,27] photonic structures. These can be produced by removing or adding material or by altering the refractive index of one or a number of elements in 1D, 2D, and 3D PCs. Introducing a quarter-wavelength space in the middle of a layered 1D sample produces a defect in the middle of the PBG. Such a defect is widely used to produce high-Q laser cavities [28].

Theoretically, three kinds of configurations may generate a defect mode in CLCs:

(1) the creation of a phase jump without any spacing layer in the CLC [29] and smetic LC structures [30],

(2) the introduction of an isotropic spacing layer in the middle of the CLC [31], and

(3) local deformation of the helix in the middle of the CLC layer [32].

The defect mode (1), i.e., a chiral twist defect, can be created by rotating one part of the CLC [29], as shown in Fig. 4.10(a). Changing the chiral twist angle from 0 to 180° tunes the defect wavelength from a high to low wavelength PBG edge. By twisting one part of the CLC by 90°, a defect mode can be generated at the center of the PBG due to the phase shift $\pi/2$ of the electromagnetic wave inside the PBG. Schmidtke *et al.* [33] and Ozaki *et al.* [34] demonstrated low-threshold defect mode lasing by using the

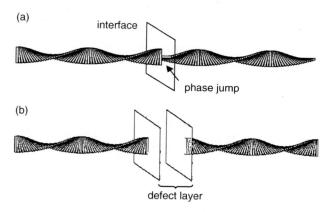

Fig. 4.10. Two kinds of defect structures in CLCs, (a) phase jump and (b) layer insertion.

phase jump in CLCs. Electrotunability of the lasing wavelength of the defect mode was also demonstrated using a layer of nematic liquid crystal (NLC) inserted into dielectric multilayers [22–24].

The defect mode (2) can be produced in a CLC structure by introducing an isotropic layer between two CLC layers in order to destroy the helical periodicity of the CLC, as shown in Fig. 4.10(b) [31]. The thickness of the isotropic defect layer which generates the phase shift, $\pi/2$, is expressed as

$$d = (2i + 1) \cdot \frac{\lambda}{(4 \cdot n)}, \quad (4.14)$$

where d is the thickness of a defect layer, i is an integer, λ the wavelength of the defect mode, n the refractive index of the defect layer.

4.2 Lasing from CLCs

Lasing from spontaneously formed photonic structures utilizing liquid crystals (LCs) has attracted much attention since its experimental demonstration by Kopp et al. [35]. Lasing has been achieved in various LC systems such as CLC [35–58], SmC* LC [59–61], and even in BP [62]. Of these, CLCs are the most extensively studied because of their ease in sample preparation. So far, various CLCs have been used, i.e., low molecular weight CLCs [35–37,39,43–45,48,49,53–58], lyotropic CLC [40], elastmer [38], photopolymerized CLCs [33,41,42,46,47], and thermotropic CLC polymer [50–52]. The performance of a variety of laser cavity structures has been demonstrated, i.e., distributed feedback (DFB) structures using simple dye-doped CLC cells [35–45,48,49,51,53–57] and defect mode structures including the insertion of both an isotropic layer [31] and an anisotropic layer [50,52,58] or a phase jump [46,47].

4.2.1 DFB mode

The DFB mode is the most fundamental lasing mode in dye-doped CLC lasers. Since the first demonstration by Kopp et al. [35], many studies have been reported. Here one of our contributions [48] will be described showing the ideal conditions for lasing based on the discussion of the section, principle of the LC dye layer, emphasizing the importance of the orientation of the dyes in CLCs. For this purpose, we used two dyes, a polymer dye and DCM (see below) with different order parameters in CLCs, and compared their lasing characteristics. We discuss herein (1) the difference between the order parameters of the polymer-dye and the DCM dye in a host NLC, (2) the comparison between the experimental fluorescence spectra and theoretically calculated fluorescence spectra in polymer-dye-doped CLCs and the DCM dye-doped CLCs, and (3) the comparison between the lasing spectra of polymer-dye-doped CLCs and DCM dye-doped CLCs.

The helical pitch (left handed) of the mixture was tuned by adjusting the amount of chiral dopant (MLC 6247, Merck Co.), 27.9, 25.4, and 23.6 w%, to the host NLC

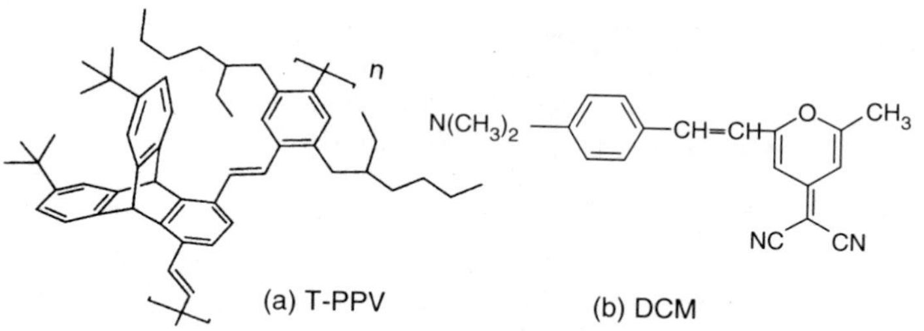

Fig. 4.11. The two dyes used, (a) triptycene-substituted PPV (T-PPV) and (b) commercial DCM.

(Zli 2293, Merck Co.). The guest fluorescent dyes used in this study are an elaborated poly(phenylene vinylene) ($M_w = 8100$ and $M_w/M_n = 2$, T-PPV(triptycene-substituted PPV)) and 4-(dicyanomethylene)-2-methyl-6-(4-dimethylaminostryl)-4H-pyran (DCM, a commercial laser dye) whose chemical structures are shown in Figs. 4.11(a) and (b). As will be shown later, T-PPV enhances the order parameter of the host though interactions of the rigid triptycene groups in each repeating unit of the polymer with the host NLC molecules, which suppress the fluctuations of the host NLC molecules [63–65]. CLC was doped with 1 wt% of either dye and was sandwiched between two glass substrates coated with unidirectionally rubbed polyimide to form a uniform helical structure with the helical axis along the normal to the substrate. The cell thickness was about 14 μm.

The emission spectra of T-PPV-doped and DCM-doped CLC cells for right- and left-circular polarization (R-CP and L-CP) are shown in Fig. 4.12(a) [48]. The R-CP

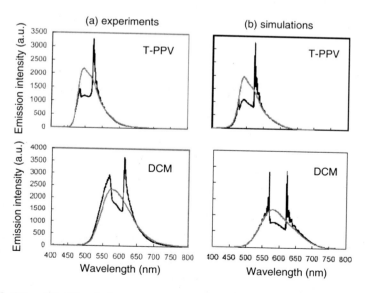

Fig. 4.12. L-CP and R-CP emission spectra of T-PPV- and DCM-doped CLCs, (a) experiments and (b) simulations.

fluorescence spectra are the same as the simple solution emissions for both samples. By contrast, the L-CP fluorescence is strongly influenced by selective reflection due to the left-handed helix structure of the CLC, and deep dips (notches) in the spectra are observed. It is also noteworthy how the fluorescence spectra are strongly enhanced at the band edges. The enhancement behavior of the T-PPV-doped and the DCM-doped cells is different; the former shows selective enhancement at the low-energy edge of the photonic gap, while the enhancement occurs at both photonic edges in the latter. This difference originates from the variations in the degree of alignment of dyes in CLCs and, as will be shown later, influences the lasing behavior.

With increasing optical pumping power, laser emission starts to occur at either or both edges of the photonic gap. Figure 4.13 shows the reflectance, fluorescence, and lasing spectra of T-PPV-doped CLCs with (a) 27.9 w% and (b) 25.4 w% chiral dopant at 24.4°C, and DCM-doped CLCs at (c) 30°C and (d) 55°C with 23.6 w% chiral dopant [48]. The relationship between the emission band and the PBG was changed by changing the chiral dopant in the T-PPV-doped cells and by changing the temperature of the DCM-doped cells. Lasing emission of the T-PPV-doped CLCs always occurs only at the

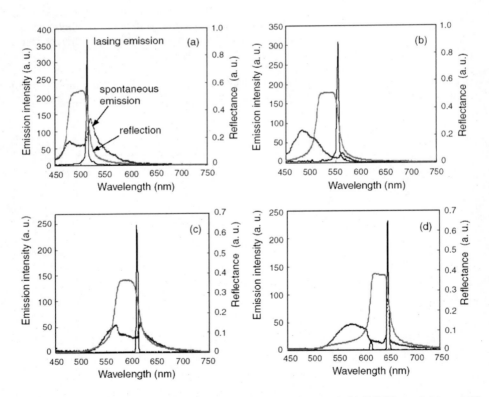

Fig. 4.13. Reflectance, fluorescence, and lasing spectra of (a) and (b) T-PPV-, and (c) and (d) DCM-doped CLCs. In (a) and (c) PBG low-energy edges are located near the emission peaks, whereas PBGs are located in the lower energy side of the emission bands for (b) and (d).

low-energy edge of the PBG. In Fig. 4.13(a), the spontaneous emission intensity at the low-energy edge of the PBG is higher than that at the high-energy edge. Hence, lasing emission at the low-energy edge is a natural consequence. In Fig. 4.13(b), however, the spontaneous emission intensity at the high-energy edge is greater than that at the low-energy edge. Hence it is surprising that lasing still occurs at the low-energy edge. Thus, lasing always occurs at the low-energy edge of the PBG in the T-PPV-doped CLC. By contrast, lasing emission of DCM-doped CLCs occurs at either or both sides of the PBG, as shown in Figs. 4.13(c) and (d), depending on the placement of the temperature-dependent selective reflection wavelengths (Fig. 4.13(c); 30°C, Fig. 4.13(d); 55°C) with respect to the fluorescence spectrum.

We will show that high alignment of the T-PPV in the LC is responsible for selective lasing only at the low-energy edge of the photonic bandgap. The emission transition moments of the monomeric units of T-PPV are well aligned to the local director of CLCs. The eigenmodes at the high- and low-energy band edges have polarizations perpendicular and parallel to the local director, respectively [66,67]. Hence lasing occurs only at the low-energy edge of the photonic gap, where an oscillation mode is along the director.

In order to clearly explain the different lasing conditions obtained in P-TTV and DCM, the orientations of these dyes in NLCs were examined. To estimate the degree of alignment of the fluorescent dyes in the host NLCs, polarized absorption and fluorescence were measured using homogeneous sample cells containing mixtures of either of the dyes and the NLC without the chiral dopant. Figure 4.14 shows polar plots of the absorbance as a function of polarization direction. The difference of the order parameters in the two dyes is very clear. For the sample cells doped with DCM dyes, as the concentration of the dye increases, the order parameter of DCM dye decreases from $S \approx 0.36$ (0.3 wt%) to $S \approx 0.21$ (0.5 wt%). By contrast, the order parameter of T-PPV increases from $S \approx 0.41$ to $S \approx 0.49$, as the concentration increases from 0.3 to 0.5 w%. We have confirmed that the order parameters determined by the fluorescence intensities give very similar values.

The enhanced fluorescence spectra at the edges of the PBG can be explained in terms of the eigenmodes for light propagating through a helical structure. It is known that the eigenmodes in CLCs are principally circularly polarized (see Fig. 4.6). If the helix of CLCs is right (left) handed, one of the eigenmodes is always L (R)-CP. The other eigenmode changes its polarization drastically and has the greatest differences in the vicinity of the photonic band; it is R (L)-CP at sufficiently long wavelengths compared with the helical pitch. On the other hand, the ellipticity becomes larger with decreasing wavelength toward the low-energy edge of the photonic bandgap. At the low-energy edge of the photonic bandgap, the polarization state of the eigenmode has a linear polarization parallel to the local director of CLCs. The polarization state at the high-energy edge is also a linear polarization, but the polarization direction is perpendicular to the local director of CLCs [66,67]. As shown in Fig. 4.12(a), enhanced fluorescence of the T-PPV occurs only at the low-energy edge of the photonic bandgap, while that of DCM occurs at both energy edges of the photonic bandgap, as shown in Fig. 4.12(a). Therefore, the difference between spontaneous emission in T-PPV and DCM is attributed to the orientation of the transition dipole moment of the dye molecule with respect to the local director of CLCs.

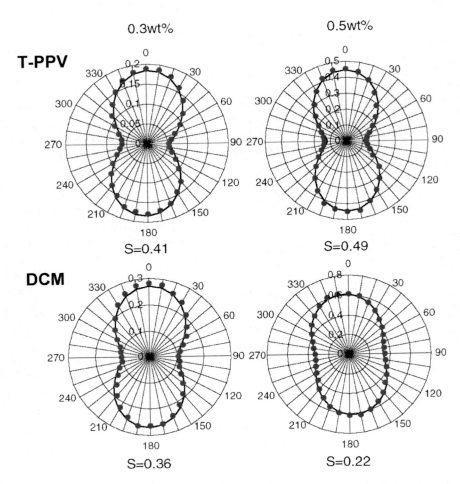

Fig. 4.14. Polar plots of absorption peak of T-PPV and DCM dyes in NLC as a function of polarizer direction. Note the opposite change of the order parameter S in T-PPV and DCM with increasing doping rate.

A simple theoretical description of spontaneous emission as a function of wavelength in terms of the order parameter S for the transition dipole moment of the dye in the CLCs is as follows [16]

$$\langle |E \cdot d|^2 \rangle = \frac{2}{3} \frac{\rho_i^2 - \frac{1}{2}}{\rho_i^2 + 1} S_{\text{dye}} + \frac{1}{3}. \tag{4.15}$$

Here, ρ_i is the ellipticity of the polarization state. Figure 4.15 shows the calculated values of $\langle |E \cdot d|^2 \rangle$ for the incidence of L-CP light as a function of wavelength, where the two profiles correspond to T-PPV with an order parameter of 0.5 and to DCM with an order parameter of 0.2. Near the edges of the photonic bandgap, a sharp variation occurs in $\langle |E \cdot d|^2 \rangle$, because the polarization state of the eigenmode with the same handedness as

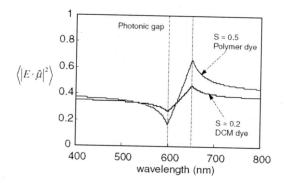

Fig. 4.15. $\langle |E \cdot d|^2 \rangle$ for the incidence of L-CPL as a function of wavelength in T-PPV- and DCM-doped CLCs.

the CLC structure is linearly polarized along the local director of the CLC. In particular, at the low-energy edge of photonic bandgap, the value of $\langle |E \cdot d|^2 \rangle$ is high because the polarization direction is parallel to the local director of the CLC. However, at the high-energy edge of the PBG, the $\langle |E \cdot d|^2 \rangle$ value is low because the polarization direction is perpendicular to the local director of the CLC. We obtained a higher $\langle |E \cdot d|^2 \rangle$ value for T-PPV than that for DCM, due to the difference of the order parameter.

In Fig. 4.12(b), the calculated fluorescence spectra are plotted for R-CP and L-CP, taking into account the $\langle |E \cdot d|^2 \rangle$ value of the fluorescent dyes (polymer dye, $S \approx 0.5$; DCM dye, $S \approx 0.2$) and the electromagnetic density of the modes (DOM) [67] of the CLC. The theoretical fluorescence spectra for T-PPV and DCM are in good agreement with experimental results. From these results, we can expect that the lasing emission of T-PPV will always occur at the lower energy edge of the photonic bandgap, and DCM lasing will occur at one or both photonic edges depending upon which side has the higher fluorescence intensity consistent with our present experimental results. In conclusion, the high alignment of laser dyes with the local CLCs director provides an ideal organization for lasing.

4.2.2 Anisotropic defect mode

We have introduced a new type of defect mode, an anisotropic defect layer, i.e., a nematic layer, into CLCs. It is interesting to note that the structure shown in Fig. 4.16 is the same

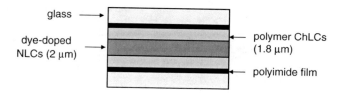

Fig. 4.16. Cell structure for anisotropic defect mode lasing.

 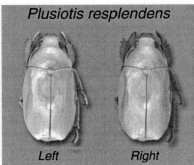

Fig. 4.17. Two types of beetles. *Plusiotis batesi* gives a selective reflection, but *Plusiotis resplendens* gives a total reflection irrespective of light polarization.

as the cuticle of one of the beetles shown in Fig. 4.17. One beetle (*Plusiotis batesi*) has a left-handed PBG structure and reflects only left-circularly polarized light, whereas the other beetle (*Plusiotis resplendens*) has a structure consisting of a nematic layer which acts as a $\lambda/2$ plate sandwiched by two left-handed (L-H) PBG structures. This structure reflects light with both circular polarizations because of the following mechanism: R-CP light with the opposite handedness to the L-H PBG helix is transmitted through the first L-H PBG layer, but then changes its polarization state to the opposite circular polarization (L-CP), which is selectively reflected by the second L-H PBG layer. This reflection directs the light back through the $\lambda/2$ plate, wherein the polarization again returns back to R-CP, to give higher reflection. Since L-CP light is totally reflected back from the first L-H PBG helix, the total reflection of natural light amounts to 100% in the ideal case. Here, we have mimicked the cuticle of a beetle to construct a sophisticated system consisting of a polymer-dye-doped nematic defect layer sandwiched by polymer CLC (PCLC) films, and have succeeded in introducing a new effect, a retardation defect mode, which gives modulation of reflectance of more than 50% and efficient lasing [50,52].

For PCLCs, mixtures of two nematic liquid crystal (NLC) polymers (Nippon Oil Corporation) were used. One of the NLC polymers contains 25% chiral units in its chemical composition, which gives right-handed helix (R-helix). By changing the mixing ratio of the two NLC polymers, the wavelength of the PBG in the PCLCs was controlled [51]. To fabricate PCLC films, solutions of the PCLC mixture were spin-cast on glass substrates with unidirectionally rubbed polyimide. After the coated PCLC films were cured for 2 min at 180°C, we obtained well-aligned R-helix PCLC film. The cholesteric helical axis of the PCLC film was normal to the substrate surface and the thickness of the film was about 1.8 μm. A pair of PCLC-coated substrates were stacked and sealed to fabricate a vacant cell with a spacer. A commercial monomeric mixture of NLCs (ZLI2293, Merck) was doped with 2 wt% of a fluorescent T-PPV[20] and introduced into the cells fabricated from the PCLC films using capillary action to form an anisotropic defect layer, as shown in Fig. 4.16.

Figure 4.18 shows the measured reflectance spectrum from such a sample cell (solid curve). The reflectance spectrum of a single PCLC film (dotted curve) is also

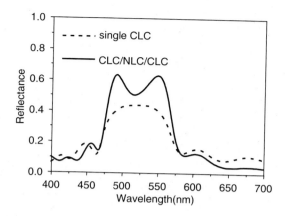

Fig. 4.18. Reflection spectra in single CLC (dotted curve) and NLC sandwiched by PCLC films (solid curve).

shown for comparison. A striking phenomenon is noted for the PCLCs with a dye-doped NLC anisotropic layer; i.e., the reflectance in the PBG region is higher than 50%, quite different from any other CLC systems. By contrast, the single PCLC film shows the expected low reflectance of about 40%. The reflectance of greater than 50% is caused by phase retardation of the NLC layer. In the ideal case, 100% reflectance would be obtained. In the present case, the thickness of the nematic layer is not controlled, so that the above effect is only partially achieved, resulting in higher reflectance than 50%, as shown in Fig. 4.18.

The other feature that should be noted is a notch of reflectance in the middle of the PBG region. This is a new retardation mode in the sense that the line shape is fairly broad in contrast to the normal defect mode, which shows a sharp, deep dip. The magnitude and shape of the reflectance can be controlled by adjusting the thickness and the anisotropy of the NLC layers.

Figure 4.19 shows the results in three cells with an NLC layer of the same thickness (2 μm) and with different mixing ratios of the component with chiral units; (a) 93 wt%, (b) 92 wt%, and (c) 87 wt%. The different chiral contents adjust the position of the reflection band (dash-dotted curve) with respect to the fluorescent emission band (broken curve). In the first example shown in Fig. 4.19(a), the reflectance spectrum was adjusted to overlap with the fluorescent emission band. The wavelength of the notch (∼500 nm) is coincident with the fluorescence peak of T-PPV. This situation promotes lasing activity, as shown by the solid curve. This result demonstrates the high efficiency of anisotropic lasing and the key differences between this approach and conventional defect mode lasing in CLCs.

The second example is a cell of PCLCs with an anisotropic NLC layer and a PBG reflectance spectrum adjusted to deviate slightly from the fluorescent emission band, as shown in Fig. 4.19(b). For this condition, two lasing modes are possible and in addition to the position at the notch of the PBG a lasing emission also occurs at the high-energy

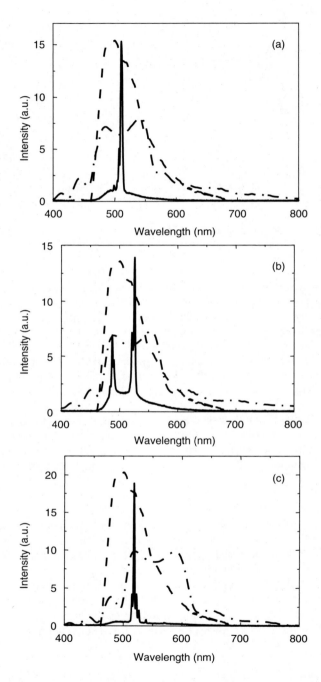

Fig. 4.19. Reflection, emission, and lasing emission spectra in three cells with different relative positions between emission and photonic bands.

edge of PBG. The lasing intensity at the notch in the middle of the PBG is more intense than at the high-energy PBG edge mode, thereby indicating that lasing at the notch of PBG is more efficient. With increasing pumping power, lasing at the notch of the PBG occurred first, followed by high-energy band edge lasing.

The third example is a cell of PCLCs with an anisotropic NLC layer and reflectance spectrum of the PBG adjusted to deviate further from the fluorescent emission band, as shown in Fig. 4.19(c). For this condition, only one sharp lasing mode is observed at the high-energy PBG edge. This result also indicates that our PCLC laser configuration with a dye-doped NLC layer provides a highly efficient lasing condition in comparison with conventional dye-doped CLC mirrorless laser assemblies. In the present case, high-energy band edge coincides with the emission peak wavelength, so that lasing occurs at the high-energy band edge. If the low-energy band edge coincides with the emission peak wavelength, we may expect lasing at the low-energy band edge. Figure 4.20 reveals a clear ring pattern, indicating the coherent laser emission in Fig 4.19(a).

It is well known that polarized fluorescent light emission of the same handedness as the cholesteric medium is suppressed inside the PBG and enhanced at the band edge of the PBG in single CLC films [69]. Hence the lasing emission at the edge of the PBG is circularly polarized with the same handedness as the CLC helix. For CLC films with 90° phase jump inside the helix, Kopp and Genack have suggested theoretical complex polarization characteristics [29]. In the present configuration with an anisotropic NLC layer, the polarization state changes after passing through the anisotropic layer. Then what is the polarization state of the lasing emission? To study the polarization characteristics of these CLC lasers, we first observed the polarization characteristics of a low molecular CLC laser [52]. As shown in Fig. 4.21(a), L-CP lasing emission with the same handedness as the CLC helix was observed at the low-energy edge of the PBG. This result is consistent with the theoretical prediction [69]. Figure 4.21(b) shows the lasing spectra of a CLC/NLC/CLC structure (solid curve) with R-C and L-C polarizers [52]. As shown in the figure, a sharp lasing emission is generated just at the notch of the PBG. A remarkable point is that lasing occurs for both L-CP and R-CP lights. We also

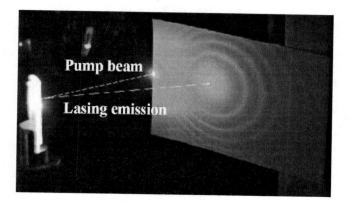

Fig. 4.20. Far-field pattern of lasing in the anisotropic defect mode.

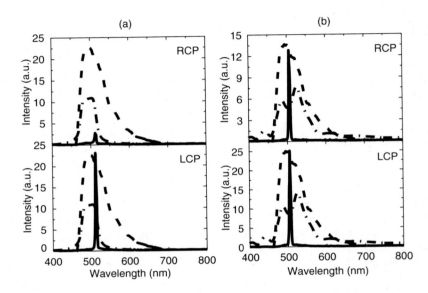

Fig. 4.21. Polarization characteristics of lasing emissions in (a) single CLC and (b) NLC sandwiched by PCLC films.

confirmed that both L-CP and R-CP lasing emissions are clearly observed even for the lasing emission from the higher edge of the PBG.

Recently, tunability of the lasing in CLC systems has been achieved by changing the chiral dopant concentration, by varying the temperature [44], by mechanical stress [38], by photoirradiation [49], or applying an electric field [43]. Here we introduce the tunability of lasing and transmittance wavelength by applying an electric field. Figure 4.22(a) shows a schematic cell structure with a polymer-dye-doped NLC layer between photopolymerized CLC films for electric tunable lasing. The thickness of the nematic layer is 50 μm. An external voltage with a 1-kHz rectangular waveform was applied to the sample cell through ITO layers to change the molecular alignment of the NLC in the anisotropic layer, as shown in Fig. 4.22(a). Figure 4.22(b) shows the shift of the lasing emissions with the applied voltage. The lasing peak shifts toward shorter wavelengths with increasing voltage due to the decrease of the optical length of the anisotropic NLC layer caused by the reorientation of the NLC molecules [43]. The range in the wavelength shift of lasing is about 24.5 nm. Two lasing modes momentarily appear near both band edges at 4.2 V, then the shorter wavelength mode of the PBG disappears and the longer wavelength mode of the PBG shifts toward the shorter wavelength.

4.2.3 Efforts to lowering the lasing threshold

Despite extensive efforts, lasing has been achieved only by excitation with pulsed laser light, but continuous wave (cw) lasing has not yet been successful. It is very important

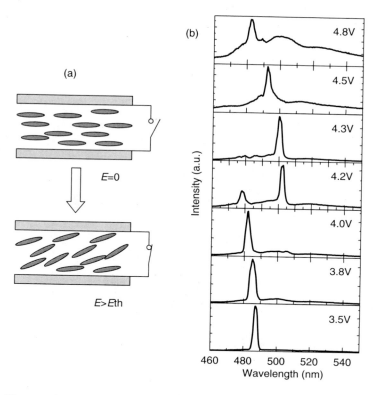

Fig. 4.22. Electrotuning of lasing emission. (a) Orientational change of NLC by applying an electric field and (b) the associated changes of the lasing peak position.

to achieve cw lasing to extend possible applications. For this purpose, the threshold input power for lasing has to be lowered. At least three efforts have been published. In the case of DFB lasing, Amemiya et al. [70] introduced two PCLC films at both surfaces; i.e., a PCLC film for reflecting a pump beam and a PCLC film for reflecting the emission, respectively. This assists the efficient use of the incident energy and contributes to amplifying the stimulated emission. As a result, the cell in which both PCLC films were introduced, achieved about 60% reduction of the lasing threshold.

An interesting attempt was made by Matsuhisa et al. [58] and Song et al. [71]. By incorporating cavity mirrors to defect mode lasing, they succeeded in lowering the threshold input energy. Although Matsuhisa et al. used dielectric multilayers as cavity mirrors, Song et al. used only CLCs. The latter is particularly interesting. Song et al. constructed a three-layered cell consisting of an L-CLC layer sandwiched by two R-CLC films. They found that the defect states emerge, and the DOS is resonantly enhanced when the defect mode coincides with the edge mode. The lasing from this enhanced mode was found to show a threshold value lowered by a factor of 3 or 4 compared with that of simple DFB lasing. The DOS shown in Fig. 4.23 clearly shows the advantage of this structure.

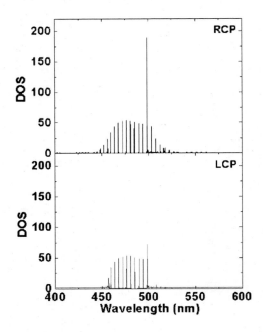

Fig. 4.23. DOS for R-CP and L-CP in L-CLC film sandwiched by two R-CLC films.

4.3 Optical Diode

In addition to the application of lasing in PCs, other optical devices have also been focused on PC applications, for example the optical diode. A diode is an essential element in most electronic circuits, permitting a current in the forward direction and blocking it in the backward direction. We can call a device an optical diode (OD) if it transmits light in one direction and blocks it in the reverse direction; i.e., non-reciprocal transmission. Optical isolators based on linear polarizers and a Faraday rotator, which provides a high extinction ratio with a low insertion loss, also show non-reciprocal transmission. Various ODs including PC-based devices [72,73] have been demonstrated. While the concept of tunability in OD operation has not been widely exploited despite such extensive studies, we have suggested a theoretically electrotunable optical heterojunction anisotropic structure consisting of an anisotropic layer sandwiched by two PCLC layers with different periodicity of the helix (hetero-PBG). This has successfully achieved novel OD effects [74,75].

4.3.1 Cell structures and principle

We have constructed two optical diode structures; i.e., two-layered [74] and three-layered [73] devices. For simplicity, let us first consider the two-layered device shown in Fig. 4.24(a). This consists of a PCLC layer and a $\lambda/2$ phase retarder layer made of a planar NLC layer. The black and gray solid (broken) lines and arrows, respectively

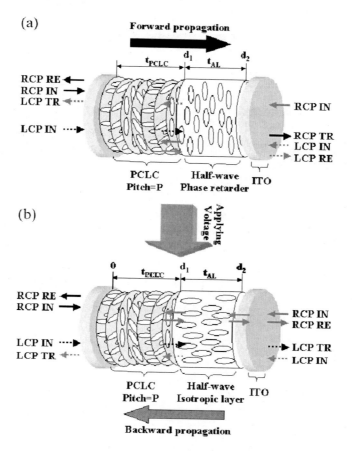

Fig. 4.24. (a) Two-layered OD device structure and OD performance and (b) electric-field induced change for electrotuning.

correspond to forward- and backward-propagating CP light with the same (opposite) handedness as the CLC's helix. For R-CP light incidence, forward-R-CP light is reflected, whereas backward-R-CP light is transmitted. For L-CP light incidence, forward-L-CP light is transmitted, whereas backward-L-CP light is reflected due to the handedness change of the incident circular polarization upon passing through the $\lambda/2$ retarder. Thus, unidirectional propagation (non-reciprocal transmission), i.e., the characteristic of the OD device, occurs.

We have also demonstrated a three-layered device, as shown in Fig. 4.25 [74]. A half-wave phase retarder layer made of planar nematic liquid crystal (NLC) is sandwiched between two PBG CLC layers with the same sense of helix handedness but possessing different pitches ($p_1 < p_2$). The principle is the same as that of the two-layered device, the only difference being the wavelength selectivity. For two different wavelength regions near p_1 and p_2, the directions with high and low transmittance are opposite, as will be shown in Section 4.3.2.

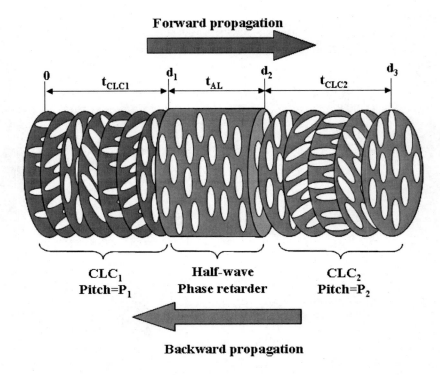

Fig. 4.25. Three-layered OD device structure.

4.3.2 Simulations

Let us consider single CLC films with pitches of p_1 and p_2 and a three-layered device. The dispersion relations for forward and backward propagation of light are readily obtained by introducing a propagation direction as well as position-dependent anisotropic dielectric tensor, as shown in Fig. 4.26(a) [74]. Two dispersion curves for L-CP light in CLC_1 (left) and CLC_2 (right), which have L-helix, are shown. PBGs due to CLC_1 and CLC_2 are shown by gray stripes. Solid and broken lines indicate the dispersions for forward and backward light propagation, respectively. Figure 4.26(b) shows the dispersion curves of the three-layered device for L-CP light (left) and R-CP light (right) [74]. Note that the handedness of the CP light refers to that of the incoming light to the device, not the outgoing light. We find that PBG positions depend on the propagation direction, exhibiting OD (non-reciprocal transmission) behavior. For the two opposite circular polarizations, the PBGs for forward and backward light propagations are interchanged. It is easy to understand that the dispersion curves shown in Fig. 4.26(b) originate from the handedness change of the incident circular polarization upon passing through the half-wave phase retarder.

We now show the transmission spectra simulated numerically by use of a 4×4 Berreman matrix [76]. As shown in Fig. 4.27 [74], the three-layered device gives different transmission spectra for forward and backward L-CP light propagations.

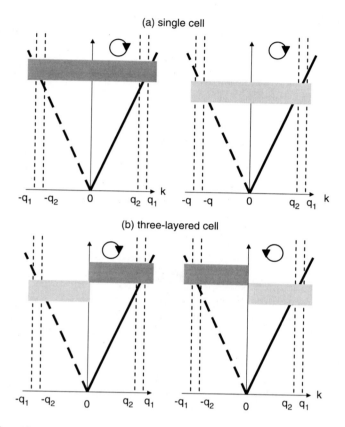

Fig. 4.26. Dispersion curves of (a) single CLC films with two different pitches and (b) the three-layered device.

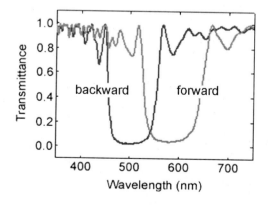

Fig. 4.27. Transmittance spectra for a three-layered OD device for forward (long wavelength band) and backward (short wavelength band) propagation light beams.

At say, 500 nm (600 nm), the film is opaque (transparent) for forward propagation, whereas it is transparent (opaque) for backward propagation, clearly demonstrating OD characteristics.

4.3.3 Experimental results

Consider a two-layered cell to explain the transmission spectra obtained experimentally [75]. Figures 4.28(a) and (b) show the transmission spectra for the forward- (black-solid (broken)) and backward- (gray-solid (broken)) propagating R-CP(L-CP) light [75]. As for the transmittance of R-CP light at 505 nm, the film is opaque for forward propagation, whereas it is transparent for backward propagation. As for the transmittance of L-CP light at 505 nm, the film is transparent for forward propagation, whereas it is opaque for backward propagation, clearly showing non-reciprocal transmittance. These spectra are those expected for the structure shown in Fig. 4.24(a). In this experiment, the extinction ratio is not very high, but it could be increased by coating both sides with antireflection films to obtain higher transmittance and by using thicker PCLC films to increase the reflectance.

We next investigated the OD action (non-reciprocal transmission) in a three-layered device [74]. The experimental setup to directly observe OD action is shown in Fig. 4.29(a). Figure 4.29(b) shows the dispersion relations of the OHAS for forward and backward propagation, respectively. The results for CP light with handedness the same as (left) and opposite to (right) that of the PCLC helix are shown. The PBG regions are indicated by black stripes, the positions of which are dependent on the direction of light propagation and circular polarization used, clearly exhibiting an OD effect.

4.3.4 Tunability

Figure 4.24(b) schematically shows the molecular orientation after reorientation of NLC molecules by an electric field. As the applied voltage gradually increases, the long axis

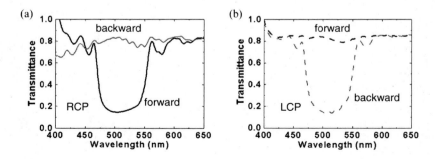

Fig. 4.28. Transmittance spectra showing OD characteristics in two-layered OD device. (a) R-CP and (b) L-CP. The thick black curves indicate forward-propagating light and gray curves indicate backward-propagating light.

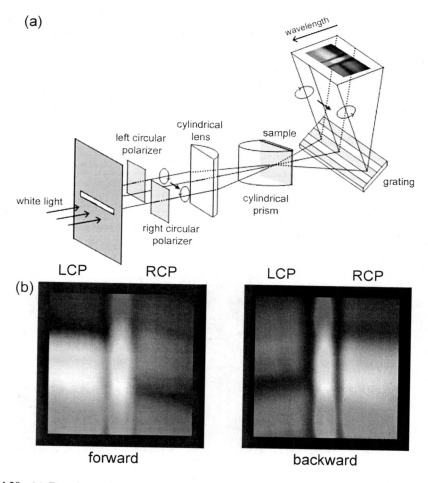

Fig. 4.29. (a) Experimental setup for visualizing PBG. (b) Visualized images for forward (left) and backward (right) propagating light.

of the NLC molecules reorients parallel to the field direction (perpendicular to the ITO surfaces) and the anisotropy of the refractive index of the NLC molecules for the light propagation along the field direction decreases after the Fredrick transition takes place, finally diminishing. For R-CP light incidence, forward-R-CP light is reflected as in Fig. 4.24(a), whereas backward-R-CP light is also reflected back contrary to Fig. 4.24(a). For L-CP light incidence, both forward and backward L-CP light beams are transmitted. After the non-reciprocal transmission property completely disappears, the PBG becomes similar to that of a single PCLC layer.

Next, in order to confirm the electrotunability of the OD device, an electric field was applied between two ITOs. Figure 4.30(a) shows the variation of the transmission spectra of the backward-propagating R-CP light under various applied voltages [75].

With gradually increasing voltage, the refractive index anisotropy Δn in the phase-retarder layer with fixed n_o, decreases, and the film becomes opaque after the Fredrick transition takes place. The film is most opaque at 5.2 V. If more than 5.2 V is applied to this device, the transmittance intensity increases then decreases repeatedly. Figure 4.30(b) shows the variation of the transmission spectra for the backward-propagating L-CP light under various applied voltages. As the applied voltage gradually increases, the film becomes increasingly transparent, with a maximum transmittance at 5.2 V [75].

Figure 4.30(c) shows the experimentally measured transmittance intensity change at a fixed wavelength (505 nm) for the present simple OD device as a function of the applied voltage. The black and gray curves are for forward and backward propagations, respectively. The solid and broken curves are for CP light the same (R-CP) as and opposite (L-CP) to the helix handedness, respectively. When the anisotropic layer behaves exactly as a $\lambda/2$ plate, the transmittance intensity is maximum or minimum for forward and backward propagation, indicating OD performance. With increasing applied voltage, the transmittance intensity changes only for the backward-propagating light, giving no OD effect at 5.5 V. Thus, electrotunability is demonstrated in the present two-layered OD device by using a reorientable anisotropic NLC layer as a tunable phase retarder.

The OD performance has an extinction ratio of 8.2 at 505 nm as can be seen from Fig. 4.30(c). Although our OD shows a low extinction ratio compared with a conventional

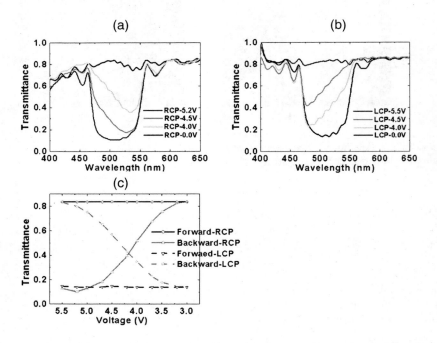

Fig. 4.30. Electrotunability of the two-layered OD device. (a) Backward-propagating R-CP light, (b) backward-propagating L-CP light, and (c) the transmittance change at 505 nm as a function of an electric field.

optical isolator which shows a high extinction ratio with a low insertion loss, we can improve our OD device by coating antireflecting multilayers at both surfaces to obtain higher transmittance, using thicker PCLC films to increase reflectance and optimizing the cell geometry.

It is also noted that the OD effect and tenability are effective for light emitted inside of the device [77]. A non-reciprocal R-CP lasing emission peak emerges at either of the band edges and diminishes upon voltage application. In contrast, two L-CP lasing emission peaks were observed at both edges of the overlapped region of two photonic bands and were shifted upon voltage application.

4.4 Concluding Remarks and Future Problems for Practical Applications

The use of organic functional materials as practical devices is still quite limited. Liquid crystals (LCs) have successfully been used for display devices. In this application, LC molecules respond dielectrically to the electric field, meaning that very small power consumption is necessary because no current flows. This is one of the great advantages of LC displays over other flat panel display devices such as plasma display panels (PDP) and organic light emitting diodes (OLED).

For laser applications, lasing by continuous wave (cw) or by charge injection will have to be developed. For this purpose, the threshold input energy must be negligible or zero. Progress has been made in this as described above. Another problem is heating of the devices. Dyes doped into CLCs absorb light and inevitably increase the temperature of the device, in turn deteriorating the organic materials such as the dyes. This is the most serious problem for this practical application.

By contrast, applications as optical diodes are more promising. Since the light wave either passes through or is reflected and no absorption takes place, damage by the incident light is not serious. At the present stage, these devices are useful only for circularly polarized light. However, their use as optical isolators may be possible. New ideas for devices applicable to natural light will open up the possibility of wider applications.

Acknowledgments

This work is supported in part by the TIT-KAIST Joint Program, the JSPS Research for the future program, the 21st century COE program, and a Grant-in-Aid for Exploratory Research (16656023) from the Ministry of Education, Sports, Culture, Science and Technology. This work is strongly supported by many collaborators outside, our staff and former and present students. We acknowledge all these collaborators. Special thanks are due to Professor T. M. Swager for supplying the polymer dye, and two students, M. H. Song and K. C. Shin who mainly conducted the experiments introduced in Sections 4.2.1 and 4.2.2, respectively. We also acknowledge Nippon Oid Corp. and Merck for supplying valuable materials.

References

[1] E. Yablonovitch, Phys. Rev. Lett. **58** (1987) 2059.

[2] S. John, Phys. Rev. Lett. **58** (1987) 2486.

[3] J.D. Joannopoulos, R.D. Meade, J.N. Winn, Photonic Crystals: Molding the Flow of Light, Princeton University Press: New Jersey, 1995.

[4] E. Yablonovitch, Phys. Rev. Lett. **67** (1991) 2295.

[5] K.M. Ho, Phys. Rev. Lett. **65** (1990) 3152.

[6] H. Kosaka, T. Kawashima, A. Tomita, M. Notomi and T. Tamamura, Phys. Rev. B **58** (1998) R10096.

[7] S. Noda, A. Chutinan and M. Imada, Nature **407** (2000) 608.

[8] E. Yablonovitch and T.J. Gmitter, Phys. Rev. Lett. **67** (1991) 3380.

[9] P.G. de Genne, The Physics of Liquid Crystals, Oxford University Press: Oxford, 1974.

[10] S. Chandrasekhar, Liquid Crystals, Cambridge University Press: Cambridge, 1992.

[11] C. Da Cruz, J.C. Rouillon, J.P. Marcerou, N. Isaert and H.T. Nguyen, Liq. Cryst. **28** (2001) 125.

[12] R.B. Meyer, L. Liebert, L. Strzelecki and P. Keller, J. Phys. (France), **36** (1975) L69.

[13] A.D.L. Chandani, T. Hagiwara, Y. Suzuki, Y. Ouchi, H. Takezoe and A. Fukuda, Jpn. J. Appl. Phys. **27** (1088) L729.

[14] A.D.L. Chandani, E. Gorecka, Y. Ouchi, H. Takezoe and A. Fukuda, Jpn. J. Appl. Phys. **28** (1989) L1265.

[15] J.W. Goodby, M.A. Wauch, S.M. Stein, E. Chin, R. Pindak and J.S. Patel, Nature **337** (1989) 449.

[16] H. Stegemeyer, Th. Blumel, K. Hiltrop, H. Onusseit and F. Porsch, Liq. Cryst. **1** (1986) 1.

[17] H. Kikuchi, M. Yokota, Y. Hisakado, H. Yang and T. Kajiyama, Nature Mater. **1** (2002) 64.

[18] H. Takezoe, Y. Ouchi, M. Hara, A. Fukuda and E. Kuze, Jpn. J. Appl. Phys. **22** (1983) 1080.

[19] H. Takezoe, K. Hashimoto, Y. Ouchi, M. Hara, A. Fukuda and E. Kuze, Mol. Cryst. Liq. Cryst. **101** (1983) 329.

[20] M. Hara, H. Takezoe, A. Fukuda, E. Kuze, Y. Kaizu and H. Kobayashi, Mol. Cryst. Liq. Cryst. **116** (1985) 253.

[21] J.P. Dowling, M. Scalora, M.J. Bloemer and C.M. Bowden, J. Appl. Phys. **74** (1994) 1896.

[22] R. Ozaki, M. Ozaki and K. Yoshino, Jpn. J. Appl. Phys. **42** (2003) L669.

[23] R. Ozaki, M. Ozaki and K. Yoshino, Appl. Phys. Lett. **82** (2003) 3593.

[24] R. Ozaki, M. Ozaki and K. Yoshino, Appl. Phys. Lett. **84** (2004) L1844.

[25] O. Painter, P.K. Lee, A. Scherer, A. Yariv, J.D. O'Brien, P.D. Dapkus and I. Kim, Science **284** (1999) 1819.

[26] S.G. Johnson and J.C. Joannopoulos, Appl. Phys. Lett. **77** (2000) 3490.

[27] K.M. Ho, C.T. Chan and C.M. Soukoulis, Phys. Lett. **65** (1990) 3152.

[28] H. Yokoyama and K. Ujihar, Eds., Spontaneous Emission and Laser Oscillation in Microcavities, CRC Press: Boca Raton, FL, 1995.

[29] V.I. Kopp and A.Z. Genack, Phys. Rev. Lett. **89** (2002) 33901.

[30] H. Hoshi, K. Ishikawa and H. Takezoe, Phys. Rev. E. **68** (2003) 020701.

[31] Y.-C. Yang. C.-S Kee, J.-E. Kim, H.Y. Park, J.-C. Lee and Y.-J. Jeon, Phys. Rev. E **60** (1999) 6852.

[32] T. Matsui, M. Ozaki and K. Yoshino, Phys. Rev. E. **69** (2004) 061715.

[33] J. Schmidtke, W. Stille and H. Finkelmann, Phys. Rev. Lett. **90** (2003) 83902.

[34] M. Ozaki, T. Matsui and K. Yoshino, Jpn. J. Appl. Phys. **42** (2003) L472.

[35] V.I. Kopp, B. Fan, H.K.M. Vithana and A.Z. Genack, Opt. Lett. **28** (1998) 1707.

[36] A. Munoz, P.P. Muhoray and B. Taheri, Opt. Lett. **26** (2001) 804.

[37] B. Taheri, A.G. Munoz, P.P. Muhoray and R. Twieg, Mol. Cryst. Liq. Cryst. **358** (2001) 73.

[38] H. Finkelmann, S.T. Kim, A. Munoz, P.P. Muhoray and B. Taheri, Adv. Mater. **14** (2001) 1069.

[39] M. Chambers, M. Fox, M. Grell and J. Hill, Adv. Func. Mater. **12** (2002) 808.

[40] P.V. Shibaev, K. Tang, A.Z. Genack, V. Kopp and M.M. Green, Macromolecules **35** (2002) 3022.

[41] J. Schmidtke, W. Still, H. Finkelmann and S.T. Kim, Adv. Mater. **14** (2002) 764.

[42] T. Matsui, R. Ozaki, K. Funamoto, M. Ozaki and K. Yoshino, Appl. Phys. Lett. **81** (2002) 3741.

[43] S. Furumi, S. Yokoyama, A. Otomo and S. Mashiko, Appl. Phys. Lett. **82** (2003) 16.

[44] F. Araoka, K.-C. Shin, Y. Takanishi, Z. Zhu, T.M. Swager, K. Ishikawa and H. Takezoe, J. Appl. Phys. **94** (2003) 279.

[45] A. Chanishvili, G. Chilaya, G. Petriashvili, R. Barberi, R. Bartolino, G. Cipparrone, A. Mazzulla and L. Oriol, Appl. Phys. Lett. **83** (2003) 5353.

[46] J. Schmidtke and W. Stille, Eur. Phys. J. B **31** (2003) 179.

[47] M. Ozaki, R. Ozaki, T. Matsui and K. Yoshino, Jpn. J. Appl. Phys. **42** (2003) L472.

[48] K.-C. Shin, F. Araoka, B. Park, Y. Takanishi, K. Ishikawa, Z. Zhu, T.M. Swager and H. Takezoe, Jpn. J. Appl. Phys. **43** (2004) 631.

[49] S. Furumi, S. Yokoyama, A. Otomo and S. Mashiko, Appl. Phys. Lett. **84** (2004) 2491.

[50] M.H. Song, B. Park, K.-C. Shin, T. Ohta, Y. Tsunoda, H. Hoshi, Y. Takanishi, K. Ishikawa, J. Watanabe, S. Nishimura, T. Yoyooka, Z. Zhu, T.M. Swager and H. Takezoe, Adv. Mater. **16** (2004) 779.

[51] T. Ohta, M.H. Song, Y. Tsunoda, T. Nagata, K.-C. Shin, F. Araoka, Y. Takanishi, K. Ishikawa, J. Watanabe, S. Nishimura, T. Toyooka and H. Takezoe, Jpn. J. Appl. Phys. **43** (2004) 6142.

[52] M.H. Song, K.-C. Shin, B. Park, Y. Takanishi, K. Ishikawa, J. Watanabe, S. Nishimura, T. Toyooka, Z. Zhu, T.M. Swager and H. Takezoe, Sci. Tech. Adv. Mater. **5** (2004) 437.

[53] K. Amemiya, K.-C. Shin, Y. Takanishi, K. Ishikawa, R. Azumi and H. Takezoe, Jpn. J. Appl. Phys. **43** (2004) 6084.

[54] T. Nagata, T. Ohta, M.-H. Song, Y. Takanishi, K. Ishikawa, J. Watanabe, T. Toyooka, S. Nishimura and H. Takezoe, Jpn. J. Appl. Phys. **43** (2004) L1220.

[55] A. Chanishvili, G. Chilaya, G. Petriashvili, R. Barberi, R. Bartolino, G. Cipparrone and A. Mazzulla, Appl. Phys. Lett. **85** (2004) 3378.

[56] A. Chanishvili, G. Chilaya, G. Petriashvili, R. Barberi, R. Bartolino, G. Cipparrone and A. Mazzulla, Adv. Mater. **16** (2004) 791.

[57] S.M. Morris, A.D. Ford, M.N. Pivnenko and H.J. Coles, J. Appl. Phys. **97** (2005) 023103.

[58] Y. Matsuhisa, R. Ozaki, M. Ozaki and K. Yoshino, Jpn. J. Appl. Phys. **44** (2005) L629.

[59] M. Ozaki, M. Kasano, D. Ganzke, W. Haase and K. Yoshino, Adv. Mater. **14** (2002) 306.

[60] M. Kasano, M. Ozaki and K. Yoshino, Appl. Phys. Lett. **82** (2003) 4026.

[61] M. Ozaki, M. Kasano, T. Kitasho, D. Ganzke, W. Haase and K. Yoshino, Adv. Mater. **15** (2003) 974.

[62] W.Y. Cao, A. Munoz, P. Palffy-Muhoray and B. Taheri, Nature Mater. **1** (2002) 111.

[63] T. Long and T.M. Swager, Adv. Mater. **13** (2001) 601.

[64] T. Long and T.M. Swager, J. Am. Chem. Soc. **124** (2002) 3826.

[65] Z. Zhu and T.M. Swager, J. Am. Chem. Soc. **124** (2002) 9670.

[66] R. Dreher and G. Meier, Phys. Rev. A **8** (1973) 1616.

[67] Y. Ouchi, H. Takezoe, A. Fukuda, E. Kuze, N. Goto and M. Koga, Jpn J. Appl. Phys, **23** (1984) L464.

[68] J.M. Bendickson, J.P. Dowling and M. Scalora, Phys. Rev. E **53** (1996) 4107.

[69] V.I. Kopp and A.Z. Genack, Proc. SPIE **3623** (1999) 71.

[70] K. Amemiya, M.H. Song, Y. Takanishi, K. Ishikawa, S. Nishimura, T. Toyooka and H. Takezoe, Jpn. J. Appl. Phys. **44** (2005) 7966.

[71] M.H. Song, N.Y. Ha, K. Amemiya, B. Park, Y. Takanishi, K. Ishikawa, J.W. Wu, S. Nishimura, T. Toyooka and H. Takezoe, Adv. Mater. **18** (2006) 193.

[72] M. Scalora, J.P. Dowling, C.M. Bowden and M.J. Bloemer. J. Appl. Phys. **76** (1994) 2023.

[73] S.F. Mingaleev and Y.S. Kivshar, J. Opt. Soc. Am. B **19** (2002) 2241.

[74] J. Hwang, M.H. Song, B. Park, S. Nishimura, T. Toyooka, J.W. Wu, Y. Takanishi, K. Ishikawa and H. Takezoe, Nature Mater. **4** (2005) 383.

[75] M.H. Song, B. Park, Y. Takanishi, K. Ishikawa, S. Nishimura, T. Toyooka and H. Takezoe, Thin Solid Films, In press.

[76] D.W. Berreman, J. Opt. Soc. Am. **62** (1972) 502.

[77] M.H. Song, B. Park, S. Nishimura, T. Toyooka, I.J. Chung, Y. Takanishi, K. Ishikawa and H. Takezoe, Adv. Func. Mater. In press.

CHAPTER 5

Nanocylinder Array Structures in Block Copolymer Thin Films

Kaori Kamata and Tomokazu Iyoda

Abstract

An overview of the microphase separation of block copolymers is described here ranging from their syntheses and basic phase behaviors to the recent progress in the fabrication of ultrafine self-organized nanostructures. Since the synthesis of block copolymers with a narrow polydispersity and a wide variety of backbones such as a rod–coil or a coil–coil block copolymer has undergone considerable development over the last 20 years, the morphology of self-organized nanostructures in block copolymer melts can be controllable from the viewpoint of polymer design and treatment conditions generating the desired phases. The self-organized nanostructure with cylindrical phase has especially attracted much interest due to the structural character of its domains, which are either continuous or noncontinuous in the solid state, i.e., the anisotropic distribution of the elements, and due to its ability to control the cylinder direction against the film. The nanocylinder array structures in block copolymer samples have also been shown to be promising for the design of electric and photonic devices.

Keywords: block copolymer, self-organized nanostructure, phase transition, nanocylinder array structure, nanostructured thin film, nanotemplating.

5.1 General Introduction

Block copolymers, which are substances composed of two or more types of polymeric segment (block) combined linearly in a macromolecular chain, have been investigated for a long time. Together with the related families of graft copolymers, they have been specific targets to form a well-established discipline in polymer science and technology. Because of the connectedness of polymer chains, phase separation leads to pattern formation on nanometer scales, defined in theoretical studies using molecular dynamic calculations. Since then, the morphology and phase behavior of block copolymers have been extensively studied experimentally by direct observations using transmission electron microscopy (TEM) and small-angle X-ray scattering (SAXS) analysis. For symmetric copolymers, in which both blocks occupy equal volume fractions, the domains generally self-organize into lamellae. The parameters that determine the morphology of a given block copolymer are the incompatibility between the blocks, polymer composition, and the volume fractions of individual blocks. By varying these parameters, spherical, cylindrical, or more complex morphologies such as a gyroid can be formed.

Recent advances in synthetic chemistry have exposed new opportunities to use judicious combinations of multiple blocks in novel macromolecular architectures to produce a seemingly unlimited number of exquisitely structured materials with tailored mechanical, optical, electric, magnetic, ionic, and other physical properties.

5.2 Synthesis of Block Copolymers

Over the last decade, there has been considerable progress in the development of synthetic strategies for preparing block copolymers of various architectures, solubilities, and functionalities. Such architectures comprise diblock, triblock, and multiblock copolymers arranged linearly or as grafts, stars, or H-shaped blocks. The solubilities vary among solvents with high cohesive-energy densities such as water and media with low cohesion energies such as silicon oil or fluorinated solvents. Additionally, the control of functionality has become an important issue, motivated by the requirements to stabilize metallic, semiconductor, ceramic, or biomaterial interfaces. The preparation of well-defined block copolymers requires a chain-growth polymerization mechanism that can be conducted in the absence of undesired transfer and termination steps. The anionic polymerization of styrene and isoprene was the first successful demonstration of this approach using AB diblock copolymers, i.e., poly(styrene-*block*-isoprene) (PS-*b*-PI) [1], although both the precedence and initial impact of this work have been questioned. Adams *et al.* [2] reported the anionic step polymerization reaction, as the method modified in this first demonstration, that involves a subsequent polymer-analogous reaction. They also mentioned that the products of such a reaction display phase separation. Zascke *et al.* demonstrated a similar synthetic route [3] for the synthesis of diblock copolymers and reviewed useful post-polymerization (polymer-analogue) reactions for the preparation of amphiphilic block copolymers [4]. Subsequently, a number of different routes to block copolymers have been reported, e.g., group transfer polymerization [5], cationic polymerization [6], combined anionic and photopolymerization [7], metathesis [8], and direct anionic polymerization [9]. Many classical synthetic routes to block copolymers such as living anionic polymerization have long been known and was summarized in an excellent review by Riess *et al.* [10]. The 'living' carbanion end of a polymer has been focused on, because it can be used to initiate the polymerization of a second monomer; thus, an AB diblock can be prepared by the sequential addition of monomers, if the 'living' carbanion can be stably generated. This was one of the original motivations for developing living anionic polymerization, and remains a key potential advantage of any living polymerization protocol.

Among the most common addition polymerization reactions based on monomers containing double carbon–carbon bonds, only anionic polymerization, which is limited to a fairly narrow range of monomers, is inherently living. Until recently, radical polymerization, which covers the broadest range of monomers, seemed to be inherently uncontrollable, due to fast chain termination of growing radicals. Thus, although very useful in the synthesis of bulk, conventional polymers, where precise control of the polymer structures has not always been the center of research subjects, radical polymerization has been hardly viewed as a tool for precisely engineering macromolecules for well-defined functional nanostructures. This outlook markedly shifted with the development of new controlled/living radical polymerization processes, such as atom transfer radical polymerization (ATRP), which were reported by Matyjaszewski group [11–13]. ATRP has enabled the preparation of new classes of amphiphilic or functional block copolymers including heteroatoms, since these techniques have the advantage of yielding narrow molecular weight distributions with a predetermined degree of polymerization, N, that depends only on the molar ratio of monomers to initiators. Through this progress in polymerization techniques, several different monomers, which greatly expand the resulting properties beyond those of prototypical styrene/diene copolymers, have been developed and the resulting block copolymers have been expected as one of the promising candidates for nanotechnology. (see Table 5.2 in Section 5.3). Modifiable monomers such as *tert*-butyl methacrylate and 2- or 4-vinylpyridine have particularly drawn attention owing to their abilities of post-polymerization chemical modification (transformation reaction), enabling the fixation of the chain length, composition, and architecture at the initial polymerization.

Side-chain liquid crystalline (LC) block copolymers have also been extensively studied, since research interests have focused on the influence of the nature and structure of mesogenic side-chain groups on the mesomorphic properties of block copolymers. The first polymerization design by Adams *et al.* involved anionic step polymerization and the production of high-molecular-weight LC block copolymers that exhibit lamellar morphology [14], although most LC monomers have been regarded as unsuitable for anionic polymerization due to the high reactivity of all typical functional groups with 'living' anions. The LC monomers studied for copolymerization with polystyrene (PS) are listed in Table 5.1, including their resulting phases and assignment method of these phases.

5.3 Self-organization and Phase Behavior of Block Copolymer Microdomains

The unique properties of block copolymer materials rely crucially on the mesoscopic self-assembly of these materials in the molten and solid states. Such self-assembly spatially produces periodic structural patterns that can exhibit considerable complexity. These patterns are commonly referred to as microphases, mesophases, or nanophases, depending on the length of the block copolymer system. The blocks making up a block copolymer generate chemical incompatibilities, which is the essence of the driving forces for the formation of microphase-segregated structures. In the simplest case of an AB diblock copolymer, there is only the issue of compatibility between dissimilar A and B blocks. Unlike binary mixtures of low-molecular-weight fluids, the entropy of mixing per unit volume of dissimilar polymers is small. Thus, even minor chemical or structural differences between A and B blocks are sufficient to produce excess free energy contributions that are usually unfavorable for mixing. The microphase separation

Table 5.1. Common liquid-crystalline blocks combined with classic blocks (C-block), studied for the synthesis and phase morphology.

C-block[*1]	LC-block	Functional group	Polymerization	Remark	Ref.
PS	PEE	**choresteryl LC**	anionic and polymer-analogue	lamellae (TEM)	[2,14]
PS	PMA[*2]		anionic and polymer-analogue	sphere and lamellae (SAXS and TEM)	[15–18]
PS	PMA	**biphenyl LC**	ATRP	cylinder (SAXS, WAXS, and AFM)	[19]
PS	PEE		anionic and polymer-analogue	sphere and cylinder (SAXS and TEM) discussion of mesogen and cylindrical phase orientation	[20]

Block A	Block B (structure)	LC type / notes	Polymerization	Morphology / notes	Refs
PMMA $-\text{CH}_2-\underset{\underset{\text{C}=\text{O}}{\|}}{\underset{\|}{\text{C}}}(\text{CH}_3)-\text{OCH}_3$	PMA $-\text{CH}_2-\underset{\underset{\text{C}=\text{O}}{\|}}{\underset{\|}{\text{C}}}(\text{CH}_3)-\text{O-L-R}$; $L=-(\text{CH}_2)_6-$; $R=$ biphenyl–OCH_3		group transfer polym. (GTR) living anionic polym.	synthesis only lamellae (SAXS and TEM) discussion of LC-layered structure in crystalline phase	[5][21]
PS	main-chain: isocyanate $-\underset{\underset{\text{O}}{\|\|}}{\text{C}}-\underset{\underset{\text{CH}_3}{\|}}{\text{N}}-(\text{CH}_2)_n$	isocyanate LC	anionic polym.	zigzag morphology (TEM) hierarchical structures	[22–24]
PS	PMA $-\text{CH}_2-\underset{\underset{\text{C}=\text{O}}{\|}}{\underset{\|}{\text{C}}}(\text{CH}_3)-\text{O-L-R}$; $L=-(\text{CH}_2)_2-$; $R=-\text{O}-\underset{\underset{\text{O}}{\|\|}}{\text{C}}-\text{CH}=\text{CH}-\text{C}_6\text{H}_5$	cinnamoyl LC	anionic and polymer-analogue	rectangularly packed cylinder (TEM and SAXS)	[25,26]
PS	PMA $-\text{CH}_2-\underset{\underset{\text{C}=\text{O}}{\|}}{\underset{\|}{\text{C}}}(\text{CH}_3)-\text{O-L-R}$; $L=-(\text{CH}_2)_n-$; $n=4$ or 6; $R=$ –C$_6$H$_4$–N=N–C$_6$H$_4$–R″	azobenzene LC; $R''=-\text{OCH}_3$	anionic polym.	synthesis only*3	[9]
				cylindrical micelle	[27]

Continued

Table 5.1. Common liquid-crystalline blocks combined with classic blocks (C-block), studied for the synthesis and phase morphology—cont'd.

C-block[*1]	LC-block	Functional group	Polymerization	Remark	Ref.
PS	PEE —CH$_2$—CH— | CH$_2$ | CH$_2$ | C(=O)—OR		R'' = —CN anionic polym.	lamellae and cylinder (TEM)	[28,29]
PEO —O—CH$_2$—CH$_2$—	PMA —CH$_2$—C(CH$_3$)— | C(=O)—O-L-R L = —(CH$_2$)$_{11}$—		R'' = —O(CH$_2$)$_3$CH$_3$	micelle	[30]
			R'' = —(CH$_2$)$_3$CH$_3$ ATRP	cylinder (TEM)	[31][*4]
PS	PMA —CH$_2$—C(CH$_3$)— | C(=O)—O-L-R L = —(CH$_2$)$_{11}$—	*coumarin* R = —O—⟨benzene⟩—COO—⟨coumarin⟩—COOC$_2$H$_5$		cylinder (TEM)	[32]

[*1] Denotations of polymers: PS polystyrene, PMMA poly(methyl methacrylate), PMA polymethacrylate, PEE poly(ethylethylene).
[*2] PMA has side-chain LC with a long hydrocarbon chain as a linker (L) and a functional LC group (R).
[*3] Although the authors showed the electron micrograph of microtome-sliced thin section, no structural feature was observed.
[*4] PEO-b-PMA with an azobenzene-based side-chain LC block copolymer shows a highly ordered hexagonal cylinder array structure, so that the Section 6 focuses on the cylindrical phase behavior and the orientation control in the thin film.

behaviors of block copolymer solutions have been theoretically studied in the early stages using a thermodynamic model on the basis of the relationship between critical micelle concentration and the incompatibility of both blocks [33,34]. In the solid states of phase-segregated structures, microscopic statistical theories on phase equilibria [35–37] and swelling equilibria [38], the illustrations of models for SAXS profiles of crystalline block copolymers [39,40] have been subsequently studied. From the experimental results of SAXS, neutron scattering, and TEM of ultrathin-sectioned samples [41–44], as well as the above theoretical predictions, the shape and size of the micelle or domain have been understood to be associated with (i) Flory-Huggins interaction parameter, χ_{AB}, (ii) the total degree of polymerization, $N = N_A + N_B$, and (iii) the volume fraction of the A component, f_A [45]. Since then, the analyses and observations of microphase separations have been systematically demonstrated using linearly combined block copolymers as the simplest AB diblock copolymers or ABA triblock copolymers consisting of, for example, PS-*b*-PI, poly(styrene-*block*-methylmethacrylate) (PS-*b*-PMMA), and poly(styrene-*block*-vinylpyridine) (PS-*b*-PVP). These studies have significantly played important roles in the elucidation of phase transitions that determine spherical, cylindrical, lamellar, or more complex gyroid phases (Fig. 5.1).

Solidification methods and their conditions, e.g., casting, dipping, and spin-coating, polymer solution (solvent and concentration), sample preparation, and sometimes complex systems such as blend systems between block copolymers and homopolymers of their constituent blocks and other blocks have been investigated extensively.

The casting solvent, polymer concentration, and blend systems with the homopolymer or block copolymer having different molecular weights or main frames have been examined for the drawing of complete phase diagrams. Most of the studies of the phase transitions of block copolymers have targeted PS-based copolymers. Table 5.2 summarizes block copolymers with PS (A-block) as the common block copolymer and their phase structures. In this Section, the self-organized nanostructures of PS-based block copolymers will be introduced with their phase transitions based on the polymer structure of the B-block, molecular weight, copolymer compositions, and other conditions of sample preparations, such as temperature, the polymer concentration of the casting solution, and the type of blend system.

5.3.1 Phase transition of self-organized structure: micelles, vesicles, spheres, cylinders, and lamellae

The self-assembly of diblock copolymers in selective solvents induces micellar aggregation with an insoluble block as the core and a soluble block as the corona [93]. Spheres are the most common morphology for block copolymer micelles. Bates investigated that four factors control polymer–polymer phase behavior in solution: choice of monomers, molecular architecture, composition, and molecular size [94]. Two representative molecular architectures, binary linear homopolymer mixtures, and diblock copolymers, exhibiting macrophage separation and microphase segregation, respectively, were mentioned in some detail therein. The unique design of shell-cross-linked Knedel-micelles consisting of block copolymers of poly(styrene-*block*-acrylic acid) (PS-*b*-PAA) [95,96] or PS-*b*-PVP [97,98] has been demonstrated by incorporating cross-linkable functional

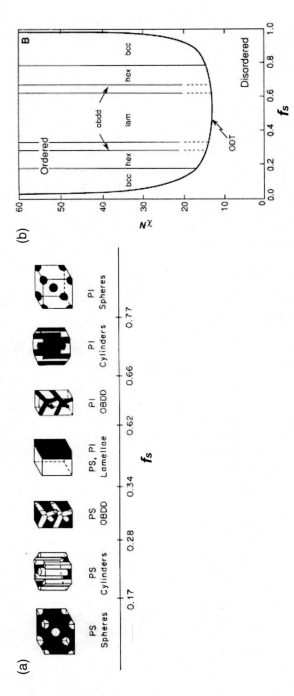

Fig. 5.1. Phase diagram and drawing of PS-*b*-PI block copolymer system.

Table 5.2. Common AB block copolymers with polystyrene as A-block and their experimental studies of their self-organized morphologies.

B block		Type of sample	Obs.*[1]	Remark	Ref.
PI polyisoprene	$-CH_2-CH=\underset{CH_3}{C}-CH_2-$	bulk	S1, CT S2	lamellar and spherical morphology	[42,46–50]
		bulk	A, T	phase transition	[51]
		film	S1, CT	phase transition	[52]
		bulk	S2, CT	salt-addition effect	[53]
		bulk	CT	effect of blend with homopolymer	[54]
		bulk		block copolymer template	[55]
				block copolymer lithography	[56,57]
PB polybutadiene	$-CH_2-CH=CH-CH_2-$	bulk	S2, CT S3	effect of blend with homopolymer	[58]
		film	A	depth profiling	[59,60]
				block copolymer lithography	[57,61,62]
PMMA polymethylmethacrylate	structure with CH_3, CH_2, $C=O$, OCH_3	film	S4	surface morphology	[63]
		film	A	surface morphology	[64–66]
		film	T	lamellar morphology	[67]
				block copolymer template	[68–73]
PVP polyvinylpyridine	$-CH_2-CH-$ (pyridine groups 2- and 4-)	film	T, X, S4	cylindrical morphology	[74]

Continued

Table 5.2. Common AB block copolymers with polystyrene as A-block and their experimental studies of their self-organized morphologies—cont'd.

B block		Type of sample	Obs.[*1]	Remark	Ref.
P4VP P2VP		film	T, A	micellar morphology	[75]
		bulk	T	supramolecular assembly	[76]
		bulk	CT	solvent and cross-linking effect	[77]
		film	T	supramolecular assembly	[78,79]
				block copolymer template	[80–84]
PAA poly(acrylic acid)	—CH$_2$–CH— | O=C–OH	particle	T	cylindrical morphology in restricted region	[85]
		film	A, T	cylindrical morphology	[86]
				block copolymer template	[87,88]
PEO Poly(ethylene oxide)	—O–CH$_2$–CH$_2$—	bulk	S1, CT	cylindrical morphology	[89]
		film	A, CT	cylindrical morphology	[90]
		bulk	S1, CT	lamellar morphology	[91]
				block copolymer template	[92]

[*1] Observation technique for obtaining direct evidence of perpendicular nanocylinder array structures, i.e., SAXS, Small-angle neutron scattering (SANS), X-ray reflectivity, TEM, cross-sectional TEM, SEM, cross-sectional SEM, AFM, and SIMS, denoted S1, S2, X, T, CT, S3, CS3, A, and S4, respectively.

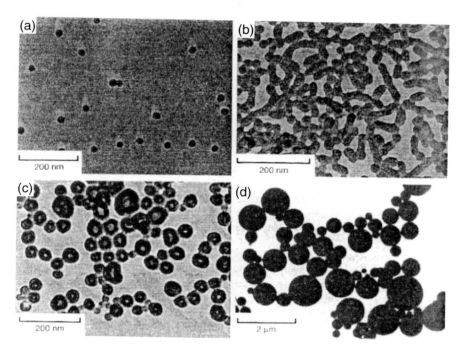

Fig. 5.2. Multiple morphologies of crew-cut aggregates from poly(styrene-acrylic acid) block copolymers (PS_m-b-PAA_n). Polymers with different molecular weights denoted m-b-n: (a) 200-b-21, (b) 200-b-15, (c) 200-b-6, and (d) 200-b-4.

groups along the peripheral block of the micelle and enables the polymerization of the exterior layer and gives stability to the micellar assembly (Fig. 5.2).

Other types of supramolecular structure for functional micelles and vesicles have recently been reported. In particular, 'crew-cut' systems [99], in which the soluble block is very short, are of great interest, since electrochemically active poly(ferrocenylsilane) is employed as the organometallic core of the micelles. The resulting water-soluble cylindrical micelles shown in Fig. 5.3 may provide a wide range of redox-active microcapsules simply by varying the micelle preparation method, copolymer composition, polymer concentration, pH and the temperature of the solution, and consequently may prove advantageous for lithographic applications.

The structural features of PS-b-PI block copolymers were first investigated by small-angle neutron scattering and TEM [51]. Wide ranges of molecular weights and compositions have been investigated, chiefly encompassing spherical, cylindrical, lamellar morphologies of PS domains [52,53]. Apart from domain separation and packing, domain size has been determined by using copolymers with fully deuterated PS blocks to improve contrast. Although domain sizes are in some agreement with the statistical thermodynamic theory, no such agreement was found for domain separation. Possible reasons for this were discussed, especially for copolymers with spherical morphology. Results also

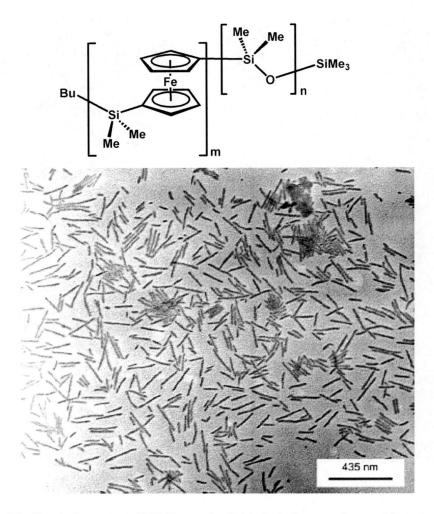

Fig. 5.3. Chemical structure and TEM image of cylindrical micelles aerosol-sprayed from aqueous solution.

indicate that the thermal annealing of the copolymers increases grain size rather than improves the packing of the domains. Representative electron micrographs are shown in Fig. 5.4 for lamellar and cylindrical PS domain morphologies. Copolymers with spherical PS domain morphology were more difficult to examine by electron microscopy primarily due to the heavy staining by osmium tetraoxide, which apparently contaminated the PS domains, thereby reducing contrast. The micrographs show that the domains are highly organized on a local scale, with cylindrical domains exhibiting perfect hexagonal close packing (*hcp*). In micrographs with a low magnification, it is clear that the copolymers have 'grains' that are oriented at different angles to the incident electron beam. This is most evident for cylindrical domains (Figs. 5.4(c–e)) but is also clearly shown by lamellar domains (Fig. 5.4(a)), which run almost orthogonal to each other in the grains.

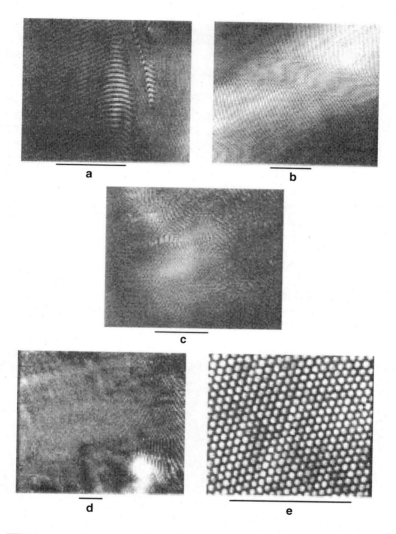

Fig. 5.4. TEM images of PS-*b*-PI diblock copolymers: (a) weight fraction of styrene, Ws = 53%, (b) Ws = 69%, (c) Ws = 36%, (d) and (e) that of deuterated PS-*b*-PI diblock copolymer with Ws = 55%. The lamellar and cylindrical morphologies were observed in (a), (b), and (c)–(e), respectively. Scale bar indicates a length of 1 μm.

The order–order transition (OOT) and order–disorder transition (ODT) in a poly(styrene-*block*-isoprene-*block*-styrene) triblock copolymer (Vector 4111, Dexco Polymers Co.), which has a weight-average molecular weight of 1.4×10^5 and a 0.183 weight fraction of PS blocks, were investigated by SAXS and rheological measurements [100]. SAXS experiments reveal that Vector 4111 undergoes an OOT at 179–185°C with (i) hexagonally packed cylindrical microdomains of PS at $T \leq 179°C$, (ii) the coexistence of cylindrical and spherical microdomains of PS at $180°C \leq T \leq 185°C$, (iii) spherical microdomains of PS in a cubic lattice at $185°C < T \leq 210°C$, and (iv) the cubic lattice of the spherical domains becomes distorted with further increase in temperature so that lattice disordering transition occurs at temperatures between 210 (onset) and 214°C (completion) and the spherical microdomain structure of PS with a liquid-like short-range order persists even up to 220°C, the highest experimental temperature employed.

TEM was employed to investigate the microdomain structures of Vector 4111 at elevated temperatures by rapidly quenching specimens, which had been annealed at a predetermined temperature, into ice water. TEM results reveal the following: (i) At 170°C, Vector 4111 has hexagonally packed cylindrical microdomains of PS. (ii) At 185°C, cylindrical and spherical microdomains of PS coexist. (iii) As temperature is increased further to 220°C, only spherical microdomains of PS persist; i.e., the order-disorder transition temperature (TODT) of Vector 4111 is higher than 220°C, as shown in Fig. 5.5. The authors found in this study that the dynamic temperature sweep experiment under isochronal conditions failed to accurately determine the TODT of Vector 4111 having a low volume fraction of PS blocks.

The components of the PS-b-PVP block copolymer are highly immiscible; at the temperatures studied, both bulk copolymer melts are predicted to be strongly segregated into cylindrical microdomains in all the theoretical treatments [101]. Surface-induced ordering in thin films of asymmetric deuterated polystyrene (dPS) and PVP diblock and triblock copolymers of comparable polymerization indices and PVP volume fractions ($f \sim 0.25$) was studied using transmission electron microscopy, atomic force microscopy, secondary ion mass spectrometry, and neutron reflectivity [74]. Both di- and triblock copolymer films were found to have cylindrical domains except for the layer adjacent to

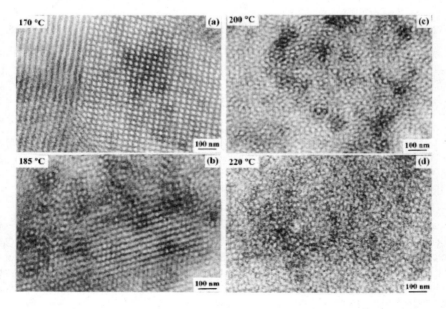

Fig. 5.5. TEM images of Vector 4111 annealed at (a) 170°C, (b) 185°C, (c) 200°C, and (d) 220°C. The morphologies were assigned as (a) cylinder, (b) and (c) cubic sphere, and (d) disordered sphere.

the silicon oxide surface, which showed lamellar domains due to the strong interaction with PVP. The spacing between adjacent cylindrical layers was found to be consistent with mean field theory predictions. In the triblock copolymer films, the cylindrical layers were oriented parallel to the silicon oxide surface, and no decay of the ordered structure was observed for at least 12 periods. The height of the holes or islands reached its equilibrium value of 21 nm. In contrast, it was far more difficult to orient microphase-segregated cylindrical domains in the diblock copolymer films parallel to the silicon oxide surface. As a result, no islands or holes were observed even after annealing. In the conclusion of this report, the difference in ordering behavior is due to the ability of the triblock copolymer to form an interconnected micelle network while the diblock copolymer formed domains that were free to move with respect to each other. This conclusion was further confirmed by diffusion measurements, in which the PS homopolymer penetrates easily into the ordered diblock copolymer films and is excluded from the ordered triblock copolymer films. Figure 5.6 shows the TEM image of a typical 50-nm thick sample, floated from the silica substrate onto a Cu microscope grid after annealing. The dark regions correspond to iodine-stained PVP blocks and the light sections to dPS blocks. The broad light and dark areas correspond to single- and double-ordered micelle layers, respectively. Inspection of the micrographs shows that the diameters of the micelle structures are the same in the single- and double-layer regions and remain unchanged under rotation of $-40°$ and $+40°$ about the axis in the sample plane as shown in the figure. Therefore, the in-plane morphology has a cylindrical symmetry as would be expected from the diblock copolymer phase diagram at this temperature. Cylinders lying perpendicularly to the rotation axis are not affected and are still difficult to resolve.

The morphological changes of poly(styrene-*block*-4-vinylpyridine) (PS-*b*-P4VP) prepared by anionic polymerization were investigated in different casting solvents, leading to three types of morphology (spherical, cylindrical, and lamellar) [77]. Subsequently, the cross-linking reaction of P4VP spherical, cylindrical, or lamellar microdomains was carried out with 1, 4-dibromobutane vapor in the solid state. The morphologies and

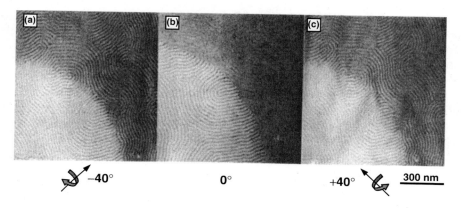

Fig. 5.6. TEM images of a 5-nm-thick triblock film, floated from the silica substrate onto a Cu grid after annealing for 24 h at 180°C. The dark and light regions correspond to double and single layers of cylinders, respectively. The sample is imaged (a) after rotation of $-40°$, (b) along the normal to the sample surface, and (c) after rotation of $+40°$.

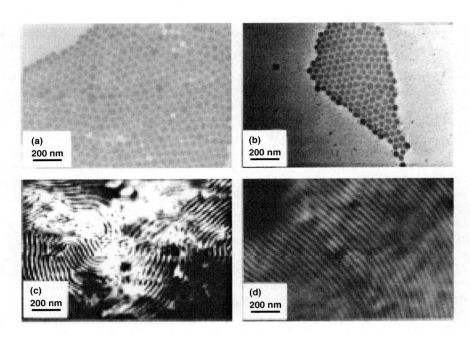

Fig. 5.7. TEM images of diblock copolymer films cast from different solvents: (a) toluene, (b) THF, (c) THF/CHCl$_3$ 5/5 (v/v), and (d) CHCl$_3$. The morphologies were assigned as (a) and (b) sphere, (c) cylinder, and (d) lamellae.

solution properties of the cross-linked products were studied by electron microscopy, SAXS, and static and dynamic light scatterings. PS-b-P4VP with a molecular weight of 27 000, a polydispersity of 1.12, and a P4VP molar fraction of 30 mol% showed P4VP spheres dispersed in a PS matrix in the case of an as-cast film from toluene solution and from tetrahydrofuran (THF) as shown in Fig. 5.7(a) and (b). The radius of P4VP spheres for the as-cast film from toluene is somewhat larger than that from THF. On the other hand, the as-cast film from CHCl$_3$ shows alternating PS/P4VP lamellae (Fig. 5.7(d)). The packing pattern obtained from the SAXS of the cast film prepared from CHCl$_3$, is identical to that of the interplanar spacing of a lamellar structure. It was found from the above results that spherical and lamellar domains were obtained by choosing the casting solvents. Then, cylindrical domains should appear in the intermediate region between spherical and lamellar microdomains. Figure 5.7(c) shows a TEM image of an as-cast film from THF/CHCl$_3$ 5/5 (v/v).

5.3.2 Polymer-structure-dependent phase structure

The morphological characteristics of star-branched block copolymers having PS-b-PI arms (30 wt% PS) have been examined by electron microscopy, SAXS, dynamic mechanical thermal analysis, and vapor sorption studies [102]. The special molecular architecture of these star-branched molecules was found to modify the nature of the

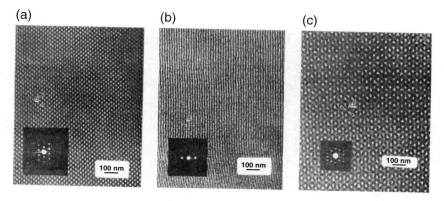

Fig. 5.8. Cylindrical morphology of 4-armed star block copolymer containing 30 wt% PS with molecular weight of PS block in each arm being 10×10^3: (a) axial view (and its optical transform) showing intercylinder spacing; (b) longitudinal view of cylinders, and (c) ordered bicontinuous morphology of the higher-arm-number and higher-molecular-weight sample.

microphase-segregated solid state. The influences of segment molecular weight and star functionality on the morphology were examined. In certain cases, branching brings about the formation of an ordered bicontinuous structure; this structure is not seen in linear materials of identical compositions where the equilibrium morphology is that of PS cylinders hexagonally packed in a polydiene matrix.

At 30 wt% PS, diblock and triblock copolymers have equilibrium morphology of PS cylinders hexagonally packed in a matrix of PI. This morphology was observed in the star block copolymers of low arm number and/or molecular weight. Figure 5.8(a) shows a micrograph and an optical transform of an axial projection of this structure for the 4/30/10 sample (4-armed star containing 30 wt% PS, the molecular weight of the PS block in each arm being 10×10^3). Figure 5.8(b) shows the longitudinal projection of cylinders and its corresponding optical transform. Both projections must be present in the film to distinguish the morphology from hexagonally packed spheres or parallel lamellae viewed edge-on. This cylindrical morphology was found for all the 2-armed (triblock) and 4-armed samples as well as for the 8-armed and 12-armed samples of the 7000-molecular-weight series. The higher-arm-number and higher-molecular-weight samples exhibit strikingly different morphologies, that is, ordered bicontinuous morphology, as shown in Fig. 5.8(c). The authors mentioned that the effect of arm molecular weight on the cylindrical to ordered bicontinuous transition should be explained by the interaction between blocks, junction placement, density maintenance, surface area-volume features of each block that determine the structure.

5.3.3 Phase structure in mixture of block copolymers

The competing long-range interactions of the two block chains in their binary mixtures were found to give intriguing effects on ordered microdomain structures. Mixtures of a block copolymer and a corresponding homopolymer have been widely studied to confirm

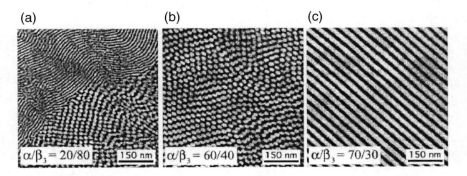

Fig. 5.9. TEM images of (PS-b-PI)$_\alpha$ (having molecular weight of 100 000 and weight fraction of styrene of 0.47) and (PS-b-PI)$_\beta$ (having molecular weight of 14 500 and weight fraction of 0.45) mixtures with composition of (a) 20/80, (b) 60/40, and (c) 70/30. The morphologies observed here were (a) between hexagonal cylinder and lamellae, (b) hexagonal cylinder, and (c) lamellae.

the interaction parameter for the interface between phases and the observation of its phase structure [58,59]. For more abrupt phase transition, a mixture of block copolymers having different molecular weights has been suggested by Hashimoto et al. [54]. One effect is a co-surfactant effect observed when the two blocks are mixed with their junctions at common interfaces, giving rise to such effects as (i) two lamellae-forming block copolymers give hexagonal cylinders and consequently increase the interface curvature, or (ii) a short symmetric block copolymer mixed with a large asymmetric block copolymer effectively decreases interface curvature, depending on the composition and molecular weight of the short block copolymer. The other effect causes two blocks to macroscopically phase-separate into different ordered domains. Figure 5.9 shows some typical morphology of the mixtures with the ratio r of the molecular weight of the large (PS-b-PI)$_\alpha$ block copolymer to that of the small (PS-b-PI)$_\beta$ block copolymer, r being equal to 6.9. In this case, macrophase separation into ordered cylinders rich in (PS-b-PI)$_\alpha$ and ordered lamellae rich in (PS-b-PI)$_\beta$ up to approximately 55 wt% of the large block copolymer (PS-b-PI)$_\alpha$, as shown in the mixture with a composition of 20/80 (Fig. 5.9(a)). Upon further increase in the composition of (PS-b-PI)$_\alpha$, the morphology of this polymer turns into an ordered single phase of cylinders and finally lamellae, as shown for the mixtures of (PS-b-PI)$_\alpha$/(PS-b-PI)$_\beta$ = 60/40 and 70/30. The results indicate lamellar-to-cylinder order-order transition with increasing composition (PS-b-PI)$_\alpha$. These effects were also theoretically studied by Shi and Noolandi [103].

5.3.4 Phase-structural control by supramolecular self-assembly

Schädler and Wiesner reported on an entirely different approach to manipulating the block copolymer phase behavior: by introducing ionic groups at one or both chain ends of a symmetric PS-b-PI diblock copolymer, the stabilization or destabilization of the microphase separation was successfully controlled [104]. In polymer matrices, these ionic groups have a strong tendency to form ion aggregates which superimposes an additional principle of self-assembly onto the system. As a first step toward the manipulation of the phase behavior, the competition between block segregation and ionic aggregation inducing changes in the lamellar spacing was examined. Schöps *et al.* reported on manipulating the phase behavior of a diblock copolymer, initiated by their study [53]. They succeeded in controlling the lamellar spacing of a symmetric PS-*b*-PI diblock copolymer by introducing ionic groups with opposite charges at chain ends. In apolar polymer matrices, these ionic groups have a strong tendency to form ion aggregates, superimposing an additional self-assembly principle onto the system. The competition between block segregation and ionic aggregation could be controlled by the addition of a low-molecular-weight salt, which leads to the observed changes in the lamellar spacing. The principle of such tandem molecular interactions in an ionically functionalized polymer was exploited in this report, i.e., one ionic group at the block junction point and another one with an opposite charge at one chain end. The addition of a low-molecular-weight salt induces a change in bulk microphase morphology, as shown in Fig. 5.10. This is similar to that observed by changing the architecture from an AB diblock copolymer to an A2B miktoarm star copolymer, but without the necessity of a new, demanding polymer synthesis.

Hierarchical morphologies were fabricated using a PS-*b*-P4VP diblock copolymer in which 4-n-$C_{19}H_{39}$-C_6H_4-OH (NDP) is hydrogen-bonded to the latter block (see Fig. 5.11) [76]. P4VP stoichiometrically complexed with NDP, denoted as P4VP(NDP)$_{1.0}$, forms a comb copolymer-like system with a microphase-segregated lamellar (smectic-like)

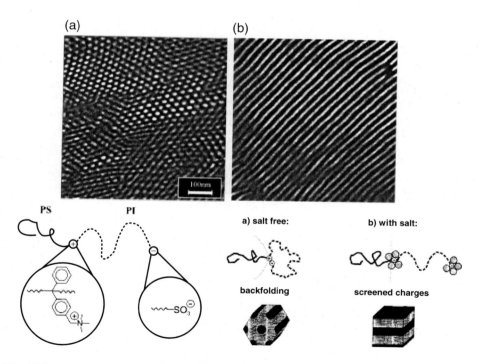

Fig. 5.10. Schematic drawing and TEM images obtained for the sample with added low-molecular-weight salt (a, lamellae) and for salt-free sample (b, cylinder).

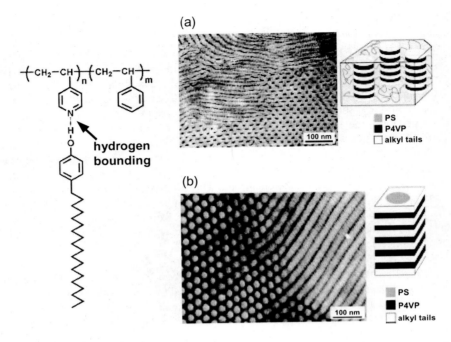

Fig. 5.11. TEM images and schematics of (a) *lamellar-within-cylindrical* structure and (b) *cylindrical-within-lamellar matrix* structure. The polymers used in (a) and (b) are PS-*b*-PVP(NDP)$_{1.0}$ with $f_{comb} = 0.27$ and 0.65, respectively.

structure with a long period of ca. 4.5 nm [105]. The diblock copolymer structure self-organizes without NDP at a larger length of 100 ± 10 nm, the actual morphology depending on the weight fraction of the P4VP block. The effective weight fraction of the latter block can be tuned by hydrogen-bonded NDP, which acts as a selective solvent.

In this study, the amount of NDP was fixed to be stoichiometric, i.e., nominally every repeat unit of P4VP was complexed with one NDP. To calculate the effective weight fractions of the P4VP(NDP)$_{1.0}$ and PS blocks, one has to take into account phenols, which are selectively hydrogen-bonded to the P4VP phase. The lengths of the P4VP blocks were thus selected to achieve different diblock copolymer regimes on the basis of the effective weight fraction f_{comb} of P4VP(NDP)$_{1.0}$ blocks, i.e., lamellar, cylindrical, and spherical phases. The expected internal organization of P4VP(NDP)$_{1.0}$ is interdigitated lamellar in all cases. Figure 5.11(a) shows a TEM image of the *lamellar-within-cylindrical* structure obtained for the sample PS-*b*-P4VP(NDP)$_{1.0}$ with $f_{comb} = 0.27$. The characterization of this polymer gave molecular weights of 34 000 and 2900 for PS and P4VP chains, respectively, and a 1.07 polydispersity. The P4VP(NDP)$_{1.0}$ domains form a hexagonal array of cylinders in the background of PS. The image also supports the expectation that an internal lamellar order exists within P4VP(NDP)$_{1.0}$ cylinders. The schematic structure is also shown in Fig. 5.11(a). In an opposite case, where PS forms the minority phase, Fig. 5.11(b) shows a TEM image for $f_{comb} = 0.65$ molecular weights of 31 900 and

13 200 for PS and P4VP, respectively. A hexagonal array of PS cylinders is now embedded in a lamellar P4VP(NDP)$_{1.0}$ matrix. The relative orientation is clearly orthogonal as can be best observed by inspecting the top right corner of the image where PS cylinders are horizontal with respect to the electron beam.

The understanding of phase behavior in a block copolymer system is particularly important for the control of the phase structure and the alignment of microdomains, where interfacial interactions fluctuate in each polymer system. The fundamental studies introduced here could surely provide several means of driving self-organized phase structures (still beset with problems such as defects and a mechanically low strength) for block copolymer engineering.

5.4 Phase-segregated Nanostructures in Block Copolymer Thin Films

The microphase separation of block copolymers has been studied so far to observe specific microdomains and their aggregates in solution and specific morphology consisting of ordered microdomains in solid, as functions of both molecular parameters, such as chemical structure, polymerization degree, volume fraction, and side chain, and processing parameters for preparing solid samples from their polymer solutions, e.g., solvents, concentration, temperature, solidification methods such as coating, casting, and reprecipitation. The feature of microphase separation has been determined from the bulk samples as the most thermodynamically stable phase. That is to say, it should be purely governed by block copolymers themselves to exclude any interfacial effects, which have potential to strongly influence the self-assembly to modify a variety of morphologies in microphase separation. Except for a few studies, most of the works have been devoted to the study of the structure of block polymers in bulk but not near a free surface or interfaces with other materials. The experimental procedures for collecting essential morphologies specific to block copolymers have also been well established for a long time, including living polymerization, annealing conditions, SAXS, and TEM observation techniques, such as staining and cross sectioning with a microtome. In the last decade, a new movement in polymer physics has come up to study the morphology of various thin films, quite different from those of the bulk. The morphology of thin films should be governed by not only molecular structures of block copolymers but also interfacial effects of substrates and air. In some cases, the latter would be predominant over the former in determining the morphology of thin films. The morphology growth of thin films would experience a restricted space along the thickness direction, quite unlike that of the bulk with isotropic free space. This is just a promising target in the microphase separation of block copolymers, which has a long history in polymer physics but is now thrown into the limelight to develop a new theory and understanding and to extend its practical use in advanced materials science, which the authors emphasize in this Chapter.

Thomas et al. utilized X-ray photoelectron spectroscopy (XPS) to investigate the influence of polymer composition and film casting solvent on the surface structure of poly(styrene-*block*-ethylene oxide) diblock copolymers (PS-*b*-PEO) [106]. Before their report, there have been no investigations of the surface properties of block copolymers,

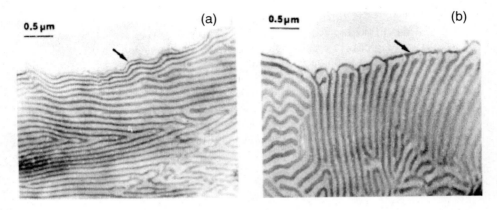

Fig. 5.12. TEM images of the ultrathin sections of PS-*b*-PI diblock copolymer show (a) the lamellar structure parallel and (b) perpendicular to the film plane. Arrow indicates PI layer darkly stained.

particularly, the surface composition and topography of such polymers at the air–copolymer interface. Hasegawa *et al.* first presented the morphology of the microdomains of an AB diblock polymer formed as a consequence of the liquid–liquid microphase separation of the constituent polymers A and B at or near the air surface in contrast to the morphology of the bulk determined by TEM observation [50]. The morphology at the surface was realized as strongly governed by surface free energy (as an additional physical factor that does not affect the morphology of the bulk) and therefore as being different from the morphology of the bulk, as shown in Fig. 5.12. Near the interface between the specimen and air, a darkly stained PI thin layer always covered the PS layer. This phenomenon was observed even for lamellae perpendicular to the free surface. The authors mentioned that these results were quite reasonable because of a thermodynamic requirement of the lowest surface free energy. It may be unnecessary to say that studies of the surface morphology of block polymers are of academic and practical importance in various types of application, particularly in nanotechnologies.

The conventional method of visualizing the self-organized nanostructures of block copolymers in the bulk is by TEM of stained and then microtome-sectioned samples. Since the surface morphologies had been studied as mentioned above, the fabrication and direct observation of self-organized nanostructures in block copolymer thin films (not in the conventional bulk samples) were then focused because the thin films with nanostructures have a large potential to be applied in nano-functionalized surfaces, patterned media, optical materials, anisotropic conducting films, separation membranes, and nano-filters among others. In addition to morphological singularity and practical potential in thin films, another motivation to draw one's interest is the successful visualization of phase-segregated surface morphology by atomic force microscopy (AFM). Van Dijk *et al.* first demonstrated the AFM imaging of PS-based block copolymer thin films, operated in the tapping mode [107,108], as shown in Fig. 5.13.

A quite clear surface morphology of phase-segregated nanostructures is visualized as the closest packed regions of 50–60 nm diameter dots and a winding striped pattern region

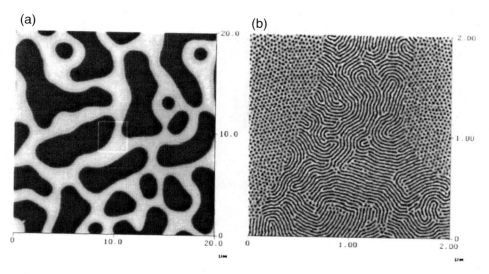

Fig. 5.13. AFM images of PS-*b*-PI block copolymer film spin-coated on silicon wafer with average thickness of 50 nm; (a) ultrafiltered large scan, (b) scan in white square in (a).

of 50–60 nm width in the right image in Fig. 5.13. The AFM large-area image on the left also shows these grain structures with several micrometers in size. The hard PS domains manifest as regions with an apparent lower thickness. This study indicates that AFM is a powerful tool for visualizing such phase-segregated surface morphology under ambient conditions, matched with analytical requirement for thin films. AFM observation only requires reasonably flat surfaces of samples, basically with no pretreatment such as metal deposition for preventing surface charging and various staining and replication techniques for enhancing contrast, usually observed in scanning electron microscopy (SEM) and TEM.

Here, a quite impressive AFM study of phase-segregated surface morphology as a function of film thickness is introduced. Both AFM observations and calculations for a given strength of the surface field were compared by Krausch *et al.* [109]. A wide variety of nanostructures with discontinuous changes in individual morphologies were observed as film thickness increases, as shown from left to right in Fig. 5.14. With increasing film thickness, both experiment and calculation exhibit a consistent sequence of various phases in thin films: a featureless film surface for the smallest thickness, isolated domains of the A phase, parallel-oriented A cylinders, and finally two layers of parallel-oriented cylinders, some of which are interestingly imaged in one frame with discontinuous boundaries. Several phase transitions occur at well-defined film thicknesses with phase boundaries denoted as white contour lines that have been calculated from the AFM height images. This combinatorial structural analysis of phase-segregated surface morphology by the smart fabrication of thin films with graded thicknesses provided a consistent and systematic conclusion concerning thickness-dependent morphological changes in thin films.

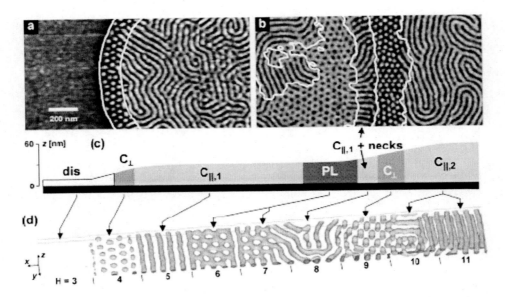

Fig. 5.14. (a)(b) Tapping-mode AFM phase images of thin SBS films on silicon substrates after annealing in chloroform vapor. The surface is covered with a homogeneous ~10-nm-thick PB layer. Bright and dark areas correspond to PS and PB microdomains, respectively, below the top PB layer. (c) Schematic height profile of phase images shown in (a) and (b). (d) Simulation of ABA block copolymer film in one large simulation box with film thickness increasing from the left to the right.

5.5 Phase-segregated Nanostructures in Thin Films Effective for Practical Use Cylindrical Phases in Block Copolymer Thin Films

Scientific interest in the microphase separation of block copolymer thin films has attracted attention in view of contrastive morphology with the case of the bulk, which means that such morphology would be controlled by the effective design and modification of their interfaces or surfaces. In particular, a thin film, commonly found in various coating materials, is one of the most significant forms of materials for practical use. The fabrication of phase-segregated nanostructures in thin films will open up a new market of polymer films with self-organized nanostructures and their related materials. When ordered nanostructures such as cylinders and lamellae found in block copolymers are considered in thin films, their orientations relative to the film should be discussed. This argument becomes less serious in the case of spherical nanostructures as one of the microphase separations, since a body-centered cubic arrangement of spherical microdomains often found in block copolymers seems spatially isotropic and has little specificity in a thin film.

Let us discuss several possible alignments of cylindrical and lamellar microdomains in thin films. To begin with, a film or membrane has been used to separate two phases or spaces and to transport molecules, ions, energy, and electrons through the film. Such transport requires running crevices or cracks or holes across films as 'channels'.

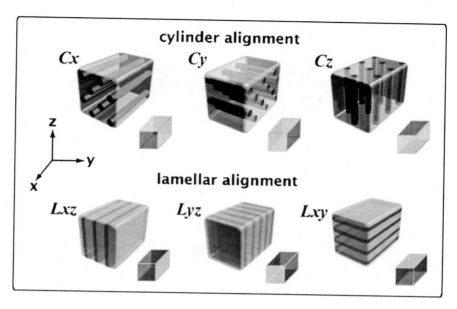

Fig. 5.15. Illustration imaging direction of cylinders and lamellae in film and direction of resulting nanochannels for transportation or anisotropic diffusion.

Size-controlled holes and molecular recognition sites should be fabricated into transport channels for selective transport. Either cylinders or layers drawn in dark gray are assumed as transport channels, so that the remaining domains colored as lightgray play an important role as frameworks of the films (Fig. 5.15). For lamellar alignment, transport channels open through configurations of the x–z (**Lxz**), y–z (**Lyz**), and x–y (**Lxy**) planes. The former two configurations include a common z-channel but have in-plane anisotropy. In this case, further in-plane orientation control of the x–z or y–z transport channel in lamellar alignment is required. No transport channel exists in the **Lxy** configuration, indicating that it is unsuitable for transport function across the film. Let us consider the cylindrical alignment. Transport channels are two-dimensionally formed along the x-direction (**Cx**), y-direction (**Cy**), and z-direction (**Cz**). Neither **Cx** nor **Cy** configurations in cylinder alignment open z-channel. Only the **Cz** configuration with a z-channel should be effective for the inherent mass transport function across the film. When these transport channels are integrated into the thin film through microphase separation, only the **Cz** configuration is effective for the transport across the film. Therefore, cylinder array structures perpendicular to the film plane should be ranked into one of the ideal ordered nanostructures through self-assembling process, as will be discussed again in our recent work described partly in Section 5.6.

Further explanation of the superiority of the cylinder array structure perpendicular to the substrate may be developed by considering a grain size of several hundreds of nanometer and centimeter-to-meter-sized substrates, as shown in Figure 5.16. In the case of 'isotropic' functions to utilize nanoporous or mesoporous materials such as catalysts, and adsorbents, the pores themselves are crucial, however, less effort has been exerted

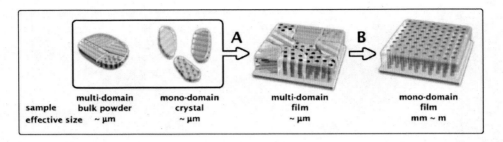

Fig. 5.16. Illustration of orientation of cylinder array structures in grains of several hundreds of nanometer onto centimeter-to-meter-sized substrates. The grains self-assemble into the microdomain structure in the film (a) and are aligned all over the film space (b).

on the precise control of grain structures. On the other hand, the transport function of the film strictly requires single-grain structures with adequate orientation of the transport channel, consistent with the argument in Fig. 5.15.

The mono-grain structures of self-organized block copolymers have been realized as the only requisite for facilitating many industrial applications. The uniform surface nanostructures of block copolymer thin films have been widely developed and recently expected as a great candidate generating nanodevices with resolutions higher than the lithographic lower limit of line width. Examples of potential applications using self-organized block copolymers particularly focused on nanotechnology were summarized in several reviews [110,111]. It should be noted that an important issue for any nanostructured film applications is the axial extension of the cylinders, i.e., it is essential that individual cylinders traverse the entire film thickness rather than form unconnected domains across the film.

Diblock copolymer thin films have significant potential for use as templates and scaffolds for the fabrication of arrays of nanostructures over large areas. A key to many future applications is the long-range ordering and precise placement of self-organized nanoscale domains. To achieve the ordering of hexagonally close-packed vertical standing cylindrical domains in thin films, the chemical modification or pretreatment of a substrate such as a polymer brush or a shearing, and the use of an external alignment field such as a flow field or an E-field have been successfully applied. The resulting minor component forming nanodomains is eliminated to transform the block copolymer thin film into a membrane and/or template. However, further alignment and positional ordering of these cylinders three-dimensionally are much more difficult to achieve since there are significantly no driving forces for the domains of close-packed columns to align macroscopically. It is still a challenging task to develop appropriate materials and technology that is a simple solution to the above difficulties and at the same time, enables the fast fabrication of well-ordered nanotemplates from block copolymer films deposited on a solid substrate. In this section, the achievements in the preparation of cylindrical morphologies perpendicular to the substrates, which should be sufficient as functional nanotemplates, are summarized and the type of block copolymer used for perpendicular cylinder array structures and the procedures are presented (see Table 5.3).

Table 5.3. Perpendicular nanocylinder array structures in block copolymer thin films.

Cylinder block	Matrix block	Substrate	Solvent	Obs.[*1]	Remark	Ref.
PMMA	PS	PS-r-PMMA coated Si	toluene	A, S2, CS2	Blend system with homopolymer Nanopore fabrication by removal of homopolymer in cylindrical domain	[112,113]
PMMA	PS	PS-r-PMMA coated Si	toluene	A, S1, CT, S2, CS2	Nanopore fabrication by chemical etching or UV cross-linking	[114–116]
PMMA	PS	PS-r-PMMA coated Si	toluene	A	Layered Au/Cr nanodot or nanoporous film templated from nanoporous PS film or PS posts in ref. [62] and [63]	[117]
PMMA	PS	PS-r-PMMA coated Si	toluene	A	CdSe quantum dot infiltration into nanopores	[118]
PMMA	PS	PS-r-PMMA coated Si	toluene	A, S2, CS2	Nanoporous SiO_2 templated with nanoporous PS film	[119]
PMMA	PS	PS-r-PMMA coated Si	toluene	A	SiO_2 nanoposts templated with cylinder arrays	[120,121]
PMMA	PS	PS-r-PMMA coated Si	toluene	A, S2	Conducted in InP-imprinted grating of 100 nm gap and 210 nm periodicity	[122]
PMMA	PS	Si, Au/Si, or Au/Kapton	toluene	S1, CS2	Co electrodeposition in nanoporous PS film	[123]
PEO	PS	Si	benzene	A, CT	One-step fabrication by only thermal treatment	[90]
PEO	PS	Si, gold, and plastic film	benzene	S1, A	Alignment by solvent evaporation	[124]
PEO	PS	Si	benzene	T, S2, CS2	Selective decoration with SiO_2 onto PEO cylinders	[125]
PEO	PS	Si	benzene	A	Terpyridine attached at the end of both blocks formed metallo-supramolecular copolymer. Nanopores were produced by Ce ion extracting	[126]
PS	PBD	Si	toluene	A	Film thickness dependence on morphology	[108]
PS	PBD	Mica	toluene	A, T	Solvent evaporation effect on morphology	[52]

Continued

Table 5.3. Perpendicular nanocylinder array structures in block copolymer thin films—cont'd.

Cylinder block	Matrix block	Substrate	Solvent	Obs.[*1]	Remark	Ref.
PS	PBD	Si (and ITO as upper electrode in sandwich cell)	toluene	CT	E-field induced perpendicular alignment inside the film in sandwich cell	[127]
P4VP/HABA[*2]	PS	Si and quartz	dioxane	A	Alignment control with casting solvent Chloroform produced parallel alignment Ni infiltration into perpendicular nanopores	[128]
P2VP	PS	Si	toluene	S1, A, CT	Hybrid film with CdSe nanoparticles	[129]
PLA-PDMA[*3]	PS	not mentioned	not used	S1, A	Nanopore fabrication by chemical etching of PLA core cylinder Modification of nanopore interior surface consisting of PDMA shell cylinder	[130][*3]
PαMS[*4]	PHOST[*4]	Si	PGMEA[*4]	S1, A	Nanopore fabrication by photoresist process producing cross-linked PHOST matrix	[131]
PFS[*5]	PS	Si	toluene	A, S1	Organometalic nanoposts by RIE of PS matrix	[132,133]
POTI[*6]	PS	Glass slide, Si, plastic film	CS_2	CT	Perpendicular cylinders in matrix of microporous structure as self-organized hierarchical structure	[134]

[*1]Observation technique for obtaining direct evidence of perpendicular nanocylinder array structures, i.e., SAXS, AFM, TEM, cross-sectional SEM, and cross-sectional SEM, denoted as S1, A, T, CT, S3, and CS3, respectively.
[*2]2-(4′-Hydroxybenzeneazo)benzoic acid was used for supramolecular assembly through hydrogen bonding with pyridine in PVP block.
[*3]Triblock copolymer composed of polylactide (PLA), polydimethylacrylamide (PDMA), and PS was used for formation of core-shell PLA@PDMA cylindrical morphology in PS matrix.
[*4]Poly(α-methylstyrene-b-4-hydroxystyrene) (PαMS-b-PHOST) and also cross-linking agent was mixed in propylene glycol methyl ether acetate (PGMEA) for the casting solution.
[*5]Poly(ferrocenylsilane) (PFS) was incorporated into the block copolymer with PS for use as an etch-resistant barrier in reactive ion etching (RIE).
[*6]Polyisoprene (PI) with oligothiophene-modified side chains (POTI) was copolymerized with PS, leading to semi-rod-coil block copolymer.

5.5.1 Solvent evaporation

Solvent evaporation is one of the strongest and highly directional fields. Kim and Libera demonstrated that the rate of solvent evaporation from solution-cast thin films of block copolymers could be used to manipulate the growth and orientation of copolymer assemblies [135,136]. Ham and Sibener [137] have also shown that evaporation-induced flow in solvent-cast block copolymer films can produce arrays of nanoscopic cylindrical domains with a high quality of in-plane orientation and lateral order. Huang et al. [52] investigated the formation of an inverted phase in a series of solution-cast poly(styrene-*block*-butadiene) (PS-*b*-PB) asymmetric diblock copolymers having nearly equal PS weight fractions (about 30 wt%) but different molecular weights (Fig. 5.17). The microstructure of the solution-cast block copolymer films resulting from different solvent evaporation rates was inspected, from which the kinetically frozen-in phase structures at qualitatively different block copolymer concentrations, and correspondingly, different effective interaction parameters can be deduced. Their result shows that there is a threshold molecular weight or range of molecular weights below which the unique inverted phase is accessible by controlling solvent evaporation rate. The formation of the inverted phase had little bearing on the chain architecture.

Minko and Stamm reported on a unique, very simple method of preparing reactive membranes and nanotemplates with nanoscopic cylindrical channels on the surface of various inorganic and polymeric substrates. Well-ordered nanostructured thin polymer films have been fabricated through the supramolecular assembly of the PS-*b*-PVP matrix and 2-(4-hydroxybenzeneazo)benzoic acid (HABA) as a side chain, consisting of cylindrical nanodomains formed by the adduct formation between the pyridine moiety and HABA surrounded by PS. The alignment of the domains has been shown to be switched upon exposure to vapors of different solvents from the parallel orientation to the perpendicular orientation to the confining surface and vice versa [128]. In this report, the regularity of nanoscopic cylindrical domains was found to be as low as that in the film produced by only annealing, while the control of cylinder alignment was successively demonstrated

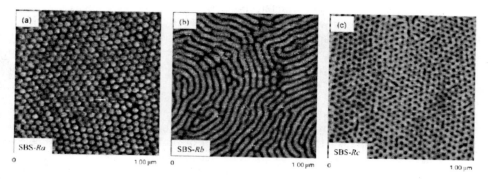

Fig. 5.17. Tapping-mode AFM phase images of SBS triblock copolymer films obtained from solution-cast at different solvent evaporation rate, R; (a) R ≈ 0.1 mL/h, (b) R ≈ 1 μmL/h, and (c) R ≈ 0.2 μmL/h. The phase structures of the resulting morphologies were (a) inverted sphere, (b) inverted cylinder and (c) normal cylinder.

in the thin film. Nearly defect-free cylinder array structures consisting of PS-*b*-PEO were achieved by controlling the rate of solvent evaporation or solvent annealing, leading to the ordering of the nanodomains induced at the surface and propagating structures over the entire film [124]. The solvent imparts mobility to the copolymer, enabling the rapid removal of defects. Thus, highly oriented, defect-free arrays of nanoscopic cylindrical domains are produced that span the entire film thickness and have a high quality of long-range lateral order. The same system was adopted for scaffolds to define an ordered array of nanometer-scale reaction vessels in which high-density arrays of silicon oxide nanostructures were exposed to silicon tetrachloride [125]. Figure 5.18 shows AFM images of the PS-*b*-PEO template and cross-sectional field-emission (FE) SEM images with an array of ∼5-nm-high silicon oxide nanostructures (∼24 nm in diameter) on the PEO microdomain. The nanotemplate formation with a block copolymer thin film was successfully demonstrated for the selective growth of silicon oxide within the PEO domain by vapor phase reaction.

5.5.2 Control of polymer-surface interaction

The control or neutralization of differential surface affinities on a block copolymer system has been highly desirable in order to elicit a specific response of a material at an interface. Little work has been performed on the control of the relative affinities of the components of an interface. Mansky *et al.* and Huang *et al.* [138,139] reported first that the interfacial energies of polymers at a solid surface can be manipulated by end-grafting statistical random copolymers onto the surface, where the chemical composition of the copolymer can be controlled by choosing the synthetic methods. This strategy for surface modification is both simple and versatile and can be extended to a wide range of polymers. The general procedure for surfaces with balanced interfacial interactions with, for example, PS and PMMA, has been reported as follows. Hydroxy-terminated random copolymers of styrene (S) and methylmethacrylate (MMA) homopolymers, prepared using nitroxide-mediated living free radical polymerization containing an S fraction of 0.58, were spin-coated onto the native oxide layer of a Si wafer. At this composition, the interactions between the random copolymer and PS or PMMA are balanced. At other compositions, there are preferential interactions with ether PS (>0.58) or (<0.58). The random copolymer film allows the hydroxyl end groups to diffuse and react with the oxide layer, anchoring the random copolymer to the substrate as polymer brush. Since then, the microdomain morphology of diblock copolymers can be oriented normal to the surface over a broad range of molecular weights. This has led to a unique means of controlling the size of nanotemplates.

The development of balanced interfacial interactions on the surface has also led one to examine the molecular weight effect on cylinder array nanostructures with perpendicular alignment to the substrate [115], although a number of studies have been carried out on cylindrical phase structures in bulk samples or in an unaligned thin film. Using the same strategy, the cylindrical microdomains of the PS-*b*-PMMA film can be aligned perpendicularly and generate an array of ordered nanoscopic pores with well-controlled sizes, orientations, and structures, combined with selective dry [114] or wet [116] etching. Very recently, this nano-scaffold, which is created by anchoring a random copolymer to a substrate to balance the interfacial interactions as well as by the flow field of solvent

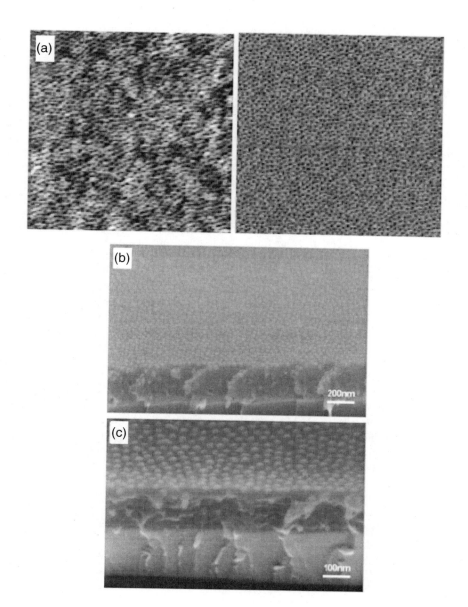

Fig. 5.18. (a) AFM topographic (left) and phase (right) images of PS-*b*-PEO cylindrical microdomains ordered on Si wafer and (b)(c) FE-SEM images after the selective growth of silicon oxide nanodots onto PEO cylindrical domains at different magnifications.

evaporation in Section 5.5.1, has been widely utilized in the block copolymer nanolithographic process. The newly developed nanolithography has been carried out by applying the infiltration of nanoparticles [118,129] or the dry etching process [117,120,121] producing ceramic posts or nanoporous films comparable in size to the block copolymer thin films. These applications of nanoscopic cylindrical structured materials using block copolymer thin films as template materials will be briefly reviewed in Section 5.5.6.

5.5.3 Blend-polymer-induced orientation

The preferential interactions of copolymer blocks with air and substrate interfaces result in an orientation of microdomains parallel to the substrate and the quantization of film thickness in terms of the lattice period of the morphology. The parallel orientation of the microdomains, however, limits the number of end uses. This limitation can be overcome by controlling interfacial interactions between the copolymer and the substrate with anchored random copolymers or by applying an external filed normal to the surface (which will be mentioned later). The limited mobility of high-molecular-weight copolymers and the order–disorder transition temperature of block copolymers restrict the range of microdomain sizes accessible, even though the orientation of microdomains normal to the film surface was achieved by a combination of the use of a random-copolymer-anchored substrate and E-field application. Numerous studies have appeared on the phase behavior of mixtures of symmetric diblock copolymers with the corresponding homopolymers [89,140–143]. Most thin-film studies have focused on symmetric, diblock copolymers having lamellar microdomains. The morphology of PS-b-PMMA thin film of asymmetric diblock copolymers having cylindrical microdomains oriented normal to the substrate was investigated as a function of the amount of added homopolymer in ref. [112]. The distribution of homopolymer and change in lattice spacing was also quantitatively demonstrated therein. The density of cylinders for the pure copolymer was found to be 7.3×10^{10} cylinders/cm^2. This decreased to 6.6×10^{10} cylinders/cm^2 by the addition of 10% PMMA homopolymer, while the uniformity of cylinder arrays was retained. The resulting decrease agrees well with that expected from the change in PMMA volume fraction induced by the addition of the homopolymer, assuming a uniform distribution of the homopolymer in cylindrical microdomains.

The orientation of cylindrical microdomains normal to the surface in a thick film was experimentally investigated by blending PS-b-PMMA with the corresponding homopolymers [113]. Figure 5.19 shows the SEM images of films of mixtures of PS-b-PMMA with MW of 71 000 and PMMA homopolymer with MW of 18 000 in which the thicknesses of the films were 90, 278, and 329 nm. All the samples were annealed and etched by UV exposure (254 nm). The volume fraction of the PMMA homopolymer based on the total amount of PMMA in the system was 0.26. The film mixtures showed a hexagonal packing of cylindrical microdomains at the surface of the film, even for a 300-nm-thick film. The FE-SEM images of the fracture surfaces of 92 and 273-nm-thick films show both the film surface and film cross section, as indicated in Fig. 5.20. The top surfaces show arrays of hexagonally packed cylindrical nanopores, whereas the cross section demonstrates that these nanopores are oriented normal to the substrate and penetrate through the entire film. The results clearly show that the simple addition of a homopolymer, i.e., increase in PMMA volume fraction, can be used to markedly promote microdomain

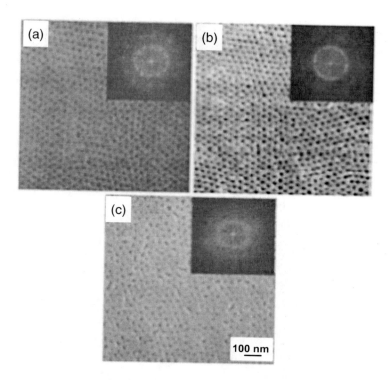

Fig. 5.19. SEM images of thin film from mixture of PS-*b*-PMMA with MW of 71 000 and PMMA homopolymer with MW of 18 000 at fixed PMMA volume fraction of 0.26 for three different film thicknesses: (a) 91 nm, (b) 278 nm, and (c) 329 nm. All the samples were annealed and treated by UV etching.

orientation without the use of external fields. Additionally, note that nanoscopic cylindrical microdomains perpendicular to the substrate even in thick films generate array nanostructures with a high aspect ratio.

5.5.4 Restricted microstructures

The success of nanotechnology lies on our ability to rationally arrange functional materials on nanoscale. This has been basically achieved by top-down lithographic processes, however, fundamental limitations associated with these techniques are rapidly becoming apparent. Not only is lithography hitting the lower limit of its length scale capability at ~20 nm, but also the process itself is often inefficiently serial in nature. Bottom-up self-assembly (categorized for block copolymer morphology) has the capacity to transcend both of these limitations because it is inherently parallel and uses molecules as building blocks to enter the nanoscale. Most hard materials, however, will not self-organize at these length scales. For this reason, hybrid systems where soft matter self-organizes into a nanoscale template that directs the assembly of hard matter have

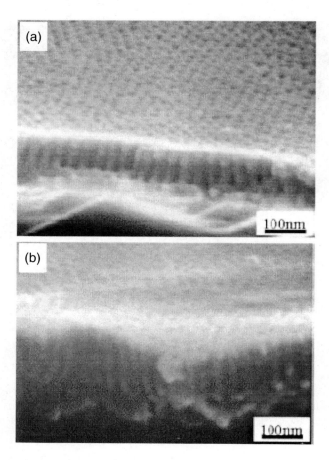

Fig. 5.20. Cross-sectional SEM images of the same samples observed in Fig. 5.17 with different thicknesses: (a) 91 nm and (b) 329 nm.

been suggested. Nevertheless, pure bottom-up methods have their own drawback: one cannot attain a long-range order. The solution to this dilemma is to adopt a hierarchical, combination top-down/bottom-up methodology. Several groups have studied the lithographically assisted self-assembly of the spherical phase. Kramer and coworkers were the first to demonstrate a graphoepitaxial strategy for growing single-crystal films of block copolymers over large areas of a substrate [144]. Ross and coworkers extended this work to more confined volumes and identified the effect of lithographic defects and the relation of channel width and polymer sphere accommodation [145]. The epitaxial self-assembly of lamellar phase on lithographically defined nanopatterned substrates was then reported using Extreme Ultraviolet (EUV) interferometric lithography through photoresists [146].

The self-assembly of the cylindrical phase structure of poly(styrene-*block*-(ethylene-*alt*-propylene) (PS-*b*-PEP) in lithographic microtroughs was examined for compliance of cylinder alignment against the width and depth of the microtroughs [147]. Silicon nitride grating substrates of various depths and periodicities were used to template the alignment of high-aspect-ratio cylindrical polymer domains. The alignment was realized to be nucleated by PS preferentially wetting the trough sidewalls and was thermally extended throughout the polymer film by defect annihilation. Since the polymer flows from the edges to the center of the troughs as the overall film flattens during annealing, the cylinders parallel to the substrate are aligned along the entire channel length, not across the trough. Similar phenomena were observed in the cylindrical phase of poly(styrene-*block-tert*-butyl methacrylate) (PS-*b*-P*t*BA) filled into a patterned silicon oxide substrate [148]. The P*t*BA cylinders parallel to the substrate in the PS matrix were found to be aligned along the patterned channel. This study also demonstrated the thermochemically induced cylinder to sphere transition in the trough, indicating that polyacrylic anhydride deprotected by *t*BA transformed the cylindrical phase to a spherical arrangement on the basis of the subsequent volume change.

The cylindrical microdomains perpendicular to the substrate patterned as microchannels have been reported very recently by the neutralized silicon wafer, the same method as mentioned in Section 5.5.2 [119]. At almost the same time, employing the combination of nano-imprint lithography and self-assembly of the PS-*b*-PMMA diblock copolymer (Fig. 5.21(a)), tailored periodic arrays of PMMA cylinders normal or parallel to the neutralized silicon surfaces were successfully formed inside the gap of imprint molds [122]. Figure 5.21(c) clearly shows that the vacant nanochannels produced by PMMA cylinder etching are aligned perpendicularly to the substrate at a film thickness smaller than one period of cylinder arrays. The number of their column rows are also realized as strictly governed by the gap width. In this work, the cylinder alignment in the thicker film was described as the cylinders were completely parallel to the substrate and packed along the channel direction. The other approach to the combination technique with 'top-down' and 'bottom-up' fabricating structures was the nanofabrication process of spatially controlled nanopores [131]. It consists of UV cross-linking of one block with a photomask, the solvent dissolution of un-cross-linked part to form micro-sized patterns on the top of substrate, and second UV irradiation for etching the other block, thereby finally yielding nanoporous channels on the substrate.

The unique cylinder arrays perpendicular to the substrate were found by Hayakawa and Horiuchi [134]. The specially designed block copolymer, PS-*b*-POTI (PI with oligothiophene-modified side chains), forms a self-organized hierarchical structure in the film. When the film was prepared under a moist air flow, a periodic microporous structure with a pore size of several microns was instantly obtained over an area of 3 cm^2 (Fig. 5.22(b)). By microtome-sectioning this microporous film, the finer self-organized nanostructure of PS-*b*-POTI with a periodicity of 25 nm was observed in the cross-sectional TEM image (Fig. 5.22(h)). Surprisingly, the POTI cylinders were realized to be perpendicular to the substrate even in the microporous film, while the polymer film with the non-microporous structure exhibited the formation of a cylindrical nanostructure parallel to the substrate.

The hierarchical structures reviewed in this Section can have applications in fields such as photonics or microfluidics, especially given the ability to precisely position engineered defects and periodicity in the structures.

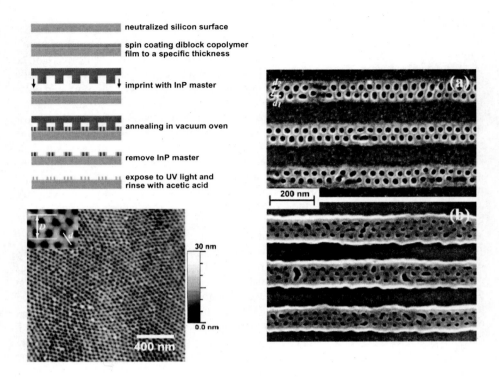

Fig. 5.21. Schematic drawing of procedures for controlled self-assembly nanolithography using block copolymer thin films and AFM image of PS-*b*-PMMA prepared onto neutralized silicon wafer with a total film thickness of 42 nm, (left). SEM images of controlled self-assembled polymer structure with imprinted InP grating of (a) 200 nm periodicity and (b) 210 nm periodicity. The PMMA cylindrical domains were removed away by oxygen plasma etching.

5.5.5 External field

For more reliable alignment of self-organized nanostructures of block copolymers, the effect of external fields such as shear field [149–151], flow field [152–154], magnetic field [19,155], and electric field [156–158] have been theoretically and experimentally studied over the last three decades. Most of these studies are focused on the self-organized morphologies of symmetric block copolymers. An important paper from the viewpoint of perpendicular cylinder arrays was reported [159]. It was mentioned therein that the external field effectively inducing the perpendicular alignment of cylindrical domains to the substrate is the electric field. As a very clear demonstration of E-field alignment for PS-*b*-PMMA cylindrical morphology, Fig. 5.23(b) shows the cross-sectional TEM image obtained by annealing under an E-field applied to the thickness direction, indicating the PMMA cylinders are oriented normal to the substrate [114]. This E-field effect on the alignment of phase-segregated nanostructures in thin films of the other conventional block copolymers such as PS-PVP, PS-PB, and PS-PI were described by cross-sectional TEM observations in ref. [127]. The E-field alignment of cylindrical nanostructures that allows

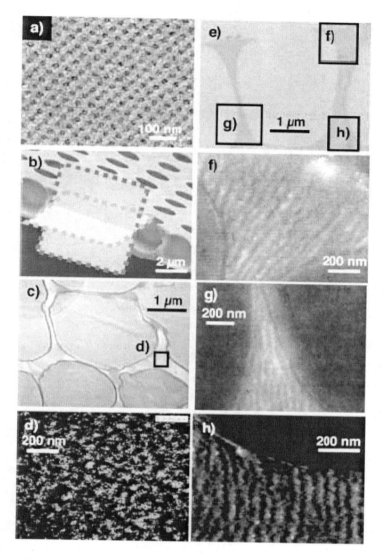

Fig. 5.22. (a) Non-microporous film. (b) Direction of cross-sectioned film. (c) Horizontally cross-sectioned film and (d) its Sulfur-distribution image to detect POTI domains (brighter part) obtained by energy-filtering TEM (EF-SEM). (e) Perpendicularly cross-sectioned film and (f) and (g) its high-magnification micrographs and (h) EF-TEM image.

the simplification of the process required to achieve the desired alignment and facile nanofabrication has been presented. As the most characteristic application of nanofabrication by E-field alignment, the completely perpendicular cylinders successively etched to form nanopores with the corresponding diameter were filled with magnetic particles by direct current electrodeposition [123]. Figure 5.24 shows a SEM image of a fracture surface of an array of Co nanowires grown within an array of nanopores formed by

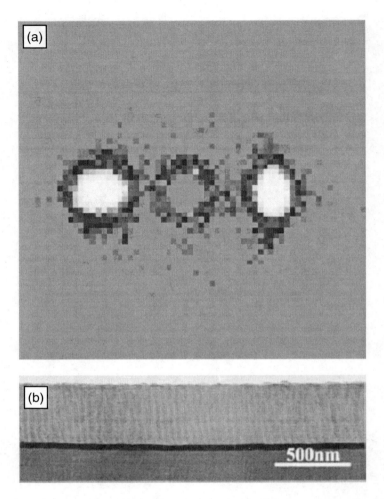

Fig. 5.23. (a) SANS pattern obtained from a PS-*b*-PMMA film after annealing under E-field. (b) Cross-sectional TEM image obtained from the same film in (a).

the selective etching of cylindrical domains in block copolymer thin films. Additionally, the control of growth of Co nanowires into nanopores is quite clear because the uniform length of Co nanowires can be realized at the termination of electrodeposition. The anisotropic magnetic properties of Co-infiltrated nanoscopic cylinder arrays were successfully demonstrated in their study. The control of the orientation of block copolymer morphologies in thin films offers a very practical means of producing nanostructures that are feasible for technological applications.

5.5.6 Template process coupled with recent advances in perpendicular cylinder arrays

Various types of combinative nanofabrication technique based on block copolymer thin films having cylindrical nanostructures perpendicular through the entire films have

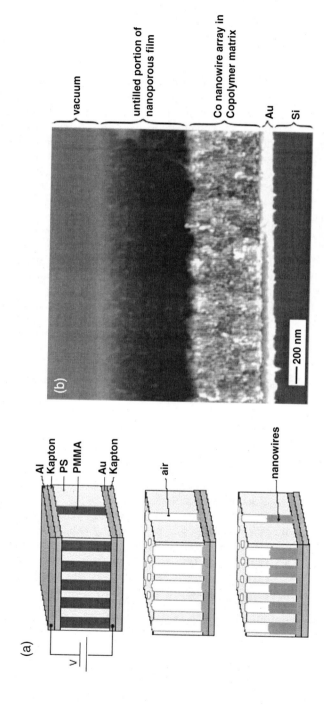

Fig. 5.24. (a) Schematic representation of high density nanowires growing within cylinder arrays in a polymer matrix. (b) Cross-sectional SEM image of a fracture surface of an array of Co nanowires grown into an array of nanopores formed from block copolymers.

emerged by the systematic modification of alignment processes, such as solvent evaporation induction, balancing of interfacial interaction (neutralization of substrate), and the use of blend systems, and external fields, as reviewed above. Here, few examples are introduced as recent advances in the nanotemplates process of cylindrical channels in block copolymer thin films.

Lin et al. reported that mixtures of PS-b-PVP diblock copolymers and either CdSe and ferritin-based nanoparticles exhibit cooperative and coupled self-assembly on the nanoscale [129]. In particular, CdSe quantum dots covered with tri-n-octylphosphine oxide (TOPO), which is a type of surfactant with long hydrocarbon chains, interestingly induces the alignment of cylindrical domains of PVP to the substrate by stable incorporation into PVP domains. The preferential interaction of PVP with the substrate and the lower surface energy of PS basically force the orientation of the cylindrical microdomains parallel to the substrate. The direct observation of the effect of TOPO-capped CdSe addition is shown in Fig. 5.25. A TEM image of a thin film from the PS-b-PVP/CdSe mixture after annealing revealed the penetration of cylindrical microdomains perpendicular to the surface and the persistence of order over very large distances, while that of a thin film from a pure PS-b-PVP block copolymer revealed that all the cylindrical microdomains are oriented parallel to the substrate. The temporary explanation for this is that cylindrical microdomains of the higher-surface-energy PVP with γ_{PVP} of 47 mN/m are coated with the lower-surface-energy hydrocarbon-capped CdSe nanoparticles with γ_{CH} of 30 mN/m, which effectively balance the surface interactions relative to the PS matrix with γ_{PS} of ~39 mN/m. This could cause an orientation of the cylindrical microdomains perpendicular to the surface in the same manner in Section 5.5.2.

The balancing interfacial interactions of blocks on the substrate surface are now reaching basic pretreatment for the fabrication of block copolymer templates with cylindrical microdomains perpendicular to the substrate. Russell's research group has dramatically established block copolymer template and lithography by combining the fundamental procedures producing alignment control of cylinders. Due to the remarkable progress in nanofabrication by block copolymer lithography, a simple route to metal nanopost arrays or nanoporous metallic films was investigated [117,120,121]. High-density arrays of chromium and layered gold/chromium nanodots and nanoholes in metal films were fabricated by evaporation onto nanoporous templates produced by the self-assembly of PS-b-PMMA diblock copolymers. The cylindrical microdomains of the asymmetric block copolymer were orientated perpendicularly to the surface by balancing the interfacial interactions of blocks with the substrate. By selectively removing either the minor or major component, nanoporous films of PS or nanoscopic posts of PS could be produced as shown in Fig. 5.26.

5.6 Nanocylindrical-structured Block Copolymer Templates

The approach using the self-organized structures in the diblock copolymer thin film becomes particularly powerful when the spatial orientation can be controlled and the resulting well-ordered alignment is utilized for the fabrication of nanocomposites with the electronic, magnetic, or photonic functional materials such as metal, metal oxide

Nanocylinder Array Structures in Block Copolymer Thin Films

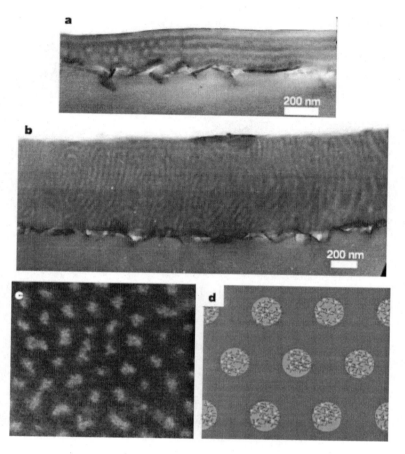

Fig. 5.25. (a) Cross-sectional TEM image of pure PS-*b*-PVP block copolymer thin film after annealing. (b)(c) Cross-sectional TEM and SEM images of thin films prepared by PS-*b*-PVP/CdSe mixtures after annealing. (d) Schematic drawing of nanoparticle aggregates in the PVP cylinders.

particles, and/or conducting organic polymers. Especially, nanocylinder array structures in block copolymer thin films have played important roles in these recent nanofabrications. Therefore, the control of the spatial orientation of nanocylinders relative to the substrate has been demonstrated by several research groups, i.e., solvent-evaporation-induced flow field, blend-induced alignment, and control by external field. However, most of the experimental studies on nanocylinder alignment have been focused on the conventional block copolymer system such as PS-*b*-PMMA or PS-*b*-PVP. Only a few studies have conducted the approach developing the control of both regularity and alignment of nanocylinder arrays in thin films using a newly designed block copolymer. The techniques of alignment should be applicable to various film shapes and thicknesses and should have no intricate steps in the preparative method. Moreover, the regularity of the nanocylinders in such conventional copolymer thin film is not sufficient for practical use and is now being improved.

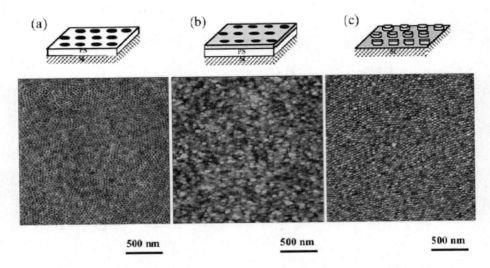

Fig. 5.26. Schematic fabrication process of Cr dot arrays (upper pictures) and height images of AFM at different preparative stages: (a) nanoscopic holes in cross-linked PS matrix, (b) evaporated Cr onto PS template, and (c) Cr nanodot arrays.

We have proposed that it should be essential to design self-organized microstructures hierarchically from the molecular level. To highlight the function of molecules made up of monomers and polymers can only be the solution to address the above problems and to develop block copolymer thin films competent for tasks of expected applications. We have designed a series of amphiphilic LC diblock copolymers, PEO_m-b-$PMA(Az)_n$, consisting of PEO as hydrophilic segment and poly(methacrylate) containing azobenzene moiety as hydrophobic liquid crystalline segment (PMA(Az)) (Fig. 5.27).

These multifunctions, i.e., amphiphilicity, LC character, and the photoisomerization of the azobenzene moiety, would be performed in individual domains specific to the phase-segregated nanostructure of the block copolymers. A series of block copolymers, PEO_m-b-$PMA(Az)_n$ was prepared by using ATRP [31]. The self-organization of PEO domains stained with RuO_4 in the PMA(Az) microdomains in the microtome-sectioned bulk sample was clearly observed as a hexagonally arranged nanocylinder array structure by TEM measurement [160]. The small grains with the individual distances between cylinders were arranged to form multigrains. The cylindrical direction in each grain was unrestricted in all directions, while the regularity of the cylinder arrangement in each grain was higher than that observed in the conventional block copolymer samples (Fig. 5.28). On the contrary, the thin film prepared onto a water surface and transferred to a TEM grid generated a highly ordered nanodot pattern with a diameter and a distance between cylinders of 5 and 17 nm, respectively (Fig. 5.29).

The microtome-slicing was carried out using 2-μm-thick films spin-coated onto a PET substrate for the identification of the phase structure. Figure 5.30 shows the perpendicular orientation of highly ordered nanocylinder array structures to the substrate, which run across the film. Additionally, small lamellar structures with a period of ca. 3 nm were

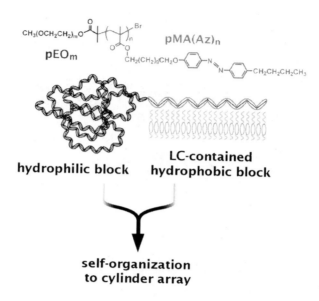

Fig. 5.27. Chemical structure of side-chain LC block copolymer, PEO-*b*-PMA(Az), newly designed for highly crystallized hexagonal cylinder array structure.

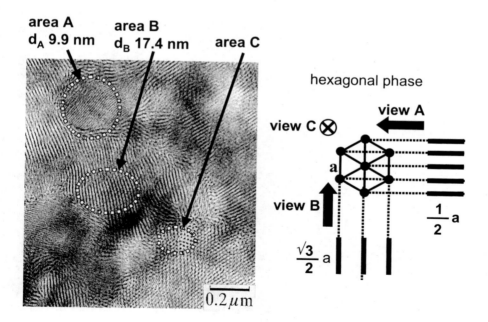

Fig. 5.28. TEM image of PEO-*b*-PMA(Az) sectioned by microtome-slicing. The image shows multigrain structure of a hexagonal cylinder phase, i.e., PEO cylinder in the PMA(Az) matrix. The PEO cylinders are well-ordered in a grain, while the orientations of grains are in three directions.

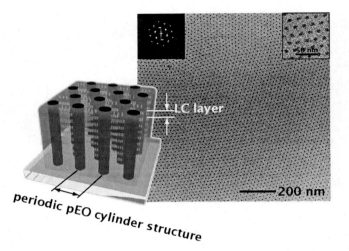

Fig. 5.29. TEM image of PEO-*b*-PMA(Az) thin film prepared on water surface and then picked up with TEM grid. The image indicates a highly ordered mono-grain structure. The FFT image of the TEM image shows the inset.

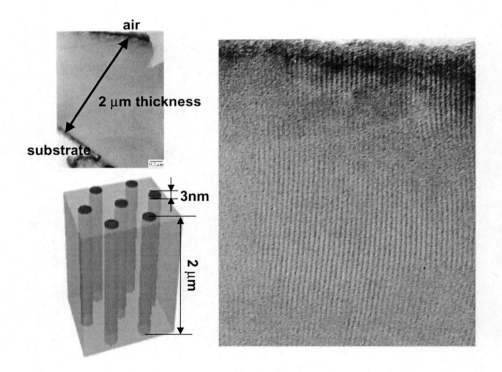

Fig. 5.30. TEM images of sectioned thin film of 3-μm thickness. The PEO cylinders are highly aligned across the film direction.

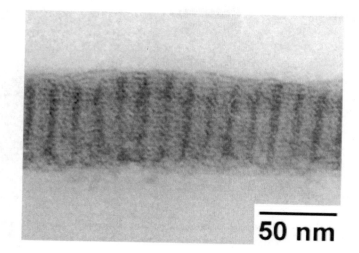

Fig. 5.31. TEM image of sectioned thin film of 80-nm thickness. An ultrafine lamellar structure with a period of ca. 3 nm was found in the PMA(Az) domains, indicating smectic layers of the azobenzene LC.

found in the PMA(Az) domains, as shown in Fig. 5.31 [161]. The 3 nm periodicity of the lamellar structures is in good agreement with the spacing of the smectic layer from SAXS and the calculated length (3.37 nm) of the fully extended conformer of the PMA(Az) side chain.

Therefore the small lamellar structures were assigned to the smectic layer of the PMA(Az) domain. The homeotropic alignment of the azobenzene side chains in the PMA(Az) domain was achieved in the smectic phase after annealing. The hierarchical structures demonstrated in this work could provide promising materials for new applications in new block copolymer nanotemplates, nanofilters, and optically functionalized materials such as holographic materials [162]. Very recently, we have succeeded in the large-area coating of PEO-*b*-PMA(Az) on a flexible PET film by the Microgravure® coating technique (Fig. 5.32). The nanocylinder array structures in the thin film of a series of PEO-*b*-PMA(Az) were successfully observed as a strong effect of different molecular weights and compositions on the diameter of cylinder and distance between cylinders all over the film surface. This first demonstration of a flexible film with very fine controllable nanostructures will play key roles in *bottom-up* block copolymer engineering.

5.7 Summary and Future Directions

The examples discussed throughout this review clearly prove the potential of block copolymers in general as building blocks or nano-objects for the fabrication of nanostructured materials on a small scale below the limitation of lithographic resolution. Although the use of block copolymer thin films for the nanotechnological applications is still in its infancy, examples of nanocylinder array structures have been considered as exciting breakthrough in a variety of applications. Of the cylindrical phase structure in block

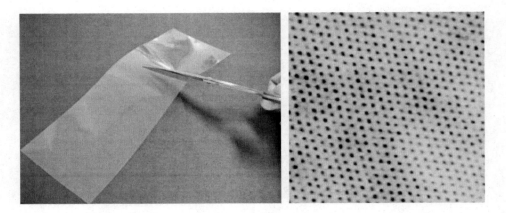

Fig. 5.32. Photograph of a large-area nanostructured film onto a roll-type PET substrate coated by microgravure technique and its AFM surface image.

copolymer thin films, a PS-based block copolymer has been studied most extensively, and nowadays, their phase behavior has been well understood. Recent work has demonstrated that the alignment control of cylindrical phase structures to produce thin films with cylindrical arrays perpendicular to the substrate holds promise for template fabrication process of semiconducting and magnetic materials, nanoparticle and nanocapsules, nanoporous materials, and photonic crystals. The future potential of this approach for new block copolymer materials lies in the achievement of large-area, highly ordered thin films as well as introduction of functional side chains that can be exploited in polymer synthesis. Only a few studies have, however, shown the large-area coating of well-aligned thin films. This is very much an ongoing activity, and we expect new developments resulting, for example, from the photo-crosslinkable side-chain block copolymers, which can facilitate the fixation of nanocylinder array structures in thin films. Finally, hierarchical ordering in LC side-chain block copolymers offers further scope for more subtle nanostructure engineering.

Acknowledgment

The authors wish to acknowledge the Japan Science and Technology Agency for their generous support of the work on nanocylindrical-structured block copolymer templates from a PEO-*b*-PMA(Az) through the division of Nanofactory and Process Monitoring for Advanced Information Processing and Communication in the Core Research for Evolutional Science and Technology program.

References

[1] M. Szwarc, M. Levy and R. Milkovich, J. Am. Chem. Soc. **78** (1956) 2656.

[2] J. Adams and W. Gronski, Makromol. Chem. Rapid Commun. **10** (1989) 553.

[3] B. Zaschke, W. Frank, H. Fischer and K. Schmutzler, Polym. Bull. **28** (1991) 243.

[4] S. Foerster and M. Antonietti, Adv. Mater. **10** (1998) 195.

[5] M. Hefft and J. Springer, Makromol. Chem. Rapid Commun. **11** (1990) 397.

[6] V. Percec and M. Lee, J. Macromol. Sci. A **29** (1992) 723.

[7] T. Kodaira and K. Mori, Makromol. Chem. **193** (1992) 1331.

[8] Z. Komiya and R.R. Shchrock, Macromolecules **26** (1993) 1393.

[9] R. Bohnert and H. Finkelmann, Macromolecular Chemistry and Physics **195** (1994) 689.

[10] G. Riess, G. Hurtrez and P. Bahadur, Encycl. Polym. Sci. Eng. **2** (1985) 324.

[11] J.-S. Wang and K. Matyjaszewski, J. Am. Chem. Soc. **117** (1995) 5614.

[12] K. Matyjaszewski and J. Xia, Chem. Rev. **101** (2001) 2921.

[13] T. Kowalewski, R.D. McCullough and K. Matyjaszewski, Eur. Phys. J. E **10** (2003) 5.

[14] M. Arnold, S. Poser, H. Fischer, W. Frank and H. Utschick, Macromol. Rapid. Comm. **15** (1994) 487.

[15] H. Fischer, S. Poser, M. Arnold and W. Frank, Macromolecules **27** (1994) 7133.

[16] H. Fischer, S. Poser and M. Arnold, Macromolecules **28** (1995) 6957.

[17] H. Fischer, S. Poser and M. Arnold, Liq. Cryst. **18** (1995) 503.

[18] B. Zaschke, W. Frank, H. Fischer, K. Schmutzler and M. Arnold, Polym. Bull. **27** (1991) 1.

[19] I.W. Hamely, V. Castelletto, Z.B. Lu, C.T. Imrie, T. Itoh and M. Al-Hussein, Macromolecules **37** (2004) 4798.

[20] J. Saenger, W. Gronski, S. Maas, B. Stuehn and B. Heck, Macromolecules **30** (1997) 6783.

[21] M. Yamada, T. Iguchi, A. Hirao, S. Nakahama and J. Watanabe, Macromolecules **28** (1995) 50.

[22] J.T. Chen, E.L. Thomas, C.K. Ober and S.S. Hqang, Macromolecules **28** (1995) 1688.

[23] M. Muthukumar, C.K. Ober and E.L. Thomas, Science **277** (1997) 1225.

[24] J.T. Chen, E.L. Thomas, C.K. Ober and G.-p. Mao, Science **273** (1996) 343.

[25] G.J. Liu, Macromol. Symp. **113** (1997) 233.

[26] G. Liu, J. Ding, L. Qiao, A. Guo, B.P. Dymov, J.T. Gleeson, T. Hashimoto and K. Saijo, Chem. A Eur. J. **5** (1999) 2740.

[27] G. Wang, X. Tong and Y. Zhao, Macromolecules **37** (2004) 8911.

[28] G. Mao, J. Wang, S.R. Clingman and C.K. Ober, Macromolecules 3048 (1997) 2556.

[29] E.L. Thomas, J.T. Chen, M.J.E. O'Rourke, C.K. Ober and G. Mao, Macromol. Symp. **117** (1997) 241.

[30] S.L. Sin, L.H. Gan, X. Hu, K.C. Tam and Y.Y. Gan, Macromolecules **38** (2005) 3943.

[31] Y. Tian, K. Watanabe, X. Kong, J. Abe and T. Iyoda, Macromolecules **35** (2002) 3739.

[32] Y. Tian, X. Kong, Y. Nagase and T. Iyoda, J. Polym. Sci., Part A: Polym. Chem. **41** (2003) 2197.

[33] E. Vanzo, J. Polym. Sci., Polym. Chem. **4** (1966) 1727.

[34] M. Moffitt, K. Khougaz and A. Eisenberg, Acc. Chem. Res. **29** (1996) 95.

[35] E. Helfand, Macromolecules **8** (1975) 552.

[36] L. Leibler, Macromolecules **13** (1980) 1602.

[37] T. Ohta and K. Kawasaki, Macromolecules **19** (1986) 2621.

[38] D.J. Meier, Appl. Polym. Symp. **24** (1974) 67.

[39] D.J. Blundell, Acta Crystallogr. A **26** (1970) 472.

[40] D.J. Blundell, Acta Crystallogr. A **26** (1970) 476.

[41] I. Goodman, Developments in Block Copolymers, Applied Science Publishers, London: New York, 1982.

[42] T. Hashimoto, K. Nagatoshi, A. Todo, H. Hasegawa and H. Kawai, Macromolecules **7** (1974) 364.

[43] T. Pakula, K. Saijo, H. Kawai and T. Hashimoto, Macromolecules **18** (1985) 1294.

[44] G. Hadziioannou and A. Skoulios, Macromolecules **15** (1982) 267.

[45] F.S. Bates, M.F. Schulz and J.H. Rosedale, Macromolecules **25** (1992) 5547.

[46] T. Hashimoto, A. Todo, H. Itoi and H. Kawai, Macromolecules **10** (1977) 377.

[47] T. Hashimoto, M. Fujimura and H. Kawai, Macromolecules **13** (1980) 1660.

[48] T. Hashimoto, M. Shibayama and H. Kawai, Macromolecules **13** (1980) 1237.

[49] G. Hadziioannou and A. Skoulios, Macromolecules **15** (1982) 258.

[50] H. Hasegawa and T. Hashimoto, Macromolecules **18** (1985) 589.

[51] R.W. Richards and J.L. Thomason, Macromolecules **16** (1983) 982.

[52] H. Huang, F. Zhang, Z. Hu, B. Du, T. He, F.K. Lee, Y. Wang and O.K.C. Tsui, Macromolecules **36** (2003) 4084.

[53] M. Schoeps, H. Leist, A. DuChesne and U. Wiesner, Macromolecules **32** (1999) 2806.

[54] T. Hashimoto, D. Yamaguchi and F. Court, Macromol. Symp. **195** (2003) 191.

[55] T. Hashimoto, K. Tsutsumi and Y. Funaki, Langmuir **13** (1997) 6869.

[56] R.R. Li, P.D. Dapkus, M.E. Thompson, W.G. Jeong, C. Harrison, P.M. Chaikin, R.A. Register and D.H. Adamson, Appl. Phys. Lett. **76** (2000) 1689.

[57] C. Harrison, M. Park, P.M. Chaikin, R.A. Register and D.H. Adamson, Journal of Vacuum Science & Technology, B: Microelectronics and Nanometer Structures **16** (1998) 544.

[58] F.S. Bates, C.V. Berney and R.E. Cohen, Macromolecules **16** (1983) 1101.

[59] A.D. Vilesov, G. Floudas, T. Pakula, E.Y. Melenevskaya, T.M. Birshtein and Y.V. Lyatskaya, Macromolecular Chemistry and Physics **195** (1994) 2317.

[60] C. Harrison, M. Park, P. Chaikin, R.A. Register, D.H. Adamson and N. Yao, Macromolecules **31** (1998) 2185.

[61] M. Park, C. Harrison, P.M. Chaikin, R.A. Register and D.H. Adamson, Science **276** (1997) 1401.

[62] P. Mansky, C.K. Harrison, P.M. Chaikin, R.A. Register and N. Yao, Appl. Phys. Lett. **68** (1996) 2586.

[63] G. Coulon, T.P. Russell, V.R. Deline and P.F. Green, Macromolecules **22** (1989) 2581.

[64] B. Collin, D. Chatenay, G. Coulon, D. Ausserre and Y. Gallot, Macromolecules **25** (1992) 1621.

[65] G. Coulon, J. Daillant, B. Collin, J.J. Benattar and Y. Gallot, Macromolecules **26** (1993) 1582.

[66] J. Hahm, W.A. Lopes, H.M. Jaeger and S.J. Sibener, J. Chem. Phys. **109** (1998) 10111.

[67] T.L. Morkved and H.M. Jaeger, Europhys. Lett. **40** (1997) 643.

[68] T.L. Morkved, P. Wiltzius, H.M. Jaeger, D.G. Grier and T.A. Witten, Appl. Phys. Lett. **64** (1994) 422.

[69] R.W. Zehner, W.A. Lopes, T.L. Morkved, H.M. Jaeger and L.R. Sita, Langmuir **14** (1998) 241.

[70] R.W. Zehner, W.A. Lopes, T.L. Morkved, H.M. Jaeger and L.R. Sita, Langmuir **14** (1998) 1942.

[71] R.W. Zehner and L.R. Sita, Langmuir **15** (1999) 6139.

[72] W.A. Lopes and H.M. Jaeger, Nature **414** (2001) 735.

[73] S. Horiuchi, T. Fujita, T. Hayakawa and Y. Nakano, Langmuir **19** (2003) 2963.

[74] Y. Liu, W. Zhao, X. Zheng, A. King, A. Singh, M.H. Rafailovich, J. Sokolov, K.H. Dai and E.J. Kramer, Macromolecules **27** (1994) 4000.

[75] Z. Li, W. Zhao, Y. Liu, M.H. Rafailovich, J. Sokolov, K. Khougaz, A. Eisenberg, R.B. Lennox and G. Krausch, J. Am. Chem. Soc. **118** (1996) 10892.

[76] J. Ruokolainen, G. Ten Brinke and O. Ikkala, Adv. Mater. **11** (1999) 777.

[77] K. Ishizu, T. Hosokawa and K. Tsubaki, European Polymer Journal **36** (2000) 1333.

[78] K. de Moel, G.O.R. Alverda van Ekenstein, H. Nijland, E. Polushkin, G. ten Brinke, R. Maki-Ontto and O. Ikkala, Chem. Mater. **13** (2001) 4580.

[79] G. ten Brinke and O. Ikkala, Macromol. Symp. **203** (2003) 103.

[80] R. Saito and K. Ishizu, Polymer **36** (1995) 4119.

[81] S. Klingelhoefer, W. Heitz, A. Greiner, S. Oestreich, S. Foerster and M. Antonietti, J. Am. Chem. Soc. **119** (1997) 10116.

[82] J.P. Spatz, P. Eibeck, S. Moessmer, M. Moeller, T. Herzog and P. Ziemann, Adv. Mater. **10** (1998) 849.

[83] J.P. Spatz, T. Herzog, S. Moessmer, P. Ziemann and M. Moeller, Adv. Mater. **11** (1999) 149.

[84] K.W. Cheng and W.K. Chan, Langmuir **21** (2005) 5247.

[85] L. Zhang, C. Bartels, Y. Yu, H. Shen and A. Eisenberg, Phys. Rev. Lett. **79** (1997) 5034.

[86] D. Bendejacq, M. Joanicot and V. Ponsinet, Eur. Phys. J. E **2005** (2005) 83.

[87] Y. Boontongkong and R.E. Cohen, Macromolecules **35** (2002) 3647.

[88] R.D. Bennett, G.Y. Xiong, Z.F. Ren and R.E. Cohen, Chem. Mater. **16** (2004) 5589.

[89] L. Zhu, B.R. Mimnaugh, Q. Ge, R.P. Quirk, S.Z.D. Cheng, E.L. Thomas, B. Lotz, B.S. Hsiao, F. Yeh and L.Liu, Polymer **42** (2001) 9121.

[90] Z. Lin, D.H. Kim, X. Wu, L. Boosahda, D. Stone, L. LaRose and T.P. Russell, Adv. Mater. **14** (2002) 1373.

[91] L. Zhu, S.Z.D. Cheng, B.H. Calhoun, Q. Ge, R.P. Quirk, E.L. Thomas, B.S. Hsiao, F. Yeh and B. Lotz, Polymer **42** (2001) 5829.

[92] S.W. Yeh, T.L. Wu, K.H. Wei, Y.S. Sun and K.S. Liang, J. Polym. Sci., Part B: Polym. Phys. **43** (2005) 1220.

[93] R. Nagarajan, M. Barry and E. Ruckenstein, Langmuir **2** (1986) 210.

[94] F.S. Bates, Science **251** (1991) 898.

[95] L. Zhang and A. Eisenberg, Science **272** (1996) 1777.

[96] L. Zhang and A. Eisenberg, Science **268** (1995) 1728.

[97] K. Bruce Thurmond II, T. Kowalewski and K.L. Wooley, J. Am. Chem. Soc. **118** (1996) 7239.

[98] K.L. Wooley, Chem. A Eur. J. **3** (1997) 13970.

[99] X.-S. Wang, M.A. Winnik and I. Manners, Macromol. Rapid Commun. **23** (2002) 210.

[100] N. Sakamoto, T. Hashimoto, C.D. Han, D. Kim and N.Y. Vaidya, Macromolecules **30** (1997) 1621.

[101] E. Helfand and Z.R. Wasserman, Macromolecules **13** (1980) 994.

[102] D.B. Alward, D.J. Kinning, E.L. Thomas and L.J. Fetters, Macromolecules **19** (1986) 215.

[103] A.-C. Shi and J. Noolandi, Macromolecules **27** (1994) 2936.

[104] V. Schadler and U. Wiesner, Macromolecules **30** (1997) 6698.

[105] J. Ruokolainen, J. Tanner and O. Ikkala, Macromolecules **31** (1997) 3532.

[106] H.R. Thomas and J.J. O'Malley, Macromolecules **12** (1979) 323.

[107] R. ven den Berg, H. de Groot, M.A. van Dijk and D.R. Denley, Polymer **35** (1994) 5778.

[108] M.A. van Dijk and R. van den Berg, Macromolecules **28** (1995) 6773.

[109] G. Krausch and R. Magerle, Adv. Mater. **14** (2002) 1579.

[110] S. Forster and T. Plantenberg, Angew. Chem., Int. Ed. **41** (2002) 688.

[111] C. Park, J. Yoon and E.L. Thomas, Polymer **44** (2003) 6725.

[112] U. Jeong, H.-C. Kim, R.L. Rodriguez, I.Y. Tsai, C.M. Stafford, J.K. Kim, C.J. Hawker and T.P. Russell, Adv. Mater. **14** (2002) 274.

[113] U. Jeong, D.Y. Ryu, D.H. Kho, J.K. Kim, J.T. Goldbach, D.H. Kim and T.P. Russell, Adv. Mater. **16** (2004) 533.

[114] T. Thurn-Albrecht, R. Steiner, J. DeRouchey, C.M. Stafford, E. Huang, M. Bal, M. Tuominen, C.J. Hawker and T.P. Russell, Adv. Mater. **12** (2000) 787.

[115] T. Xu, H.-C. Kim, J. DeRouchey, C. Seney, C. Levesque, P. Martin, C.M. Stafford and T.P. Russell, Polymer **42** (2001) 9091.

[116] T. Xu, J. Stevens, J. Villa, J.T. Boldbach, K.W. Guarini, C.T. Black, C.J. Hawker and T.P. Russell, Adv. Funct. Mater. **13** (2003) 698.

[117] K. Shin, K.A. Leach, J.T. Goldbach, D.H. Kim, J.Y. Jho, M. Tuominen, C.J. Hawker and T.P. Russell, Nano Lett. **2** (2002) 933.

[118] Q. Zhang, T. Xu, D. Butterfield, M.J. Misner, M. Du, D.Y. Ryu, T. Emrick and T.P. Russell, Nano Lett. **5** (2005) 357.

[119] X. Yang, S. Xiao, C. Liu, K. Pelhos and K. Minor, Journal of Vacuum Science & Technology, B: Microelectronics and Nanometer Structures – Processing, Measurement, and Phenomena **22** (2004) 3331.

[120] B.J. Melde, S.L. Burkett, T. Xu, J.T. Goldbach, T.P. Russell and C.J. Hawker, Chem. Mater. **17** (2005) 4743.

[121] H.-C. Kim, X. Jia, C.M. Stafford, D.H. Kim, T.J. McCarthy, M. Tuominen, C.J. Hawker and T.P. Russell, Adv. Mater. **13** (2001) 795.

[122] H.-W. Li and W.T.S. Huck, Nano Lett. **4** (2004) 1633.

[123] T. Thurn-Albrecht, J. Schotter, G.A. Kastle, N. Emley, T. Shibauchi, L. Krusin-Elbaum, K. Guarini, C.T. Black, M.T. Tuominen and T.P. Russell, Science (Washington, D. C.) **290** (2000) 2126.

[124] S.H. Kim, M.J. Misner, T. Xu, M. Kimura and T.P. Russell, Advanced Materials **16** (2004) 226.

[125] D.H. Kim, X. Jia, Z. Lin, K.W. Guarini and T.P. Russell, Advanced Materials **16** (2004) 702.

[126] C.-A. Fustin, B.G.G. Lohmeijer, A.-S. Duwez, A.M. Jonas, U.S. Schubert and J.-F. Gohy, Adv. Mater. **17** (2005) 1162.

[127] H. Xiang, Y. Lin and T.P. Russell, Macromolecules **37** (2004) 5358.

[128] A. Sidorenko, I. Tokarev, S. Minko and M. Stamm, Journal of the American Chemical Society **125** (2003) 12211.

[129] Y. Lin, A. Boeker, J. He, K. Sill, H. Xiang, C. Abetz, X. Li, J. Wang, T. Emrick, S. Long, Q. Wang, A. Balazs and T.P. Russell, Nature (London, United Kingdom) **434** (2005) 55.

[130] J. Rzayev and M.A. Hillmyer, J. Am. Chem. Soc. ASAP (2005)

[131] M. Li, K. Douki, K. Goto, X. Li, C. Coenjarts, D.M. Smilgies and C.K. Ober, Chem. Mater. **16** (2004) 3800.

[132] R.G.H. Lammertink, M.A. Hempenius, J.E. Van Den Enk, V.Z.H. Chan, E.L. Thomas and G.J. Vancso, Adv. Mater. **12** (2000) 98.

[133] R.G.H. Lammertink, M.A. Hempenius, G.J. Vancso, K. Shin, M.H. Rafailovich and J. Sokolov, Macromolecules **34** (2001) 942.

[134] T. Hayakawa and S. Horiuchi, Angew. Chem. Int. Ed. **42** (2003) 2285.

[135] G. Kim and M. Libera, Macromolecules **31** (1998) 2569.

[136] G. Kim and M. Libera, Macromolecules **31** (1998) 2670.

[137] J. Hahm and S.J. Sibener, Langmuir **16** (2000) 4766.

[138] P. Mansky, P. Liu, T.P. Russell, E. Huang and C.J. Hawker, Science **275** (1997) 1458.

[139] E. Huang, L. Rockford, T.P. Tussell and C.J. Hawker, Nature (London) **395** (1998) 757.

[140] S. Koizumi, H. Hasegawa and T. Hashimoto, Macromolecules **27** (1994) 6532.

[141] S. Koizumi, H. Hasegawa and T. Hashimoto, Macromolecules **27** (1994) 4371.

[142] L. Kane, D.A. Norman, S.A. White, M.W. Matsen, M.M. Satkowski, S.D. Smith and R.J. Spontak, Macromol. Rapid Commun. **22** (2001) 281.

[143] J. Peng, X. Gao, Y. Wei, H. Wang, B. Li and Y. Han, J. Chem. Phys. **122** (2005) 114706/1.

[144] R.A. Segalman, H. Yokoyama and E.J. Kramer, Adv. Mater. **13** (2001) 1152.

[145] J.Y. Cheng, C.A. Ross, E.L. Thomas, H.I. Smith and G.J. Vancso, Appl. Phys. Lett. **81** (2002) 3657.

[146] S.O. Kim, H.H. Solak, M.P. Stoykovich, N.J. Ferrier, J.J. de Pablo and P.F. Nealey, Nature **424** (2003) 411.

[147] D. Sundrani, S.B. Darling and S.J. Sibener, Langmuir **20** (2004) 5091.

[148] Y.-H. La, E.W. Edwards, S.M. Park and P.F. Nealey, Nano Lett. **5** (2005) 1379.

[149] F.A. Morrison, J.W. Mays, M. Muthukumar, A.I. Nakatani and C.C. Han, Macromolecules **26** (1993) 5271.

[150] F.S. Bates, K.A. Koppi, M. Tirrell, K. Almdal and K. Mortensen, Macromolecules **27** (1994) 5934.

[151] C.L. Jackson, K.A. Barnes, F.A. Morrison, J.W. Mays, A.I. Nakatani and C.C. Han, Macromolecules **28** (1995) 713.

[152] S. Sakurai, J. Sakamoto, M. Shibayama and S. Nomura, Macromolecules **26** (1993) 3351.

[153] R.J. Albalak and E.L. Thomas, J. Polym. Sci., Part B: Polym. Phys. **31** (1993) 37.

[154] Z.-R. Chen, J.A. Kornfield, S.D. Smith, J.T. Grothaus and M.M. Satkowski, Science **277** (1997) 1248.

[155] C. Osuji, P.J. Ferreira, G. Mao, C.K. Ober, J.B.V. Sande and E.L. Thomas, Macromolecules **37** (2004) 9903.

[156] E. Gurovich, Phys. Rev. Lett. **74** (1995) 482.

[157] T.L. Morkved, M. Lu, A.M. Urbas, E.E. Ehrichs, H.M. Jaeger, P. Mansky and T.P. Russell, Science **273** (1996) 931.

[158] A. Boker, H. Elbs, H. Haensel, A. Knoll, S. Ludwigs, H. Zettl, A.V. Zvelondovsky, G.J.A. Sevink, V. Urban, V. Abetz, A.H.E. Muller and G. Krausch, Macromolecules **36** (2003) 8978.

[159] T. Thurn-Albrecht, J. DeRouchey, H.M. Jaeger and T.P. Russell, Macromolecules **33** (2000) 3250.

[160] K. Watanabe, Y. Tian, H. Yoshida, S. Asaoka and T. Iyoda, Trans. Mater. Res. Soc. Jpn. **28** (2003) 553.

[161] K. Watanabe, H. Yoshida, K. Kamata and T. Iyoda, Trans. Mater. Res. Soc. Jpn. **30** (2005) 377.

[162] H. Yu, K. Okano, A. Shishido, T. Ikeda, K. Kamata, M. Komura and T. Iyoda, Adv. Mater. **17** (2005) 2184.

CHAPTER 6

Nano-Size Charge Inhomogeneity in Organic Metals

Takehiko Mori

Abstract

An organic charge-transfer salt called θ-phase shows a giant nonlinear conductance, and when a DC voltage is applied to this material, an AC voltage is generated. Thus the charge-transfer salt works as a new type of organic device like a thyristor in a DC–AC inverter. Charge-order and the nano-size charge inhomogeneity is responsible for the giant nonlinear conductance. In order to understand the mechanism of the organic thyristor, beginning with naming of the θ-phase, we review the universal phase diagram of the θ-phase, and the mechanism of the metal-insulator transitions. After surveying various experimental evidences of charge order, theoretical background that predicts non-stripe charge order is discussed.

Keywords: **organic thyristor, charge order, nonlinear conductance, charge-transfer salts, metal-insulator transitions, strong correlation.**

6.1 Introduction

Over the past two decades, there has been remarkable progress in the development of organic electronics [1]. Flat-panel displays based on organic light-emitting diodes (OLED) are now emerging as commercial products [2]. A number of industrial laboratories are working hard to develop organic field-effect transistors (OFET) in view of potential applications to large-area, low-cost, and flexible devices for flat-panel displays, electronic paper, and chemical sensors [3]. These applications are based on the use of organic semiconductors basically in the form of thin films. More conducting organics are realized using doped conducting polymers and charge-transfer salts, and particularly in the latter case, even superconductivity has been attained [4].

Organic metals and superconductors have attracted academic attention mainly due to their characteristic properties based on their low dimensionality and simple Fermi surfaces [5]. Very recently, however, giant nonlinear conductance has been discovered by Terasaki et al. in a charge-transfer salt of BEDT-TTF (bis(ethylenedithio)tetrathiafulvalene) (Figs. 6.1 and 6.2) [6], and it has been used to construct an organic thyristor (Fig. 6.1) [7]. When a DC bias is applied to a circuit shown in the inset in Fig. 6.1, including such a charge-transfer salt, the AC voltage as shown in Fig. 6.1(b) appears between the terminals of the series resistance. It is noteworthy that a bulk organic crystal works as an electronic device, and the structure is much simpler than an OFET and even an OLED. Here only two usual electrical contacts are pasted onto an organic charge-transfer crystal. When the applied DC voltage is gradually increased, the resistance of the charge-transfer

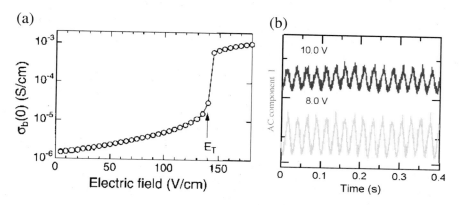

Fig. 6.1. (a) Giant nonlinear DC conductance of an organic charge-transfer salt, θ-(BEDT-TTF)$_2$CsZn(SCN)$_4$, measured perpendicular to the conducting layer [6]. When an appropriate DC bias (for example 8 V and 10 V) is applied to the inset circuit including such a charge-transfer salt the organic crystal of θ-(BEDT-TTF)$_2$CsCo(SCN)$_4$ works as a bulk device, and an AC voltage appears between the terminals of the series resistance as shown in (b).

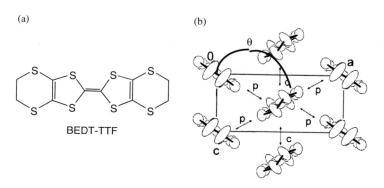

Fig. 6.2. (a) Molecular structure of BEDT-TTF, and (b) the molecular arrangement of BEDT-TTF molecules in a θ-phase charge-transfer salt.

salt abruptly drops to a definite threshold value. At the same time the current increases, and the voltage drop at the series resistance increases. This pushes down the voltage on the sample below the threshold value, and the sample oscillates between the two resistance values, generating the AC current. The origin of the giant nonlinear conductance is based on the competition of two charge-order patterns, where charge order is an inhomogeneous distribution of electrons on the originally equivalent molecules. There are molecular-scale or unit-cell scale patterns of charge-rich and charge-poor sites, whereas, in addition, there should be larger scale domains of different charge-order patterns. This is the true origin of the nonlinear conduction, a phenomenon that has recently become interesting as an example of intrinsic inhomogeneity [8]. Behind the discovery of the organic thyristor, there are a large number of intensive studies of organic charge order. In this chapter, the experimental evidence of organic charge order is first historically surveyed with the emphasis on the two-dimensional θ-phase system. Second, the naive theoretical aspects are described. Here the concept of non-stripe charge order, which is unique in nearly triangular organic θ-phase systems, is crucial to the appearance of giant nonlinearity. Finally, a brief interpretation of the organic thyristor effect and the future prospects are provided.

6.2 Universal Phase Diagram of the θ-Phase

Charge order in organic charge-transfer salts has been most extensively studied in the so-called θ-phase charge-transfer salts. It is a tradition of BEDT-TTF salts that different crystal structures are distinguished by Greek letters, and the θ-phase is a two-dimensional structure as shown in Fig. 6.2(b), where the molecular planes of the adjacent stacks are tilted alternately in opposite directions. This structure bears a close resemblance to the herringbone structure in single-component polyacenes [9]. Although the herringbone structure is the most popular structure in single-component polyacenes, the dihedral angles of polyacenes are much larger (closer to vertical direction) than those of BEDT-TTF θ-salts (110°–130°). Here the unit cell contains only two donor molecules, and there are only two kinds of intermolecular interactions, the interaction c along the stacking axis, and the interaction p in the diagonal direction. Although the molecular planes are tilted in the opposite directions alternately, the c and p interactions make a uniform two-dimensional network. In the limit of $p = c$, this pattern is reduced to a triangular lattice. On the other hand, the $c = 0$ limit is a square lattice.

The history of θ-phase charge-transfer salts dates back to the dawn of the organic superconductors [4]. After the discovery of the first organic superconductor [10], it soon became clear that BEDT-TTF forms many salts of the same composition, but different crystal structures. Kobayashi distinguished these phases by Greek letters, e.g., α-(BEDT-TTF)$_2$I$_3$ and β-(BEDT-TTF)$_2$I$_3$ (Fig. 6.3) [11]. Successively, a Russian group reported several iodine salts; although these phases had different compositions, they distinguished them by Greek letters like γ-(BEDT-TTF)$_3$(I$_3$)$_{2.5}$ [12], δ-(BEDT- as θ-TTF)I$_3$(TCE) [13], ε-(BEDT-TTF)I$_3$(I$_8$)$_{0.5}$ [14], ζ-(BEDT-TTF)$_2$I$_2$I$_8$ [15], and η-(BEDT-TTF)I$_3$ [16]. Subsequently Kobayashi discovered two organic superconductors with the same composition as the first α and β phases, and named them as θ-(BEDT-TTF)$_2$I$_3$ and κ-(BEDT-TTF)$_2$I$_3$ [17,18]. Accordingly, the θ-phase is actually the third among the 2:1 composition I$_3$ salts.

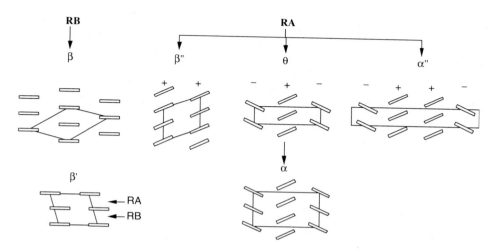

Fig. 6.3. Various crystal structures of BEDT-TTF salts, distinguished by Greek letters. The crystal is viewed perpendicular to the BEDT-TTF conducting layers, then viewed parallel to the long axes of the BEDT-TTF molecules. RB denotes the usual ring-over-bond stacking, and RA denotes ring-over-atom stacking, where the molecule shifts along the molecular long axis [11].

Following these triiodide salts, several other θ-salts with different anions were reported [19]. It is a characteristic feature of the θ-salts that the unit cell of the conducting plane is a rectangle (Fig. 6.2(b)), so that the overall space group is monoclinic or orthorhombic. On the other hand, α-(BEDT-TTF)$_2$I$_3$ has a similar herringbone-like structure, but the unit cell is tilted and has a triclinic space group. Another indication that distinguishes the θ-phase from the α-phase is the periodicity along the stacking axis; the θ-phase has uniform stacking along c, whereas the α-phase has two-fold periodicity. In this respect, θ-(BEDT-TTF)$_2$I$_3$ is exceptional because the I$_3$ anion has a two-fold periodicity along the donor stacking axis.

In the early history of organic superconductors, the θ-phase has been a less interesting system because some θ-phase salts are insulators, others like θ-(BEDT-TTF)$_2$I$_3$ are metals, and there is no unified impression of the θ-phase. Only after the preparation of θ-(BEDT-TTF)$_2$MM'(SCN)$_4$ salts [M = Cs, Rb, Tl and M' = Zn, Co] the θ-phase has begun to attract considerable attention [20–22]. Since the thiocyanate compounds constitute a series, it has become clear that the metal–insulator (M–I) transition temperature, T_{MI}, changes as a function of the dihedral angle (θ) between the molecular planes of the adjacent stacks, or alternately, as a function of the axis ratio. For example, θ-(BEDT-TTF)$_2$RbM'(SCN)$_4$ [M' = Zn and Co] with θ = 111° undergoes an M–I transition at 190 K (Fig. 6.4(a)), whereas θ-(BEDT-TTF)$_2$CsM'(SCN)$_4$ with θ = 104° exhibits a transition at 20 K (Fig. 6.4(b)). Since these properties are independent of the second metal atom, M' (Zn or Co), from now on, we shall call the above salts simply the Rb salts and the Cs salts, respectively. Accordingly, all the θ-phase salts are located on a single phase diagram as shown in Fig. 6.5, where the dihedral angle is taken as the horizontal axis. This plot is sometimes called the universal phase diagram of the

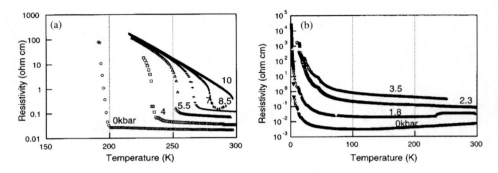

Fig. 6.4. Temperature dependence of electrical resistivity of (a) θ-(BEDT-TTF)$_2$RbZn(SCN)$_4$ and (b) θ-(BEDT-TTF)$_2$CsZn(SCN)$_4$ under hydrostatic pressure [22].

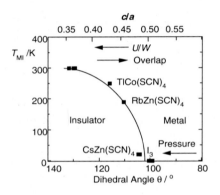

Fig. 6.5. Universal phase diagram of the θ-phase.

θ-phase [22]. It is characteristic of the θ-phase that T_{MI} increases with the application of hydrostatic pressure (Fig. 6.4). Under hydrostatic pressure the salts move to the left in the universal phase diagram (Fig. 6.5). Because the decreasing order of the alkali ion radius Cs > Rb > Tl produces the same effect, chemical pressure works in the same way as hydrostatic pressure.

The mechanism of the universal phase diagram is explained by the change of the band width W. The transfer integrals t_p and t_c calculated on the basis of molecular orbital calculations are plotted in Fig. 6.6(a). At θ = 180°, t_c is overwhelmingly large, because this corresponds to a face-to-face arrangement. At θ = 180°, t_p takes a small value, but below 140°, it increases significantly, and attains a maximum around 80°. Also, t_c changes sign twice around θ = 135° and 105°, but in the θ range of the actual θ-phase, the absolute value of t_c is not much larger than t_p. As far as $t_c < |t_p|$, the bandwidth, W, of the θ-phase is given by $8|t_p|$, and W is determined singly by t_p. The actual θ-phase takes an angle between 100° and 130°, and in this region t_c does not much exceed t_p.

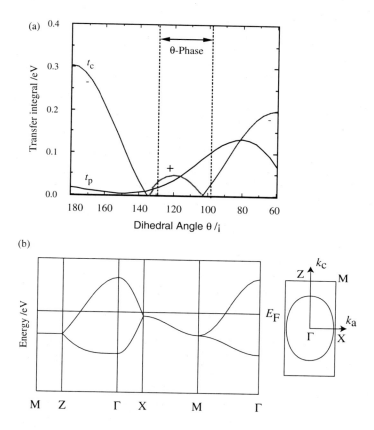

Fig. 6.6. (a) Transfer integrals, t_p and t_c, between HOMO of two tilted BEDT-TTF molecules in the θ-phase, calculated as functions of the dihedral angle, θ. (b) Tight-binding band structure and the Fermi surface in the θ-phase, where $t_p = 0.02$ eV and $t_p = 0.10$ eV.

Accordingly the bandwidth is almost solely determined by t_p, and rapidly increases with decreasing θ. The tight-binding band structure is that of a typical two-dimensional metal band (Fig. 6.6(b)).

Figure 6.7 shows another model calculation. Two BEDT-TTF molecules are placed on the same plane, and the two BEDT-TTF molecules are rotated in opposite directions around their inner S-S lines running parallel to the molecular long axes. As the dihedral angle θ decreases, the overlap increases. This is convincing because, as shown in the inset, the comparatively weak side-by-side (π-like) overlap of the p-orbitals at θ = 180° is replaced by the direct (σ-like) overlap of the shaded lobes. The overlap integral monotonously increases from θ = 180 to 90° and attains a maximum around 80°. This situation is not strictly the same as the actual interaction in the θ-phase, where the two molecules are shifted in the vertical direction. However this calculation shows the same tendency as Fig. 6.6, and more clearly demonstrates that the θ dependence of the overlap arises from the direct overlap of the π-lobes.

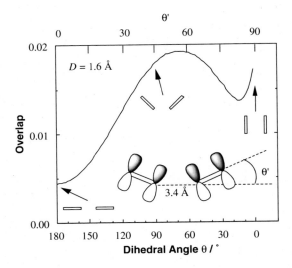

Fig. 6.7. Overlap integrals of two inclined BEDT-TTF molecules on the same horizontal level, as a function of the dihedral angle θ [19].

Accordingly, the universal phase diagram of the θ-phase is understood in view of the change of W competing with the correlated electrons. If W is sufficiently large at the right side of the phase diagram (small θ), the system retains metallic conductivity down to low temperatures. When W is relatively small at the left of the phase diagram (large θ), the system undergoes an M–I transition at a high temperature.

Although hydrostatic pressure increases T_{MI}, a uniaxial strain technique has recently been developed [23], where the sample is encased in epoxy resin, and pressurized in a conventional clamp cell. Figure 6.8 shows the temperature dependence of electrical resistivity of θ-(BEDT-TTF)$_2$CsZn(SCN)$_4$ under uniaxial strain [24]. When the uniaxial strain is applied along the c axis, the dihedral angle increases, and T_{MI} rises, whereas

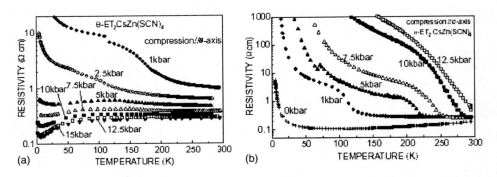

Fig. 6.8. Temperature dependence of electrical resistivity of θ-(BEDT-TTF)$_2$CsZn(SCN)$_4$ under uniaxial strain [24].

pressure along the b axis decreases the dihedral angle, and a drop in T_{MI} is observed. A similar uniaxial strain effect has been observed in the Rb salt [25]. Thus we can move to both the left and right in the universal phase diagram. This is the direct evidence that the dihedral angle is the controlling factor of the universal phase diagram.

Figure 6.9 shows the temperature dependence of the magnetic susceptibility and the lattice constants [22]. The magnetic susceptibility of the Rb salt is perfectly continuous at T_{MI} (Fig. 6.9(a)). Thus the M–I transition is not a Peierls-type transition. The magnetic susceptibility, however, drops to zero below 20 K; this transition is attributed to the spin Peierls transition, where the system becomes a band insulator in a spin singlet state, with doubling the lattice periodicity. As shown in Fig. 6.9(b), the b axis exhibits a characteristic expansion upon cooling, with an abrupt jump around 200 K. At the same time the a axis shrinks considerably in the direction where the dihedral angle increases to form the insulating phase. The system moves to the left in the universal phase diagram. The change of the lattice constants under hydrostatic pressure has not been investigated, but is usually the same as that at low temperatures. Therefore the dihedral angle is expected to increase under hydrostatic pressure. This is the reason why T_{MI} exhibits an anomalous increase under hydrostatic pressure. From a naive viewpoint, pressure usually increases W, and decreases T_{MI}, but this does not apply to the θ-phase. The phase transitions of the Rb salts and the Cs salts are quite different; the Rb salt shows a steep increase of the resistivity at T_{MI} (Fig. 6.4), whereas the transition of the Cs salts is very gradual. This has been further demonstrated by the investigation of the alloyed system, θ-(BEDT-TTF)$_2$Rb$_{1-x}$Cs$_x$Zn(SCN)$_4$ [22], where the conductance behavior discontinuously jumps from Rb type to the Cs type at $x = 0.25$. The abrupt jump of the lattice constants at T_{MI} indicates the first order nature of the phase transition. Clearer evidence of the first order transition is the fact that, when the compound is rapidly cooled, the transition is wiped out, and the Rb salt behaves like the Cs salt [27].

As shown in Fig. 6.9(c), relatively strong 2c X-ray reflections appear below $T_{MI} = 200$ K. To understand the mechanism of the M–I transition, we must investigate the insulating phase more in detail.

6.3 Charge Order

There are several different mechanisms, shown in Fig. 6.10, when a conductor becomes insulating due to the correlation of electrons, or the Coulomb repulsion between the conduction electrons. If the molecules are dimerized as depicted in Fig. 6.10(b), each dimer has one electron. In this case, the electron cannot move unless we assume a transition state in which two electrons occupy a dimer at the same time. Such a situation is mathematically represented by the Hubbard model [28],

$$H = \sum_{i \neq j} t_{ij} a_i^\dagger a_j + U \sum_i n_{i\uparrow} n_{i\downarrow} \tag{6.1}$$

where t is the transfer integral, and U is the energy loss at the doubly occupied state, called the on-site Coulomb repulsion. $c_i, c_j\, a_i^\dagger, a_j$, and n_i are creation, annihilation, and

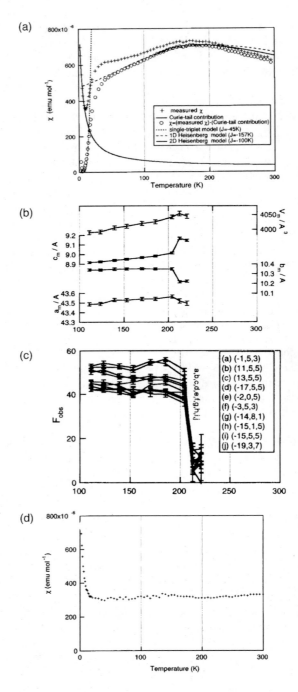

Fig. 6.9. (a) Temperature dependence of the static magnetic susceptibility of θ-(BEDT-TTF)$_2$RbZn(SCN)$_4$ [22]. (b) Temperature dependence of the lattice constants and (c) $c*/2$ reflections in θ-(BEDT-TTF)$_2$RbCo(SCN)$_4$ [26]. (d) Temperature dependence of the static magnetic susceptibility of θ-(BEDT-TTF)$_2$CsZn(SCN)$_4$ [22].

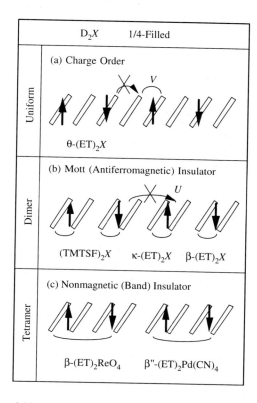

Fig. 6.10. Different mechanisms of correlated insulators.

number operators on site i. If U is sufficiently large, the system is expected to be an insulator called a Mott insulator [29]. If we assume a naive mean-field approximation, for a sufficiently large U, the energy band splits in two; the lower band, called the lower Hubbard band, for the first (singly occupied) electrons, and the upper Hubbard band for the second (doubly occupied) electrons. The splitting of these bands is given by U, and if U is larger than the bandwidth W, an energy gap, called the Hubbard gap, appears between the lower and the upper Hubbard bands.

If the energy band is half-filled, the lower Hubbard band is entirely occupied, and the system becomes a Mott insulator. It is noteworthy that this insulating state appears only at half-filling. Usually organic conductors of 2:1 composition are quarter-filled, but effective half-filling is realized in a dimerized system [30]. Figure 6.10 schematically depicts a one-dimensional system, but the same logic applies to two-dimensional systems, and mean-field calculations based on the parameters of indivisual BEDT-TTF compounds have been extensively carried out by the Fukuyama group [31]. In two-dimensional systems, whether or not a system is dimerized is not immediately obvious from the crystal structure, but a criterion based on the crystal symmetry has been proposed [32].

As shown in Fig. 6.10(a), a uniform system without dimerization, does not become a Mott insulator because it is not half-filled in any sense. The θ-phase falls into this category. An insulating state in the uniform quarter-filled system is explained if we take the off-site Coulomb repulsion into account, which is the repulsion when two electrons come to the neighboring molecules. In this case, the electrons prefer an alternating arrangement such as $D^+D^0D^+D^0$ because successive D^+D^+ arrangements give rise to an energy loss V. This is a phenomenon which has sometimes been called charge separation or charge disproportionation. Mathematically this is based on the so-called extended Hubbard model including V.

$$H = \sum_{i \neq j} t_{ij} a_i^\dagger a_j + U \sum_i n_{i\uparrow} n_{i\downarrow} + V \sum_{i \neq j} n_i n_j. \tag{6.2}$$

The periodicity of the dimerized Mott insulator does not change in the insulating phase, but the periodicity of the charge order phase is different from the metallic phase. Therefore, the two-fold periodicity in the θ-phase Rb salt (Fig. 6.9(c)) is evidence of charge order. In one dimension, a charge order state corresponds to the $4k_F$ charge density wave (CDW) state. A charge-ordered state is essentially paramagnetic. By contrast, the usual $2k_F$ CDW state (Peierls state) is nonmagnetic. The CDW state depends on the nesting of the Fermi surface and the dimensionality, whereas the charge order is, in principle, independent of the dimensionality. In practice, the charge inhomogeneity is fractional, i.e., $D^{0.5+\delta}D^{0.5-\delta}D^{0.5+\delta}D^{0.5-\delta}$.

Charge order is a phenomenon that has been extensively studied in inorganic compounds such as transition metal oxides [33]. Charge order in organic conductors was first suggested in the one-dimensional system $(DIDCNQI)_2Ag$ (DIDCNQI: 2,5-diiodo-N,N'-dicyanoquinonediimine) [34], and then in several other one-dimensional organic conductors [35]. One of the most important organic charge-order systems is TMTTF (tetramethylthetrathiafulvalene) salts, in which giant peaks of dielectric constants have been reported [36], together with a ^{13}C-NMR anomaly [37]. Charge order in the θ-phase Rb salt was first indicated by the ^{13}C-NMR measurements of the Kanoda group [38] (Fig. 6.11). The spectra change drastically below T_{MI}; the two sharp peaks are ascribed to the Pake doublet coming from the two adjacent ^{13}C atoms, but a very broad band overlapping with these sharp peaks is evidence of charge order. The charge distribution has been estimated as +0.35:+0.65. The same conclusion was reached from independent ^{13}C-NMR measurements by the Takahashi group, carried out almost at the same time [39].

The charge-order patterns were investigated by Seo [40], who included V in the mean-field calculations of the Fukuyama group. Seo explored several stripe patterns, designated vertical, diagonal, and horizontal in Fig. 6.12. The vertical pattern does not give the $2c$ periodicity, and the observed periodicity indicates horizontal order [20]. The horizontal charge distribution has been verified by X-ray investigation [41], where the charge distribution was found to be +0.2:+0.8.

Charge order in the θ-phase compounds was also supported by analysis of the reflectance spectra [42]. Tajima has carried out similar mean-field calculations to those of Seo, and obtained the optical conductivity spectra shown in Fig. 6.13. The conductivity spectrum

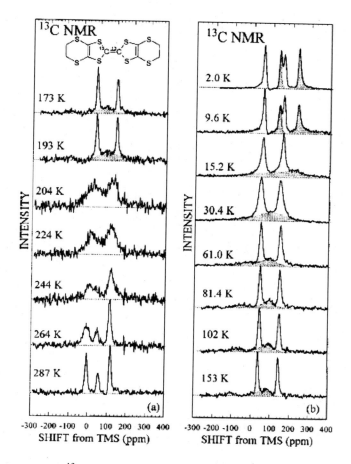

Fig. 6.11. ^{13}C-NMR spectra of θ-(BEDT-TTF)$_2$RbZn(SCN)$_4$ [38].

of the normal metal is Drude-like, while there appears a sharp peak in the vertical charge order. The horizontal pattern, however, leads to a broad band consisting of several peaks. The observed spectrum of the Rb salt is very broad below T_{MI}, and is attributed to horizontal charge order (Fig. 6.14). The spectrum of the high-temperature phase as well as that of the Cs salt is more metal-like. Although Tajima has ascribed them to a vertical phase, they are probably associated with some kind of non-stripe charge order. The C=C stretching band around 1500 cm^{-1} observed in the Raman spectrum also shows splitting below T_{MI} (Fig. 6.15(a)) [43]. These results were analyzed in view of the horizontal charge order, and the charge distribution was estimated to be +0.15:+0.75.

More recently, a type of charge inhomogeneity has been observed in the "metallic" phase of the Rb salt above 190 K. Watanabe et al. have observed satellite spots of the Rb salt at 270 K which appear at $3a \times 4c$ positions, together with diffuse lines running approximately in the $a^* + c^*$ direction [44] (Fig. 6.16). The broad NMR spectra above

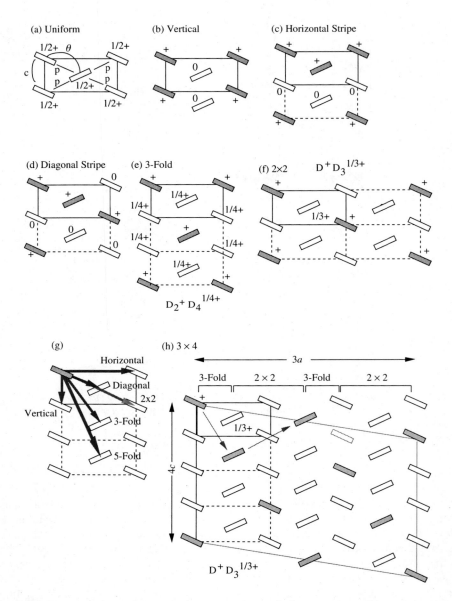

Fig. 6.12. Charge ordered patterns of the θ-phase. (a) Uniform distribution and the definition of the interactions. (b) Vertical, (c) horizontal, and (d) diagonal stripes. (e) 3-fold pattern and (f) 2 × 2 pattern are non-stripe patterns. For a more extended view of the 3-fold pattern, see Figs. 6.21 and 6.25. (g) These patterns are defined by the vectors connecting charge-rich molecules. (h) 3 × 4 pattern that explains Watanabe's modulation. This is a hybrid of 3-fold and 2 × 2 patterns.

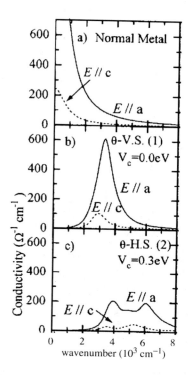

Fig. 6.13. Optical conductivity spectra calculated for (a) uniform, (b) vertical, and (c) horizontal patterns [42].

T_{MI} (Fig. 6.11) have been attributed to charge disproportionation because the second moment exhibits the same angular dependence as the Knight shift (Fig. 6.17) [45].

The amount of charge disproportionation has been estimated to be +0.3:+0.7. The observation of a large dielectric constant also supports this conclusion [46]. The temperature dependence of the resistivity above 190 K does not show a definite metallic decrease, but a very gradual increase (Fig. 6.4(a)), attributable to the existence of some kind of charge order.

In contrast with the Rb salt, the origin of the low-temperature insulating phase of the Cs salt is still ambiguous. The magnetic susceptibility is flat down to low temperatures (Fig. 6.9(d)) [22,47]. Nakamura has concluded charge disproportionation of +0.43:+0.57 below 20 K from the rotation of the ESR g-tensor [47]. The Raman spectra, however, do not show any splitting over the whole temperature range, though small broken symmetry has been observed even at room temperature (Fig. 6.15(b)) [48]. X-ray measurements show the development of $3a \times 3c$ spots (q_1 modulation) below 120 K, together with weak $2c$ diffuse spots (q_2 modulation) [49]. The intensity of the q_1 modulation does not change below 80 K, but the q_2 modulation rapidly grows below 50 K. Under pressure, the q_1 modulation becomes dominant. ^{13}C-NMR shows a remarkable broadening below

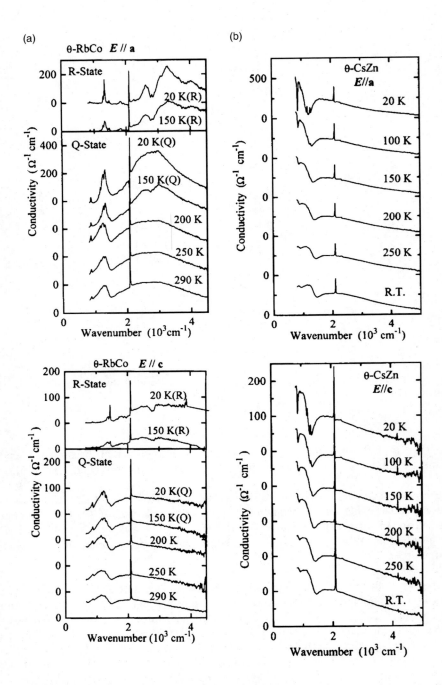

Fig. 6.14. Temperature dependence of the optical conductivity spectra of (a) θ-(BEDT-TTF)$_2$RbCo(SCN)$_4$, and (b) θ-(BEDT-TTF)$_2$CsZn(SCN)$_4$ [42].

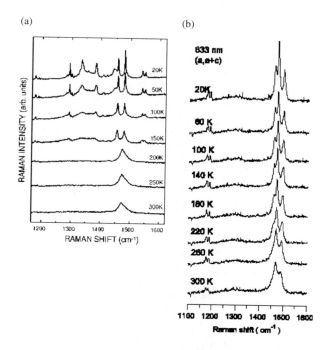

Fig. 6.15. Temperature dependence of the Raman spectra around the C=C stretching region for (a) θ-(BEDT-TTF)$_2$RbZn(SCN)$_4$ [43] and (b) θ-(BEDT-TTF)$_2$CsZn(SCN)$_4$ [48]; $(a, a+c)$ polarization, excited by 633 nm laser.

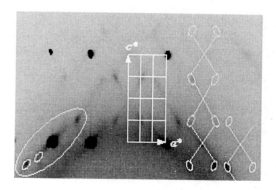

Fig. 6.16. X-ray oscillation photograph of θ-(BEDT-TTF)$_2$RbZn(SCN)$_4$ observed at 270 K.

170 K [50]. Since the second moment exhibits the same angular dependence as the Knight shift, as in Fig. 6.16, this broadening has been ascribed to charge order.

All these characteristic properties of the θ-phase are based on the uniform molecular arrangement of this phase (Fig. 6.10(a)), where the role of V is very important. In fact, the θ-phase is the only uniform molecular arrangement among the two-dimensional

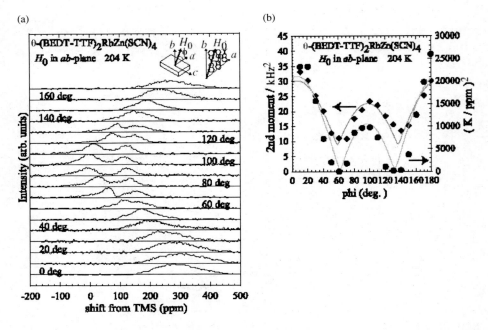

Fig. 6.17. (a) Angular dependence of ^{13}C-NMR line shape in the ab plane, and (b) the second moment and the Knight shift observed in the metallic phase (204 K) of θ-(BEDT-TTF)$_2$RbZn(SCN)$_4$ [45].

molecular conductors. To investigate the mystery of these fluctuating charge-order states, we have attempted quantum chemical calculations of V for various molecular geometries [51]. Using these results, we will show that non-stripe charge order is important in the θ-phase compounds [51].

6.4 Theoretical Background and Estimation of V

Estimation of V Figure 6.18 is the calculated intermolecular Coulomb repulsion V,

$$V = \int \phi_1^*(r_1)\phi_1(r_1)\frac{1}{r_{12}} \int \phi_2^*(r_2)\phi_2(r_2) d\tau_1 d\tau_2 = \sum_{ij} \frac{c_i^2 c_j^2}{R_{ij}}, \qquad (6.3)$$

for various geometries of the θ-phase BEDT-TTF molecules, where $\phi_1(r)$ and $\phi_2(r)$ are the highest occupied molecular orbitals (HOMO) of the BEDT-TTF molecules, and c_i are the coefficients of the HOMO. Here the point charge approximation represented as the final form in Eq. 6.3 is used according to [51]. The value of V_c at θ = 180° is large; this corresponds to the face-to-face stack (Fig. 6.6(a)). As θ decreases, V_c decreases, and V_p increases. In the actual θ range of the θ-phase, V_c is still larger than V_p, but near the metallic limit (θ = 90°) the V_c curve crosses the V_p curve. As shown in Fig. 6.6,

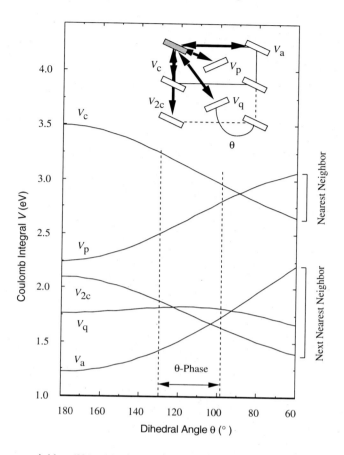

Fig. 6.18. Nearest neighbor (V_c and V_p) and second neighbor (V_a and V_q) intermolecular Coulomb integrals as functions of the dihedral angle θ, calculated from the molecular orbital calculations for the same molecular geometries as Fig. 6.5(a).

the p interaction is predominant as a transfer, but V_c is more important as the Coulomb repulsion. To compare Fig. 6.18 with the experimental results, it should be noted that U calculated from the same molecular orbital is 6.49 eV [51]. The calculated U and V values are bare values without screening. Because U of BEDT-TTF is experimentally evaluated to be about 1 eV, the V values in Fig. 6.18 are to be divided by the above U value.

Phase diagram. Elaborate calculation is necessary to investigate the extended Hubbard model represented by Eq. 6.2. However, if the first term is neglected ($t = 0$), the remaining U and V terms give simple analytic forms under the mean-field approximation. This is called the static limit or the atomic limit. In the Hartree–Fock theory, the electron numbers are approximated as

$$n_i n_j = n_i \langle n_j \rangle + \langle n_i \rangle n_j - \langle n_i \rangle \langle n_j \rangle. \tag{6.4}$$

Then Eq. 6.2 is diagonalized, and can be solved after iteration, which determines n_i self-consistently. Here, however, the lowest order term of the Hartree–Fock approximation is treated, and Eq. 6.4 is further approximated to

$$n_i n_j = \langle n_i \rangle \langle n_j \rangle \tag{6.5}$$

In the second place, the spin non-polarized case is considered. The number operators in the U term (Eq. 6.2) are restricted to up or down spins, but the numbers of up (or down) spin electrons are assumed to be half of the total electrons.

$$\langle n_i\uparrow \rangle = \langle n_i\downarrow \rangle = (1/2)\langle n_i \rangle \tag{6.6}$$

With these approximations, the potential energy terms of Eq. 6.2 are represented by

$$E = (1/4)U\Sigma \langle n_i \rangle \langle n_i \rangle + \Sigma V \langle n_i \rangle \langle n_j \rangle. \tag{6.7}$$

In the third place, complete charge separation is assumed, where the charges are separated to D^0 and D^+. This restriction is, however, easily removed in the later discussion. According to these approximations, the potential energy, E of the uniform state per unit cell is,

$$\text{Uniform:} \quad E = 2 \times \frac{1}{4}U \times \frac{1}{2} \times \frac{1}{2} + 4V_p \frac{1}{2} \times \frac{1}{2} + 2V_c \frac{1}{2} \times \frac{1}{2} = \frac{U}{8} + V_p + \frac{V_c}{2} \tag{6.8}$$

where the factor 2 before U is the number of molecules in a unit cell (Fig. 6.12(a)), and 4 and 2 before the V terms are the number of interactions in a unit cell. The factor, 1/4, in the U term comes from the spin non-polarization (Eqs. 6.6 and 6.7). Similarly obtained classical energies are listed in Table 6.1.

The phase diagram without considering the second nearest V' is depicted in Fig. 6.19. By comparing the energies of charge order phases, the phase boundaries are obtained. For example, the boundary of the uniform and the vertical phases are estimated as:

$$\text{Uniform} < \text{Vertical:} \quad U/8 + V_p + V_c/2 < U/4 + V_c \rightarrow 8V_p - 4V_c < U \tag{6.9}$$

The phase diagram is summarized in Fig. 6.19. If V' is neglected, the horizontal phase has the same energy as the diagonal phase (Table 6.1). Then the horizontal is designated in Fig. 6.19. When U is sufficiently large, the uniform state is stable. This is reasonable, because the charge-ordered states increase the U term. If V is comparatively large, charge-ordered states appear. When V_p is larger than V_c, the vertical stripe is stable. The vertical stripe eliminates the V_p term, because the charge on every other chain is zero. On the other hand, the vertical stripe enhances the V_c term, but this is unimportant because $V_p > V_c$. On the contrary, the condition, $V_p < V_c$, leads to the diagonal or horizontal stripe. From this, we can extract a simple general rule: when charge ordering occurs, the charge is arranged in the direction of the smallest V.

The 3-fold phase appears near the center of the phase diagram. If $V_p = V_c$, the network of the θ-phase is reduced to a simple triangular lattice, and the diagonal stripe is equivalent to the vertical stripe. This happens on the diagonal line (B point) in Fig. 6.19. In this case,

Table 6.1. Ground state energies of charge order states per unit cell (two molecules) [52].

Uniform	$D^{1/2+}$	$E = \dfrac{U}{8} + \dfrac{V_c}{2} + V_p + \dfrac{V_a}{2} + V_q$
Vertical	D^+D^0	$E = \dfrac{U}{4} + V_c + V_a$
Diagonal	D^+D^0	$E = \dfrac{U}{4} + V_p + V_q$
Horizontal	D^+D^0	$E = \dfrac{U}{4} + V_p + V_a + V_q$
3-Fold	$D^+D_2^{1/4+}$	$E = \dfrac{3}{16}U + \dfrac{3}{8}V_c + \dfrac{3}{4}V_p + \dfrac{3}{4}V_a + \dfrac{3}{2}V_q$
5-Fold	$D^+D_4^{3/8+}$	$E = \dfrac{5}{32}U + \dfrac{15}{32}V_c + \dfrac{15}{16}V_p + \dfrac{5}{8}V_a + \dfrac{15}{16}V_q$
2 × 2	$D^+D_3^{1/3+}$	$E = \dfrac{U}{6} + \dfrac{4}{9}V_c + \dfrac{8}{9}V_p + \dfrac{4}{9}V_a + \dfrac{8}{9}V_q$
4 × 3	$D^+D_3^{1/3+}$	$E = \dfrac{U}{6} + \dfrac{4}{9}V_c + \dfrac{8}{9}V_p + \dfrac{4}{9}V_a + \dfrac{10}{9}V_q$
4 × 1	D^+D^0	$E = \dfrac{U}{4} + \dfrac{V_c}{2} + \dfrac{3}{2}V_p + \dfrac{V_a}{2} + \dfrac{V_q}{2}$
3 × 1	$D^+D_2^{1/4+}$	$E = \dfrac{3}{16}U + \dfrac{3}{8}V_c + \dfrac{9}{8}V_p + \dfrac{3}{8}V_a + \dfrac{3}{4}V_q$

there is considerable competition or fluctuation between the different stripe patterns, and in the moderate U region, the 3-fold pattern of Fig. 6.12(e) becomes more stable than the stripes. This pattern has a 3-fold repeating unit along c, so it is called the 3-fold state. Here, the corner and the center molecules are charge-rich, and the other molecules are charge-poor. The repeating unit then contains six molecules such as $D_2^+D_4^{1/4+}$. We can invert the pattern of the charge-rich and charge-poor molecules like $D_2^0D_4^{3/4+}$, but these patterns have the same energy in the present approximation. The 3-fold pattern has been found in actual compounds such as (BEDT-TTF)$_3$CuBr$_4$ and (BEDT-TTF)$_3$CoCl$_4$ [53]. Both compounds are not quarter-filled but 2/3-filled, and have charge-poor centered patterns like $D^+D^+D^0$.

Since the non-stripe 3-fold pattern has a stable region, more complicated patterns are investigated. When the periodicity of the c axis is 5-fold instead of 3-fold, a 5-fold phase is obtained (Fig. 6.12). The energy of this phase is listed in Table 6.1. Another possible pattern is two-fold periodicity along both the c and a axes (Fig. 6.12(f)). This phase will be called 2 × 2-fold phase. These phases do not become the most stable phase in any regions. On the boundary between the uniform and the 3-fold phases, however, these more complicated phases have the same energy as the stable phases. There is a possibility that these complicated phases actually appear on account of the slight change in the boundary conditions and of a finite temperature effect, because the energies of these complicated phases are close to that of the 3-fold phase, and are located slightly above the 3-fold phase. Then we will treat the 3-fold phase as a representative of non-stripe phases.

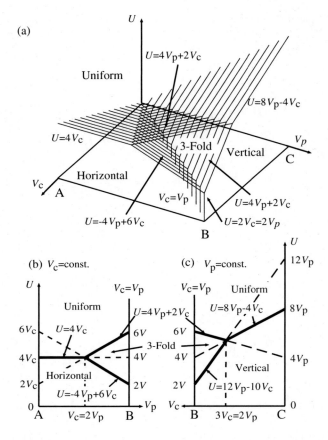

Fig. 6.19. Phase diagram of charge-ordered patterns of the θ-phase without considering V' [52]. (b) and (c) are cross sections along AB and BC in (a).

As shown in Fig. 6.12(g), these charge-order patterns are characterized by a vector connecting the charge-rich molecules. Similarly to the 3-fold and 5-fold patterns, the c-axis vector represents the vertical stripe, and the diagonal vector generates the diagonal stripe. From this figure, it is obvious that we have accounted for all possible patterns represented by simple vectors. Watanabe's modulation observed in the metallic state of the Rb salt has $4a \times 3c$ periodicity (Fig. 6.15). Figure 6.12(h) is a candidate that explains this modulation. The charge-rich molecules are generated alternately by the same rules as the 3-fold and the 2×2 patterns. The ground state energy is in between those of the 3-fold and the 2×2 phases (Table 6.1). This phase has no stable region at $T = 0$ K.

Since V_p and V_c have been given as functions of the dihedral angle, θ, as shown in Fig. 6.18, combining the results of Figs. 6.18 and 6.19, we can plot a phase diagram with respect to θ and U (Fig. 6.20). For the usual values of θ, a wide area above the stripe phases is occupied by the 3-fold phase. This means that the difference between V_p and

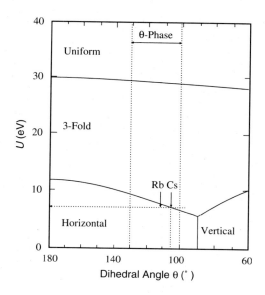

Fig. 6.20. Phase diagram of the θ-phase without including the second nearest V', constructed from Figs. 6.18 and 6.19 [52].

V_c is not large, and the actual system is located near point B in Fig. 6.19. Consequently a direct transition from the stripe state to the uniform phase is unlikely. By contrast, the square lattice of cuprate superconductors corresponds to point C in Fig. 6.19, where the stripe phase is the only possibility. If we assume that the Cs salt is located near the boundary between the horizontal stripe and the 3-fold phase, U becomes ~ 7 eV. This value is consistent with the calculated $U = 6.49$ eV; since we have used bare V values to construct Fig. 6.20, the corresponding U value is also a bare value obtained from the same calculation.

Second nearest V. To examine the phase diagram including the second nearest V', the number of the parameters is too large. If we assume that all the nearest neighbor V's are the same, namely $V_c = V_p = V$, the network is reduced to a triangular lattice. To include the second nearest V' in a triangular lattice, we may include $V_a = V_q = V'$ and neglect V_{2c}, because V_{2c} corresponds to a kind of third nearest interaction in the triangular lattice. The resulting phase diagram is depicted in Fig. 6.21, which is plotted for the horizontal V/U axis and the vertical V'/U axis.

This result is similar to the phase diagram of the triangular Ising model including the second J_2 [54]; the phase diagram of the Ising model is usually plotted for J_1/H and J_2/H, where J_1, J_2, and H act like V, V', and U, respectively. The negative J corresponds to the positive V, so that the directions of the axes are opposite. Although similar phases appear in both phase diagrams, the positions of the stable regions are a little different because the condition of charge conservation is strictly applied to the present case, while the number of up-spins is not restricted in the Ising model.

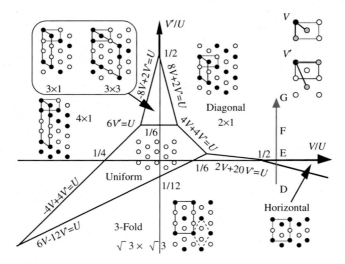

Fig. 6.21. Phase diagram of the 'triangular' θ-phase plotted with respect to V and V' [52]. The molecules are represented by the circles, and the closed and open circles stand for charge.

The ferromagnetic phase corresponds to the uniform phase, which is stable when V/U and V'/U are small (around the origin) and in the negative V and V' region (positive J_1 and J_2 region). The so-called 2×1 phase, which is the most stable phase in the positive V and V' region, is identical to the diagonal stripe; in the triangular lattice, the diagonal and the vertical stripes are of the same pattern.

The 3-fold phase is called the $\sqrt{3} \times \sqrt{3}$ phase in the Ising model. This phase has a large stable area in the negative V' region. The positive part of the horizontal axis ($V'/U = 0$) is equivalent to point B in Fig. 6.19; with increasing V (decreasing U), one passes from the uniform phase to the 3-fold phase, and successively to the stripe phase. Figure 6.21 shows that with an increase of positive V', the stripe phase is largely stabilized compared with the 3-fold phase.

In addition to these phases, the 4×1 phase has a large stable area in the negative V region. The 3×1 phase and the 3×3 phase have the same energy, and have a stable region between the 2×1 and the 4×1 phases. In the triangular Ising model, the 2×2 phase has a stable area between the uniform (ferromagnetic) and the diagonal (2×1) phases [54]. In the present model, the 2×2 phase has the same energy only on the boundary of these phases ($4V + 4V' = U$).

It is noteworthy that the horizontal phase is stable only in the negative V' region. This is obvious from the equations of the diagonal and the horizontal phases (Table 6.1). The energy of the horizontal phase is obtained by adding V_a to the energy of the diagonal phase. Thus the horizontal phase is more stable than the diagonal phase only when $V_a < 0$.

Clay et al. have shown from the exact diagonalization that the horizontal phase is stable because it consists of the 11001100 pattern rather than the 1010 pattern in the diagonal directions [55]. The 11001100 pattern corresponds to the $2k_F$ modulation, and is more stable than the 1010 type $4k_F$ modulation when t is sufficiently large. The present static-limit calculation does not include the effect of t. The above phase diagram, however, shows that negative V' has the same effect as t. The actual θ-phase is considered to be located near point E, in the slightly negative V' region.

Despite the relatively large calculated values of the second and the third V' (Fig. 6.18), the observation of the horizontal stripe in the real system suggests V' is less important, and even effectively negative. This arises from the significant influence of screening, and from the effect of t. Nonetheless, the phase diagram including V' (Fig. 6.21) is important in understanding the position of the real systems.

Fractional charge. In these calculations, full charge separation such as D^+D^0 is investigated, but in the actual systems, the charge separation is fractional such as $D^{0.5+\delta}D^{0.5-\delta}$. In this case, the charges are not 1 and 0 but represented as fractional numbers, n_1 and n_2 (Fig. 6.22(a)). From the charge conservation, $n_1 + n_2 = 1$; n_1 is taken as n, and the number of variables is reduced to one. The potential energy of the vertical stripe (Fig. 6.23(a)) up to the nearest neighbor is then,

$$\text{Vertical:} \quad E = \frac{1}{4}U\left(n_1^2 + n_2^2\right) + 4V_p n_1 \times n_2 + V_c\left(n_1^2 + n_2^2\right)$$

$$= \left(\frac{U}{2} + 2V_p - 4V_c\left(n - \frac{1}{2}\right)\right)^2 + \frac{U}{8} + V_p + \frac{V_c}{2}. \quad (6.10)$$

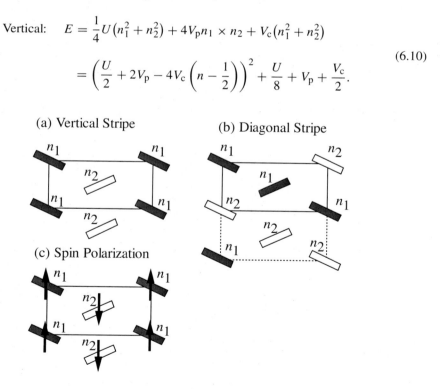

Fig. 6.22. Stripe patterns of charge order states with fractional charges. (a) vertical and (b) diagonal stripes. (c) Spin polarization coexisting with vertical charge order.

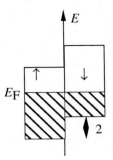

Fig. 6.23. Energy difference 2D of up and down spin energy bands in the Stoner model.

When $n = 1/2$, this equation correctly gives the energy of the uniform state (Table 6.1). When n is 0 or 1, this equation reduces to that in Table 6.1. Equation 6.10 is quadratic with respect to $n - 1/2$, so that the energy minimum depends on the sign of the coefficient of the $(n - 1/2)^2$ term. If $U/2 + 2V_c - 4V_p$ is positive, $n = 1/2$ is the energy minimum. When $U/2 + 2V_c - 4V_p$ becomes negative, n deviates from $1/2$, resulting in charge separation. The boundary $U/2 + 2V_c - 4V_p = 0$ is the same as Fig. 6.19(b). The energy of the horizontal or diagonal charge order is similarly obtained (Fig. 6.22(b)).

$$\text{Horizontal} = \text{Diagonal:} \quad E = \left(\frac{U}{2} - 2V_c\right)\left(n - \frac{1}{2}\right)^2 + \frac{U}{8} + V_p + \frac{V_c}{2}. \quad (6.11)$$

The sign of the first term gives the same boundary between the horizontal phase and the uniform phase as shown in Fig. 6.19(b). Therefore, the phase diagrams which we have obtained for complete charge separation are valid even for the fractional charge separation.

In Eqs. 6.10 and 6.11, once the charge order takes place, the charge separates as much as possible to a degree that is limited by other unconsidered factors. This happens because we have assumed Eq. 6.5. If we assume Eq. 6.4 or perturbatively estimate the higher order terms, the potential energy is expressed as

$$E = A\left(n - \frac{1}{2}\right)^2 + B\left(n - \frac{1}{2}\right)^4 + . \quad (6.12)$$

This is the Ginzburg–Laudau expansion of the potential energy with respect to the order parameter, $n - 1/2$. When $A > 0$, $n = 1/2$ gives the energy minimum. When $A < 0$, $n - 1/2$ becomes non-zero. In this case, n does not necessarily jump to the maximum and minimum values. As long as B is non-zero, $n - 1/2$ which gives the energy minimum, grows in proportion to $\sqrt{|A|/B}$. Then a second order phase transition takes place. What we have done in the above discussion is to estimate A, the sign of which gives the correct phase boundary.

Spin polarization. The influence of spin polarization is similarly treated. So far we have not considered spin polarization, that is, we have assumed Eq. 6.6. Here we consider that the numbers of electrons with upward and downward spins are different.

$$n_1\uparrow + n_1\downarrow = n_1, \quad n_2\uparrow + n_2\downarrow = n_2 \tag{6.13}$$

The degree of spin polarization is expressed by S_z,

$$\begin{aligned} n_1\uparrow = n_1(1/2 + S_z), & \quad n_1\downarrow = n_1(1/2 - S_z) \\ n_2\uparrow = n_2(1/2 - S_z), & \quad n_2\downarrow = n_2(1/2 - S_z) \end{aligned} \tag{6.14}$$

The potential energy at the atomic limit is obtained as in the preceding paragraph. For instance, in the vertical stripe (Fig. 6.22(c)),

$$\begin{aligned} \text{Vertical:} \quad E &= \frac{U}{4}\left(n_1^2\left(\frac{1}{2} - S_z\right)\left(\frac{1}{2} + S_z\right) + n_2^2\left(\frac{1}{2} - S_z\right)\left(\frac{1}{2} + S_z\right)\right) \\ &\quad + 4V_p n_1 \times n_2 + V_c(n_1^2 + n_2^2) \\ &= \left(2U\left(\frac{1}{4} - S_z^2\right) + 2V_c - 4V_p\right)\left(n - \frac{1}{2}\right)^2 + \frac{U}{8} + V_p + \frac{V_c}{2}, \end{aligned} \tag{6.15}$$

where we have used $n_1 + n_2 = 1$, and n_1 is set as n. The only difference between this equation and Eq. 6.11 is that $4(1/4 - S_z^2)$ is multiplied by the U term. If there is no spin polarization ($S_z = 0$), Eq. 6.11 is recovered. If complete spin polarization takes place ($|S_z| = 1/2$), the U term becomes zero. This is reasonable because no double occupancy occurs under complete spin polarization. If U is set to zero in Fig. 6.18, charge-ordered states are always stable. In the actual system, however, the spin polarization competes with the kinetic energy. This will be taken into account in the following.

It is a characteristic of the quarter-filled band that the spin polarization, S_z, and the charge order, $n - 1/2$, are entirely independent. In a half-filled band, S_z is restricted by the charge order. For example, under complete charge separation like D^0D^{2+}, S_z must be zero. This is the reason that such a simple treatment as Eq. 6.15 is possible in the quarter-filled band.

In order to discuss the restricting factor of spin polarization, we must consider the influence of t. We shall treat the Hubbard model, which contains the t and U terms but not the V term in Eq. 6.1, within the mean-field approximation. We assume that the energy band is not affected by U; then, as shown in Fig. 6.23, the energy bands of up and down spin electrons shift mutually by 2Δ. This model is known as the Stoner model [56]. According to the usual notation of the Stoner model, the magnetization, m, is,

$$m = n_\uparrow - n_\downarrow = \int_{E_F - \Delta}^{E_F + \Delta} D(E)dE = 2D\Delta, \tag{6.16}$$

where $D = D(E_F)$ is the density of states at the Fermi level, E_F. The magnetization, m, is connected to S_z like $S_z n = m/2$, as obtained from Eqs. 6.14 and 6.16. The loss of the kinetic energy induced by the band shift as in Fig. 6.23 is,

$$\Delta E = \int_0^{E_F+\Delta} ED(E)dE + \int_0^{E_F-\Delta} ED(E)dE - 2\int_0^{E_F} ED(E)dE$$

$$= \int_{E_F-\Delta}^{E_F+\Delta} ED(E)dE = D\Delta^2 = \frac{m^2}{4D} = \frac{Wm^2}{4}, \qquad (6.17)$$

where Eq. 6.16 is used, and $W = 1/D$ is the bandwidth. This applies exactly when the density of states is constant with respect to E, like the two-dimensional free-electron model (Fig. 6.23), but the following discussion applies generally if we use $1/D$ instead of W. The U term is readily obtained from the definition of m (Eq. 6.16) and $n = n\uparrow + n\downarrow$,

$$E = Un_\uparrow n_\downarrow = \frac{U}{4}(n^2 - m^2). \qquad (6.18)$$

Summation of Eqs. 6.17 and 6.18 gives the total energy.

$$\Delta E = \frac{Wm^2}{4} + \frac{U}{4}(n^2 - m^2) = \frac{W-U}{4}m^2 - \frac{U}{4}n^2. \qquad (6.19)$$

The second term is constant as long as the total electron number, n, is unchanged. The first term is quadratic with respect to m, affording the A term in the Ginzburg–Landau expansion (Eq. 6.12). When $W > U$, the coefficient of the m^2 term is positive, resulting in no spin polarization ($m = 0$). When $W < U$, the m^2 term is negative, giving rise to spin polarization. Strictly speaking, the Stoner model predicts a transition from a paramagnetic metal to a ferromagnetic metal. An antiferromagnetic state becomes stable above a limit U_{AF} which is different from the Stoner condition, $U_F = 1/D = W$. In the quarter-filling, when the Hartree–Fock approximation is adopted, U_F is usually smaller than U_{AF}, namely the ferromagnetic state is more stable than the antiferromagnetic state [57]. Thus we shall take U_F to be the condition of the spin polarization.

Finite temperature. Charge orders at finite temperatures have been investigated under the static approximation. In order to examine the unevenly charge separated phases generally, we consider a situation in which one of the n charge-rich sites is converted to a charge-poor site. Accordingly the repeating unit $2n$ consists of $n - 1$ charge-rich sites and $n + 1$ charge-poor sites. Each of the $N(n - 1)/n$ charge-rich sites has $(1+s)/2$ charges, and each of the $N(n + 1)/n$ charge-poor sites has $1/2 - [(n - 1)/(n + 1)s/2]$ charges, where s is the degree of charge transfer; $s = 0$ represents the uniform distribution, and $s = 1$ is the complete charge separation like D^+D^0. According to the standard theory of mixed crystals [58], the entropy S of such a state is given by,

$$S = -\left(1 - \frac{1}{n}\right)\left[\left(\frac{1+s}{2}\right)\ln\left(\frac{1+s}{2}\right) + \left(\frac{1-s}{2}\right)\ln\left(\frac{1-s}{2}\right)\right]$$
$$- \left(1 + \frac{1}{n}\right)\left[\left(\frac{1}{2} - \left(\frac{n-1}{n+1}\right)\left(\frac{s}{2}\right)\right)\ln\left(\frac{1}{2} - \left(\frac{n-1}{n+1}\right)\left(\frac{s}{2}\right)\right)\right. \quad (6.20)$$
$$\left. + \left(\frac{1}{2} + \left(\frac{n-1}{n+1}\right)\left(\frac{s}{2}\right)\right)\ln\left(\frac{1}{2} + \left(\frac{n-1}{n+1}\right)\left(\frac{s}{2}\right)\right)\right].$$

This equation affords the entropy of all the charge order phases listed in Table 6.1. In particular, the evenly ordered 1:1 phase corresponds to infinite n and $S = 0$. The uniform phase ($s = 0$) gives $S = 2k_B \ln 2$. The degree of charge transfer s is determined by minimizing the free energy $F = E - TS$. The internal energy E at the ground state is given by $E = E_U + (E_i - E_U)s^2$, where E_U and E_i are respectively the ground-state energies of the uniform phase and the charge order state listed in Table 6.1. This gives the temperature dependence of s as shown in Fig. 6.24(c). Finally the entropy S is obtained from Eq. 6.20, and the temperature dependence of the free energy F is calculated as shown in Fig. 6.24, where Figs. 6.24(a) and 6.24(b) respectively correspond to points E and F in Fig. 6.21.

The phase with the lowest free energy is the stable phase. In Fig. 6.24(b), the diagonal phase is most stable, and merges to the uniform line $F = E_U - 2k_B \ln 2$ at

$$T_{c2} = \left(\frac{E_U - E}{k_B}\right)\left(\frac{n+1}{n-1}\right) \quad (6.21)$$

where the order parameter s drops to zero. This transition is second order.

At point E (Fig. 6.24(a)), the 3-fold phase is most stable over a wide temperature range, but the horizontal phase becomes the most stable below $T_{c1} = 0.02$ eV $= 200$ K. The transition from the 3-fold phase to the horizontal phase is first order because these two phases have different order parameters. This transition resembles the actual 190 K transition of the Rb salt, though the high temperature phase is not identified with the 3-fold phase.

For the evenly charge-ordered patterns such as the D^+D^0 stripes, around $T = 0$ K the entropy S is zero from Eq. 6.20, $s = 1$ (Fig. 6.24(c)), and the free energy starts horizontally (Figs. 6.24(a) and (b)). By contrast, the unevenly separated patterns like the 3-fold phase $D^+D_3^{1/3+}$ have finite entropy even at $T = 0$ K, and the free energy has a gradient. This is the reason that the uneven phases become more stable than the stripe phases with increasing temperature.

Finite-temperature phase diagrams are shown in Fig. 6.25. Figure 6.25(a) shows a phase diagram along the horizontal line ($V' = 0$) in Fig. 6.21. At $T = 0$ K, the ground-state changes from the uniform phase to the 3-fold phase and to the horizontal phase.

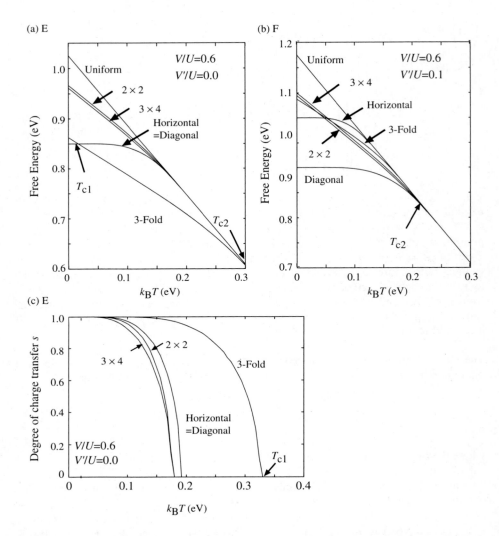

Fig. 6.24. Finite-temperature free energies of various charge-ordered patterns (a) at point E and (b) at point F in Fig. 6.21. (c) Temperature dependence of the order parameter s at point E.

With increasing temperature, the uniform phase becomes relatively stable because the uniform state has the largest entropy.

At point E, which has been shown in Fig. 6.24(a), a first-order transition from the horizontal phase to the 3-fold phase takes place.

Figure 6.25(b) is along the vertical line D to G in Fig. 6.21. Around point E and in the slightly negative V' region, the horizontal phase is stable at relatively low temperatures, and the 3-fold phase becomes stable at high temperatures.

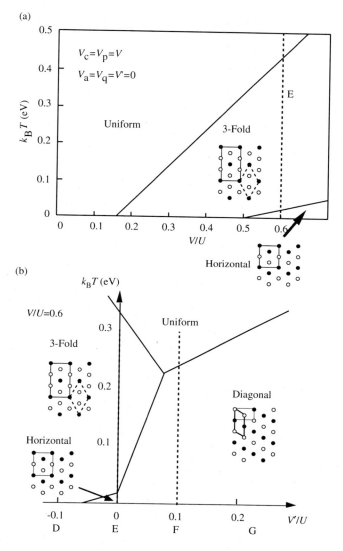

Fig. 6.25. Finite-temperature phase diagrams (a) along the horizontal axis, and (b) along the D to G line in Fig. 6.21. In (a), the horizontal phase is degenerated with the diagonal phase.

On the basis of the calculated V_c and V_p values obtained in Fig. 6.18, the free energy is compared as in Fig. 6.24; Fig. 6.26 is the resulting finite-temperature phase diagram plotted with respect to θ. The border between the horizontal and the 3-fold phases seems to reproduce the universal phase diagram. The transition from the 3-fold phase to the uniform phase is expected to occur at a very high temperature ($T > 1500$ K).

The free energy of the 3×4 phase (Fig. 6.12(h)) is plotted in Figs. 6.24(a) and (b). Since the 3×4 phase is a hybrid of the 3-fold phase and the 2×2 phase, the ground-state

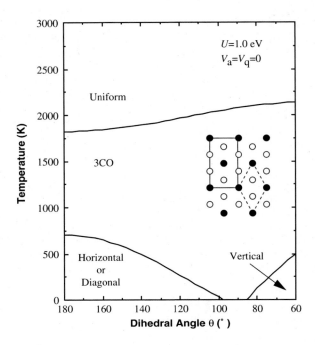

Fig. 6.26. Finite-temperature phase diagram plotted against θ without considering V'. The V_c and V_p values are taken from Fig. 6.17, and rescaled to $U = 1.0$ eV.

energy is close to these two. The charge distribution of the 3×4 phase $D^+D_3^{1/3+}$ is the same as the 2×2 phase, so that the entropy (the gradient of the free energy) is the same as that of the 2×2 phase. The 3×4 phase does not become the most stable one even at finite temperatures.

However, in Fig. 6.24(b), the 3-fold, the 2×2, and the 3×4 phases have very close free energies. If the lowest-lying diagonal phase is destabilized for some reason (probably by the effect of t), a transition from the low-temperature horizontal phase to the mixture of these three phases may take place. In this context, it is noteworthy that a Devil's staircase of many subphases has been found in the triangular Ising model [59]. Such subphases will appear around the 'tricritical' point of Fig. 6.25(b) among the diagonal, 3-fold, and uniform phases. Although the present model (Fig. 6.24(b)) does not prove the stability of such subphases, it is probable that the inclusion of t stabilizes these subphases.

The stability of more complicated patterns, which are constructed by changing some charge-rich sites of stripe phases to charge-poor sites, has been examined. The 2×2 phase is the simplest example of such 'defected' stripes, where half of the shaded sites of the diagonal phase (Figs. 6.12(d)) are converted to the open sites (Fig. 6.12(f)). The free energy of these defected patterns is located between those of the mother patterns, and does not become the most stable one.

In the uneven phases, for example $D^+D_3^{1/3+}$, the charge-poor sites $D^{1/3+}$ have fractional charge distribution, and this seems to be consistent with the metallic or flat temperature dependence of the resistivity. Flat resistivity has been universally observed in organic conductors. A famous example is α-(BEDT-TTF)$_2$I$_3$, in which broadening of the NMR line shape has been observed in the metallic state [60]. It is likely that some kind of charge order is inherent in these flat-resistivity compounds. The same is true for many one-dimensional compounds such as (TMTTF)$_2$PF$_6$ [36,37], (DMDCNQI)$_2$Li$_{1-x}$Cu$_x$ [61], and (DIDCNQI)$_2$Ag [34] (TMTTF: tetramethyltetrathiafulvalene, and DMDC-NQI:dimethyldicyanoquinonediimine).

6.5 Discussion

The θ-phase is the simplest molecular arrangement of the BEDT-TTF charge-transfer salts, and this is the only arrangement that realizes uniform two-dimensional network. In the region of the dihedral angle in the usual BEDT-TTF compounds, the diagonal transfer t_p is much larger than t_c (Fig. 6.6(a)). Even if $t_c = 0$, the remaining t_p forms a square lattice and two-dimensional band structure as shown in Fig. 6.6(c). Such an ideal two-dimensional Fermi surface has been verified by observation of the Shubnikov-de Haas oscillations [62] and angle-dependent magnetoresistance oscillations in θ–(BEDT-TTF)$_2$I$_3$ [63]. The off-site Coulomb repulsion is, however, most important in the vertical direction V_c rather than the diagonal direction V_p (Fig. 6.16), and this is the reason that the charge of the Rb salt below 190 K orders not in the vertical direction but in the horizontal direction (Fig. 6.12). Horizontal charge order and diagonal charge order are degenerated up to the nearest neighbor V. The former is associated with large transfer, similar to the $2k_F$ transition in one dimension, and the large t is the origin of the horizontal charge order. This is also related to the spin Peierls state in the Rb salt observed below 20 K (Fig. 6.9(a)). When we consider the second nearest V', the horizontal phase appears only in the region of negative V' (Fig. 6.21). Although the calculated second nearest V' is relatively large (Fig. 6.18), the actual system behaves as if V' is negative. This is ascribed to over-screening on account of large t. The horizontal charge order in the Rb salt is in consistent with many experimental results such as X-ray diffraction (Fig. 6.9(c)) [20,41], optical reflectance (Fig. 6.14) [42], Raman spectra (Fig. 6.15(a)) [43], and ^{13}C-NMR (Fig. 6.11) [38,39].

In contrast with the well-understood horizontal charge order, the high-temperature state of the Rb salt is still mysterious. The existence of some kind of charge order or fluctuation of charge order is certain in view of the observations of X-ray diffraction (Fig. 6.17) [44] and ^{13}C-NMR (Fig. 6.11) [45]. The peak structure of the optical conductivity (Fig. 6.14(b)) is obviously different from the simple Drude edge [64], and suggests some kind of charge order. The theoretical consideration indicates non-stripe charge order. The θ-phase makes an approximately triangular lattice with respect to V, and the frustration in the triangular lattice is the essential origin of the non-stripe phase (Fig. 6.19). Destruction of the stripe phase due to triangular frustration has also been demonstrated by an exact diagonalization calculation as [65], but only a metallic phase instead of a non-stripe phase has been obtained. Contrary to the 1:1 charge-rich and poor sites in the stripe states, the non-stripe phases have uneven compositions such as 2:1 and 3:1, giving rise to larger entropy. Accordingly, if the zero-temperature energy of the non-stripe phase is located slightly above the stripe phase, the non-stripe phase becomes more stable as shown in Fig. 6.24(a). This accounts for the first-order transition of the Rb salt at 190 K. The finite-temperature phase diagram in Fig. 6.26 reproduces the universal phase diagram of the θ-phase (Fig. 6.5). The phase diagrams plotted with respect to U (Figs. 6.19 and 6.20) closely resemble the finite-temperature phase diagram in Fig. 6.26 because large U prefers more uniformly distributed states with large entropy. However, there is a large remaining issue; the only non-stripe pattern that is verified to be stable by calculation is the 3-fold pattern, but the observed charge order has three- or four-fold periodicity not only along the c axis but also along the a axis [44]. Another remaining issue is that the NMR broadening is most remarkable between 200 K and 250 K (Fig. 6.11), but the room temperature spectrum contains sharp peaks. It seems that the non-stripe phase gradually transforms to a normal metal phase without transition. The same is true for the Cs salt, in which the broadening disappears above 170 K [50]. This disagrees with the static limit calculations shown in Figs. 6.20 and 6.26, where the non-stripe phase continues to several thousand K, and a definite second-order transition takes place at this temperature. This discrepancy may be due to the influence of t.

The most mysterious finding is the 20 K transition in the Cs phase. This transition is very gradual, and obviously very different from the abrupt first-order transition of the Rb salt (Fig. 6.4), although the universal phase diagram proposes a universal boundary between the metallic and the insulating phases. The 20 K transition seems to be not a true phase transition but a crossover. X-ray diffraction [49], optical reflectance (Fig. 6.14) [42], and ^{13}C-NMR measurements (Fig. 6.17) [50] suggest that some kind of charge order develops even in the "metallic" phase. Since X-ray diffraction detects both the two-fold (q_2) and three-fold (q_1) modulations, two patterns of charge order coexist in this state [49].

The nonlinear conduction in this Cs salt reminds us of the nonlinear conduction observed in low-dimensional materials [66]. Such nonlinear conduction is based on collective motion of charge or spin density waves. That of the Cs salt is, however, observed even in the direction perpendicular to the conducting layer, and is most prominent in this direction (Fig. 6.1(a)). Thus we cannot simply attribute it to the collective motion of the charge order.

Nonlinear conduction has been realized in various semiconductor devices such as thyristor and transferred electron devices [67]. A semiconductor thyristor consists of P–N–P–N junctions, and when the forward bias exceeds a certain breakover voltage, the device goes into an on state, and conducts a large current. This device is understood as two connected PNP and NPN transistors. On the other hand, a transferred electron device utilizes two close energy levels. Since the ground state is more conducting than the excited state, when the bias exceeds a certain limit, a major part of the charge carriers are excited to the upper state, and the resistance increases significantly.

The characteristics of the Cs salt are phenomenologically interpreted in terms of such a two-level model. Since the low bias state is more resistive than the high bias state, the ground state is more resistive, and the excited state is more conducting. This is reasonable if we assume that the ground state is the horizontal charge order and the excited state

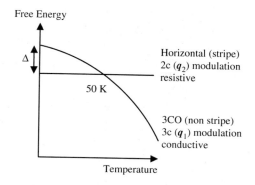

Fig. 6.27. Schematic energy diagram of the Cs salt.

is the non-stripe state (Fig. 6.27). By analogy with Fig. 6.24(a), the non-stripe state is assumed to fall, and to become the most stable state above 50 K. This is consistent with the X-ray result that the intensity of the $2c$ (q_2) modulation drops steeply above 50 K, but the three-fold (q_1) modulation remains unchanged [49]. Since the entropy of the 3-fold state is $0.75k_B$ from Eq. 6.20 at $T = 0$ K, the energy difference of these two states Δ is approximately 0.75×50 K $= 37.5$ K. When a charge carrier is excited up to this energy, this carrier enters the high conducting state. This brings about a relation, $\Delta = eE_T l_F$, where E_T is the threshold electrical field of the nonlinear conduction, l_F is the mean free path of the charge carrier, and e is the electron charge. The observed value $E_T = 300$ V/cm leads to $l_F = 0.1$ μm. The conductivity value 20 S/cm in the conduction layer together with the carrier density 1.25×10^{21} cm^{-3} gives $\mu = 0.1$ cm^2/Vs and $\tau = 5 \times 10^{-9}$ s from $\sigma = ne\mu$, and $\mu = e\tau/m$. If the band width W is assumed to be 1 eV, the corresponding Fermi velocity is $v_F = W/m(\pi/a) = 34$ m/s, and this leads to $l_F = v_F \tau = 0.2$ μm, in excellent agreement with the estimation from the threshold field.

However, the perpendicular conductivity affords a much shorter mean free path. The non-stripe modulation (q_1) appears below 120 K in the X-ray diffraction pattern [49] and below 170 K in the NMR spectrum [50], and the charge order is not robust at 50 K. This is probably the reason the 50 K crossover from the non-stripe order to the horizontal order is not a first-order transition. However, since this was originally a first-order transition, there should be domains of the non-stripe and the horizontal phases. The scale of the above mean free path, 0.1 μm may be related to the domain size of such charge inhomogeneity. A similar size of inhomogeneity has been found in the boundary of the Mott insulating phase and the metallic phase in the κ-phase organic sucperconductors, which has been investigated by microscopic infrared measurements [68].

6.6 Summary

We have reviewed giant nonlinear conduction and its use as an organic thyristor, together with a variety of experimental evidence of charge order and the theoretical background to the θ-phase charge-transfer salts. Since this finding is quite new, we cannot discuss the entire scope of this phenomenon. There are many issues remaining about the mechanism of the 20 K transition of the Cs salt as well as the high temperature state of the Rb salt. The mechanism of giant nonlinear conduction must be examined more detail in the future.

Practical applications of the organic thyristor are not certain in the present stage. However, it constitutes a new electronic device that does not use PN junctions or the field effect, and may open up a new direction in organic electronics.

Acknowledgments

This work was partly supported by the 21st century COE program and a Grant-in-Aid for Scientific Research on Priority Areas of Molecular Conductors (No. 15073211) from the Ministry of Education, Culture, Sports, Science and Technology, Japan. The author is grateful to I. Terasaki (Waseda University) for the results of the organic thyristor and many valuable discussions, and to H. Mori (Institute for Solid State Physics, the University of Tokyo) for the continuous collaboration on the θ-phase. The author thanks Y. Nogami (Okayama University), M. Watanabe (Tohoku University), and K. Yakushi (Institute for Molecular Science) for providing their results prior to publication.

References

[1] S.A. Jenekhe, Ed., Special Issue on Organic Electronics, Chem. Mater. **16** (2004) 4381.

[2] R.H. Friend, R.W. Gymer, A.B. Holmes, J.H. Burroughes, R.N. Marks, C. Taliani, D.D.C. Bradley, D.A. Dos Santos, J.L. Brédas, M. Lögdlund, and W.R. Salaneck, Nature **397** (1999) 121; L.J. Rotheberg and A.J. Lovinger, J. Mater. Res. 11 (1996) 3174; A.J. Heeger, Solid State Commun. **107** (1998) 673; J.R. Sheats, H. Antoniadis, M. Hueschen, W. Leonard, J. Miller, R. Moon, D. Roitman, A. Stocking, Science **273** (1996) 884.

[3] C.D. Dimitrakopoulos and P.R.L. Malenfant, Adv. Mater. **14** (2002) 99; G. Horowitz, Adv. Mater. **10** (1998) 365; H.E. Kats, Z. Bao and S.L. Gilat, Acc. Chem. Res. **34** (2001) 359; C.R. Kagan and P. Andry, Ed., Thin-Film Transistors; Marcel Dekker: New York, 2003.

[4] P. Batail, Ed., Thematic Issue on Molecular Conductors, Chem. Rev. **104** (2004) 4887; T. Ishiguro, K. Yamaji and G. Saito, Organic Superconductors, 2nd Ed. Springer: Berlin, 1998; J.M. Williams, J.R. Ferraro, R.J. Thorn, K.D. Carlson, U. Geiser, H.H. Wang, A.M. Kini, and M.-H. Whangbo, Organic Superconductors, Prentice-Hall: New Jersey 1992; J. Yamada and T. Sugimoto, TTF Chemistry, Fundamentals and Applications of Tetrathiafulvalenes, Kodansha-Springer: Tokyo, 2004.

[5] J. Wosnitza, Fermi Surfaces of Low-Dimensional Organic Metals and Superconductors, Springer: Berlin, 1996; J. Wosnitza, Int. J. Mod. Phys. B **7** (1993) 2707.

[6] K. Inagaki, I. Terasaki, H. Mori and T. Mori, J. Phys. Soc. Jpn. **73** (2004) 3364.

[7] F. Sawano, I. Terasaki, H. Mori, T. Mori, M. Watanabe, N. Ikeda, Y. Nogami and Y. Noda, Nature **437** (2005) 522.

[8] R. Kumai, Y. Okamoto and Y. Tokura, Science **284** (1999) 1645; J. Burgy, M. Mayr, V. Martin-Mayor, A. Moreo and E. Dagotto, Phys. Rev. Lett. **87** (2001) 277202.

[9] G.R. Desiraju and A. Gavezzotti, Acta Crystallogr. B **45** (1989) 473.

[10] D. Jérome, A. Mazaud, M. Riboult and K. Bechgaard, J. Phys. **41** (1980) L95.

[11] T. Mori, A. Kobayashi, Y. Sasaki, H. Kobayashi, G. Saito and H. Inokuchi, Chem. Lett., 1984, 957; T. Mori, Bull. Chem. Soc. Jpn. **71** (1998) 2509.

[12] R.P. Shivaeba, V.F. Kaminskii and E.B. Yagubskii, Mol. Cryst. Liq. Cryst. **119** (1985) 361.

[13] R.P. Shivaeba, R.M. Lobkovskaya, V.F. Kaminskii, S.V. Lindeman and E.B. Yagubskii, Soc. Phys. Crystallogr. **31** (1986) 546.

[14] R.P. Shivaeba, R.M. Lobkovskaya, E.B. Yagubskii and E.E. Kortyuchenko, Soc. Phys. Crystallogr. **31** (1986) 267.

[15] R.P. Shivaeba, R.M. Lobkovskaya, E.B. Yagubskii and E.E. Kortyuchenko, Soc. Phys. Crystallogr. **31** (1986) 657.

[16] R.P. Shivaeba, R.M. Lobkovskaya, E.B. Yagubskii and E.E. Laukhina, Soc. Phys. Crystallogr. **31** (1986) 530.

[17] H. Kobayashi, R. Kato, A. Kobayashi, Y. Nishio, K. Kajita and W. Sasaki, Chem. Lett. **15** (1987) 789 and 833.

[18] R. Kato, H. Kobayashi, A. Kobayashi, S. Moriyama, Y. Nishio, K. Kajita and W. Sasaki, Chem. Lett. **16** (1987) 459 and 507.

[19] T. Mori, H. Mori and S. Tanaka, Bull. Chem. Soc. Jpn. **72** (1999) 179.

[20] H. Mori, S. Tanaka, T. Mori and Y. Maruyama, Bull. Chem. Soc. Jpn. **68** (1995) 1136.

[21] H. Mori, S. Tanaka and T. Mori, Mol. Cryst. Liq. Cryst. **284** (1996) 15.

[22] H. Mori, S. Tanaka and T. Mori, Phys. Rev. B **57** (1998) 12023; H. Mori, T. Okano, S. Tanaka, M. Tamura, Y. Nishio, K. Kajita and T. Mori, J. Phys. Soc. Jpn. **69** (2000) 12023.

[23] S. Kagoshima and R. Kondo, Chem. Rev. **104** (2004) 5593.

[24] R. Kondo, S. Kagoshima, M. Chusho, H. Hoshino, T. Mori, H. Mori and S. Tanaka, Curr. Appl. Phys. **2** (2002) 483.

[25] K. Iwashita, H.M. Yamamoto, H. Yoshino, D. Graf, K. Storr, I. Rutel, J.S. Brooks, T. Takahashi and K. Murata, Synth. Met. **133–134** (2003) 153.

[26] H. Mori, S. Tanaka, T. Mori, A. Kobayashi and H. Kobayashi, Bull. Chem. Soc. Jpn. **71** (1998) 797.

[27] T. Nakamura, W. Minagawa, R. Kinami, Y. Konishi and T. Takahashi, Synth. Met. **103** (1999) 1898.

[28] A. Montorsi, Ed., The Hubbard Model, World Scientific, Singapore, 1992.

[29] N.F. Mott, Metal Insulator Transitions 2nd Ed., Taylor and Francis, London, 1990.

[30] K. Kanoda, Hyperfine Interact. **104** (1997) 235.

[31] H. Kino and H. Fukuyama, J. Phys. Soc. Jpn. **64** (1995) 1877; H. Kino and H. Fukuyama, J. Phys. Soc. Jpn. **64** (1995) 2726; H. Kino and H. Fukuyama, J. Phys. Soc. Jpn. **64** (1995) 4523; H. Kino and H. Fukuyama, J. Phys. Soc. Jpn. **65** (1996) 2158; H. Seo and

H. Fukuyama, J. Phys. Soc. Jpn. **66** (1997) 3352; H. Seo and H. Fukuyama, J. Phys. Soc. Jpn. **67** (1998) 3352.

[32] T. Mori, Bull. Chem. Soc. Jpn. **72** (1999) 2011.

[33] M. Imada, A. Fujimori and Y. Tokura, Rev. Mod. Phys. **70** (1998) 1039, Chap. IV E; V.J. Emery, S.A. Kivelson and J.M. Tranquada, Proc. Natl. Acad. Sci USA. **96** (1999) 8814.

[34] K. Hiraki and K. Kanoda, Phys. Rev. B **54**, 17276 (1996); K. Hiraki and K. Kanoda, Phys. Rev. Lett. **80** (1998) 4737.

[35] T. Yamamoto, H. Tajima, J. Yamaura, S. Aonuma and R. Kato, J. Phys. Soc. Jpn. **68** (1999) 4737; J. Dong, K. Yakushi, K. Takimiya and T. Otsubo, J. Phys. Soc. Jpn. **67** (1998) 971.

[36] F. Nad, P. Monceau, C. Carcel and J.M. Fabre, Phys. Rev. B **62** (2000) 1753; P. Monceau, F. Nad and S. Brazovskii, Phys. Rev. Lett. **86** (2001) 4080; for a review, see D. Jerome, Chem. Rev. **104** (2004) 5565.

[37] D.S. Chow, F. Zamborszky, B. Alavi, D.J. Tantillo, A. Baur, C.A. Merlic and S.E. Brown, Phys. Rev. Lett. **85** (2000) 5565.

[38] K. Miyagawa, A. Kawamoto and K. Kanoda, Phys. Rev. B **62** (2000) R7679.

[39] R. Chiba, H. Yamamoto, K. Hiraki, T. Takahashi and T. Nakamura,.Phys. Chem. Solids **62** (2001) 389; R. Chiba, H. Yamamoto, K. Hiraki, T. Takahashi and T. Nakamura: Synth. Met. **120** (2001) 919.

[40] H Seo, J. Phys. Soc. Jpn. **69** (2000) 805.

[41] M. Watanabe, Y. Noda, Y. Nogami and H. Mori, J. Phys. Soc. Jpn. **73** (2004) 116; M. Watanabe, Y. Noda, Y. Nogami and H. Mori, J. Phys. Soc. Jpn. **74** (2005) 2011; M. Watanabe, Y. Noda, Y. Nogami, H. Mori and S. Tanaka, Synth. Met. **133–134** (2003) 283.

[42] H. Tajima, S. Kyoden, H. Mori and S. Tanaka, Phys. Rev. B **62** (2000) 9378.

[43] K. Yamamoto, K. Yakushi, K. Miyagawa, K. Kanoda and A. Kawamoto, Phys. Rev. B **65** (2002) 085110; K. Suzuki, K. Yamamoto and K. Yakushi, Phys. Rev. B **69** (2004) 085114.

[44] M. Watanabe, Y. Noda, Y. Nogami and H. Mori: Synth. Met. **135–136** (2003) 665.

[45] R. Chiba, K. Hiraki, T. Takahashi, H.M. Yamamoto and T. Nakamura, Phys. Rev. Lett. **93** (2004) 216405.

[46] K. Inagaki, I. Terasaki and H. Mori, Physica B **329–333** (2003) 1162.

[47] T. Nakamura, W. Minagawa, R. Kinami and T. Takahashi, J. Phys. Soc. Jpn. **69** (2000) 504.

[48] K. Suzuki, K. Yamamoto, K. Yakushi and A. Kawamoto, J. Phys. Soc. Jpn. **74** (2005) 2631.

[49] M. Watanabe, Y. Nogami, K. Oshima, H. Mori and S. Tanaka, J. Phys. Soc. Jpn. **69** (2000) 2654; Y. Nogami, J.P. Pouget, M. Watanabe, K. Oshima, H. Mori, S. Tanaka and T. Mori, Synth. Met. **103** (1999) 1911; Y. Nogami, Private Communication.

[50] R. Chiba, K. Hiraki, T. Takahashi, H.M. Yamamoto and T. Nakamura, Synth. Met. **133–134** (2003) 305.

[51] T. Mori, Bull. Chem. Soc. Jpn. **73** (2000) 2243.

[52] T. Mori, J. Phys. Soc. Jpn. **72** (2003) 1469.

[53] T. Mori, F. Sakai, G. Saito and H. Inokuchi, Chem. Lett. **16** (1987) 927; H. Mori, M. Kamiya, M. Haemori, H. Suzuki, S. Tanaka, Y. Nishio, K. Kajita and H. Moriyama, J. Am. Chem. Soc. **124** (2002) 1251.

[54] B.D. Metcalf, Phys. Lett. **46A** (1974) 325; Y. Tanaka and N. Uryu, J. Phys. Soc. Jpn. **39** (1975) 825.

[55] R.T. Clay, S. Mazumdar and D.K. Campbell: J. Phys. Soc. Jpn. **71** (2002) 1816.

[56] E.C. Stoner, J. Phys. Radium, **12** (1931) 372.

[57] D.R. Penn, Phys. Rev. **142, 350** (1966); J.E. Hirsch, Phys. Rev. B **31** (1985) 4403.

[58] For example, R. Kubo, Statistical Mechanics, Kyoritsu, Tokyo (1952) Chap. 6D.

[59] K. Nakanishi, J. Phys. Soc. Jpn. **52** (1983) 2449; K. Nakanishi and H. Shiba, J. Phys. Soc. Jpn. **51** (1982) 2089.

[60] Y. Takano, K. Hiraki, H.M. Yamamoto, T. Nakamura and T. Takahashi: J. Phys. Chem. Solids **62** (2001) 393; Y. Takano, K. Hiraki, H.M. Yamamoto, T. Nakamura and T. Takahashi: Synth. Metals **120** (2001) 1081.

[61] T. Yamamoto, H. Tajima, J. Yamaura, R. Kato, M. Uruichi and K. Yakushi, J. Phys. Soc. Jpn. **71** (2002) 1956.

[62] M. Tokumoto, A.G. Swanson, J.S. Brooks, M. Tamura, H. Tajima and H. Kuroda, Solid State Commun. **75** (1994) 439; M. Tokumoto, A.G. Swanson, J.S. Brooks, C.C. Agosta, S.T. Hannahs, N. Kinoshita, H. Anzai, M. Tamura, H. Tajima and H. Kuroda, A. Ugawa and K. Yakushi, Physica B **164** (1993) 508.

[63] K. Kajita, Y. Nishio, T. Takahashi, W. Sasaki, R. Kato, H. Kobayashi, A. Kobayashi, Solid State Commun. **70** (1989) 1189.

[64] H. Kuroda, K. Yakushi, H. Tajima, A. Ugawa, M. Tamura, Y. Okawa, A. Kobayashi, R. Kato and H. Kobayashi, Synth. Met. **27** (1988) A491.

[65] J. Merino, H. Seo and M. Ogata, Phys. Rev. B **71** (2005) 125111.

[66] G. Grüner, Density Waves in Solids, Addison-Wesley, Massachusetts, 1994; G. Grüner, Rev. Mod. Phys. **60** (1988) 1129; Grüner and A. Zettl, Rev. Mod. Phys. **66** (1994) 1.

[67] S. Sze, Semiconductor Devices: Physics and Technology, Wiley: New York, 1985.

[68] T. Sasaki, N. Yoneyama, N. Kobayashi, Y. Ikemoto and H. Limura, Phys. Rev. Lett. **92** (2004) 227001; T. Nishi, S. Kimura, T. Takahashi, T. Ito, H.J. Im, Y.S. Kwon, K. Miyagawa, H. Taniguchi, A. Kawamoto and K. Kanoda, Solid State Commun. **134** (2005) 189.

PART III

Nanostructure Design for New Functions

CHAPTER 7

Size Control of Nanostructures by Quantum Confinement

Hiroyuki Hirayama

Abstract

Conduction electrons are confined in nanostructures and quantized. The energy levels shift and the number of quantized states below the Fermi level increase with structure size. This causes size-dependent oscillation of electronic energy. As a consequence, electronic energy has local minimums at specific *magic* sizes. This phenomenon is applicable to the automatic control of the size of nanostructures.

Here, we review recent progress in the study of electron quantum confinement and its effect on the thickness control of ultrathin metal layers of nanometer height on semiconductor and metal substrate surfaces. As a prominent application of the quantum well states (QWSs) in the nanostructures, we mention the tailoring of chemical reactivity toward the synthesis of new catalysts at the nanostructure surfaces.

Keywords: **nanostructure, quantum confinement, surface.**

7.1 Introductory Remarks

7.1.1 Fermi wavelength: scale of nanostructures

With the recent progress in materials science, the size of materials has reached the nano- or subnanometer scale. In these regions, the structure size is comparable with the Fermi wavelength of the conduction electrons. For example, metal has typically a Fermi wave vector on the order of 10^8 cm^{-1} [1]. That is, the wavelength of the metal is approximately 0.6 nm. Semiconductors have smaller Fermi wave vectors on the order of 10^6 cm^{-1} [2]. Thus, their Fermi wavelengths are approximately 60 nm. In state-of-the-art nanotechnology, materials can be scaled down to these Fermi wavelengths.

In nanostructures of Fermi wavelength size, the quantum confinement of conduction electrons becomes significant. A small change in structure size has a marked influence on the quantized electron states. This suggests the possibility of the size control of nanomaterials via quantum confinement of the conduction electrons.

7.1.2 Manifestation of quantum well states by rising over thermal disturbance

As stated above, quantum well states (QWSs) are a remarkable feature of nanoscale materials. However, it is necessary to manifest the quantized feature that the energy splitting between QWSs is sufficiently large to overcome thermal disturbance on the order of $k_B T$ in the electron distribution.

Here, we estimate the effect of thermal disturbance on QWSs in a simple one-dimensional quantum well with infinite potential barriers. For simplicity, a conduction electron is assumed to have a static electron mass (m) in a well, and move freely between the confinement potential barriers (Fig. 7.1). As described in an elementary textbook of quantum physics [3], electrons are confined and quantized to have discrete energy levels of

$$E_n = \frac{\hbar^2}{2m}\left(\frac{n\pi}{d}\right)^2,$$

where $\hbar$ denotes the reduced Planck's constant, d the width of the well, and n the quantum number. The quantized condition is given by '$n \times$ (half wavelength of the conduction electron) equals well width' (see the illustration). The energy separation between QWSs is on the order of

$$\Delta = \frac{\hbar^2}{2m}\left(\frac{\pi}{d}\right)^2.$$

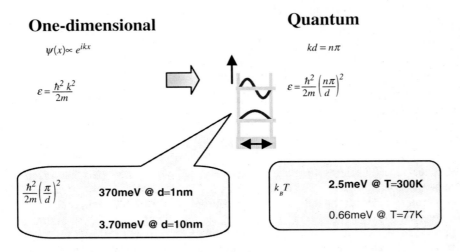

Fig. 7.1. Relationship between size and quantized state of nanometer-scale materials.

For $d = 1$, 10, and 100 nm, Δ is 370, 3.70, and 0.037 meV, respectively. On the other hand, $k_B T$ is 2.5, 0.66, and 0.034 meV at $T = 300$, 77, and 4 K, respectively. Thus, the quantized nature becomes significant even at room temperature ($T = 300$ K) in a structure of 1 nm size. Meanwhile, no quantized feature is observable even at liquid He temperature ($T = 4$ K) in a structure of 100 nm size.

The above analysis demonstrates that the size of the structure is crucial in making the intrinsic quantized nature of nanomaterials overt. Even with the recent progress in nanolithographic techniques, the smallest size of the fabricated structure remains on the order of 0.1 μm, i.e., 100 nm. This causes the so-called artificial atoms, which are fabricated by conventional semiconductor device techniques, to reveal their quantum nature only on the milli Kelvin (mK) scale. Meanwhile, in the root of so-called bottom-up nanotechnology, a structure of 1 nm size or less can be easily fabricated. Therefore, the nanostructure by the bottom-up process is more realistic for realizing versatile nanotech devices which operate at room temperature. However, unfortunately, the control of the size and position of nanostructures is difficult in the bottom-up process. This is because the structure formation is conventionally governed by statistic nucleation processes in the bottom-up strategy. As a promising solution to this problem, size control via quantized electronic states is considered. In the following sections, we briefly review the progress of this new solution. First, we review the formation of QWSs in thin films of nanometer height on substrate surfaces. Then, some examples of the height control of flat-top nanofilms via the quantized electronic states are shown. Finally, we present the control of surface catalytic reactions via QWSs in nanoscale thin films as a promising example of technical applications.

7.2 Quantum Well States in Surface Nanostructures

In this section, we present some examples of electron quantization of nanoscale structures on solid surfaces. The theoretical background and experimental evidence of electron quantization along the surface-normal direction, by taking Ag thin films of nanoscale height on Si(111) substrates as a typical example are explained. The quantization along the surface-parallel direction is also described in small-sized islands and artificially built-up quantum corrals.

7.2.1 Quantum well states in thin films of nanoscale height: quantization of bulk electronic states along surface-normal direction

For many combinations of metals on semiconductor surfaces, metal films have been reported to grow in a layer-by-layer fashion. This results in the formation of flat-top metal films of nanoscale height. The electronic structures of the nanoscale-height films have been investigated conventionally by ultraviolet photoelectron spectroscopy (UPS) [4]. UPS studies have revealed that, in the very initial stage of growth, films have an electronic structure characteristic of their surface. However, the films obtain an intrinsic bulk-band structure after the growth of several monolayer (ML) thicknesses [5]. Then, in nanoscale thin films, the electrons of the bulk-band structure are confined between the film surface and interface, and are quantized along the surface-normal direction.

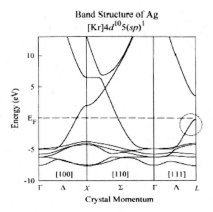

Fig. 7.2. Band dispersion of Ag crystal (after Chiang [5]).

The characteristic bulk-band dispersion gives a variety of quantizations, different from that of the simple particle-in-a-box model described in elementary textbooks of quantum physics. For example, consider the Ag films grown on Si(111) surfaces. On Si(111) 7 × 7 surfaces, {111}-oriented Ag films grow epitaxially [6]. Bulk band dispersion of Ag is shown in Fig. 7.2 [4]. Along the Γ–L direction (i.e., //{111} direction), the Ag sp-band conduction electrons have a downward parabolic band dispersion in the occupied state. The top of the downward dispersion is at the L-point just below the Fermi level (E_F). A band gap of approximately 4 eV separates the occupied state band from the empty state band. Because of the band gap, in the energy range of $E_F \pm 3$ eV, the QWSs appear only below the Fermi level.

To discuss the quantum confinement of the Ag sp-band electron along the {111} direction, the band dispersion should be correctly taken into calculation. As the simplest model, we approximate the band dispersion along Γ–L by the two-band model [7], in which the Γ–L band dispersion of Ag is regarded as a result of the two interacting plane waves e^{ikz} and $e^{i(k-G)z}$. Here, k is the wave vector, z is the position of the electron in the Ag film along {111}, and G is the reciprocal lattice vector along Γ–L. Referring to the energy at the bottom of the sp-band (V_0 at approximately −5 eV), these two waves degenerate at approximately an energy equal to E_F at the L-point. However, the crystal potential (V_G) lifts off the degeneracy and causes the band gap of a width of $2V_G$. Finally, the wave function of the sp-electrons with the Γ–L band dispersion is described as

$$\psi(z) \sim e^{ikz} + \frac{V_G}{E - V_0 - (\hbar^2/2m)(G/2)^2} e^{i(k-G)z},$$

where m is the static mass of the electrons, E the energy of the electrons, and $\hbar$ is the reduced Planck's constant.

The QWSs are solved as a one-dimensional confinement of this wave function between the potential wall at the surface and interface of the Ag film. For a given potential width (i.e., the thickness of the Ag film), this results in discrete energy levels.

a simple, free-electron-like dispersion starting from the Γ-point. The characteristic shape of the wave-function and the well-width dependence are due to the specific bulk-band dispersion of Ag along the Γ–L direction.

The formation of the QWSs in the Ag films was experimentally verified by UPS [10–14]. Figure 7.5 shows the UPS spectra of the Ag films grown on the Si(111) 7 × 7 surface at room temperature [10]. The spectra of the Ag films grown on the hydrogen-terminated Si(111) surfaces at 300°C [13], and of the Ag films deposited on the Si(111) 7 × 7 surfaces at liquid nitrogen temperature and subsequently annealed slowly to room temperature [11] are shown in Figs. 7.6 and 7.7, respectively.

In these three figures, QWS peaks were observed in the UPS spectra as peaks. The peaks in Fig. 7.5 appeared at the theoretically expected energy, and shifted toward the Fermi level (E_F) with increasing film thickness as theoretically predicted in Fig. 7.4. However, as indicated by triangles in Fig. 7.5, the QWSs had very weak peaks for the Ag films grown on the Si(111) 7 × 7 surface at room temperature. In contrast, the peaks of the QWSs were observed distinctly for the Ag films grown on the H-terminated Si(111) surfaces at 300°C (Fig. 7.6) and for the Ag films grown by low-temperature deposition followed by annealing to room temperature (Fig. 7.7). These differences are attributed to the local thickness distribution of the Ag films. Ion scattering [6] and scanning probe microscopy studies [8,9,15] indicated that the Ag films show distribution

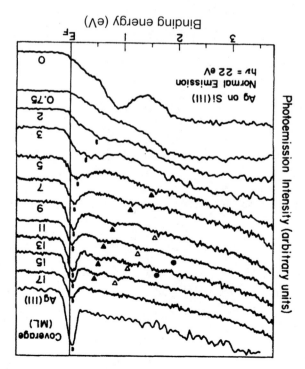

Fig. 7.5. UPS of Ag films grown on the Si(111)7 × 7 surfaces at room temperature (after Wachs et al. [10]).

Size Control of Nanostructures by Quantum Confinement 269

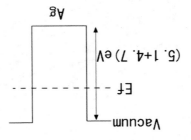

Fig. 7.3. A model of confinement potential in the Ag film.

Figure 7.3 shows the simplest model of the confinement potential; the Ag sp-band electrons are confined by the potential walls of the height of the sum of the sp-band Fermi energy ($V_0 = 5.1$ eV) and work function (4.7 eV) at the surface side of the Ag film [8,9]. For simplicity, the potential at the interface side is also assumed to have the same height as in the model in Fig. 7.3. In a more realistic model potential (shown in the inset of Fig. 7.4), we can solve the problem of the quantization of the sp-band electrons of above wave function, numerically. The numerically obtained QWS energy levels and wave functions are displayed in Fig. 7.4 [9]. In this figure, the numerically calculated energy levels of the QWSs are indicated by solid lines as functions of Ag film thickness. The experimentally observed QWS energies are indicated by solid circles. This simple calculation reproduces the experimental results well [10].

In the Ag band dispersion along the Γ–L line, the wave vector k is conventionally referred to L; i.e. $\tilde{k} = k - (G/2)$. The quantum number $n = 1,2,3$ is referred to the Bohr–Sommerfeld quantized condition for this $\tilde{k}$ [8]. The inset shows the wave function of the $n = 1$ (the upper curve), 2 (the middle curve), and 3 (the lower curve) QWSs. As shown in the inset, the wavefunction is a product of high-frequency modulation and a smooth envelope function due to the interference of the waves $e^{i\tilde{k}z}$ and $e^{i(\tilde{k}-G)z}$. Furthermore, note that each QWS increases in energy with Ag film thickness (i.e., quantum well width). This trend is just opposite to what is expected for the QWSs of the electrons with

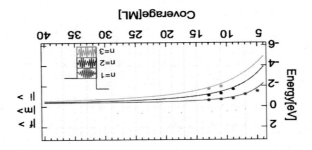

Fig. 7.4. Numerically calculated QWSs and their wavefunctions by two-band model. Reprinted with permission from M. Watai & Hirayama, Phys. Rev. B 72, 085435 (2005). Copyright (2005) by the American physics society [9] (after Watai and Hirayama [9]).

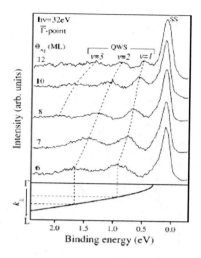

Fig. 7.6. UPS of Ag films grown on the H-terminated Si(111) surfaces at 300°C (after Arranz et al. [13]).

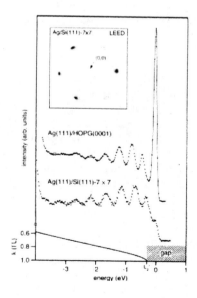

Fig. 7.7. UPS of Ag films grown by low temperature deposition and subsequent annealing to room temperature (after Neuhold and Horn [11]).

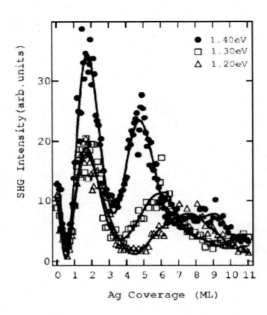

Fig. 7.8. Optical SHG intensity oscillation during the Ag growth on the Si(111) 7 × 7 surfaces at room temperature (after Hirayama et al. [15]).

of their local heights for the films grown on the 7 × 7 surfaces at room temperature. Meanwhile, the Ag film is very uniform in height in the latter two growth methods. The QWS peaks were broadened and weakened by the distribution of the local height in the growth on the 7 × 7 surface at room temperature, but not in the latter two growth methods.

In the QWSs in the Ag films, the distance between QWSs is on the order of 0.1 eV. It is sufficiently larger than the thermal energy $k_B T$. Thus, the thermal disturbance is not the cause of the weak peaks in the growth on the 7 × 7 surfaces at room temperature. In fact, the quantum confinement nature was observed clearly by optical second-harmonic generation (SHG) at room temperature. Figure 7.8 shows an example of the SHG intensity oscillation observed during the growth of the Ag films on the Si(111) 7 × 7 surfaces at room temperature [15]. SHG is a surface sensitive monitoring tool for growth on surfaces [16,17]. In the oscillation, except for the first peak due to the local plasmon resonance, the peaks shifted to the lower coverage side with increasing pump photon energy of the SHG signal. This shows that these peaks are due to the two-photon resonant transition from the QWSs of the Ag films below E_F to the Ag–Si interface state above E_F [15]. The clear peaks revealed the quantum confinement nature of the Ag electrons. The SHG study demonstrates that the QWSs survive even at room temperature in nanothick Ag films.

The QWSs along the surface-normal direction are due to the confinement of the bulk-band electrons between the film surface and interface. Here, the main problem is the origin and nature of the confinement potential at the interfaces. As the origin of the confinement

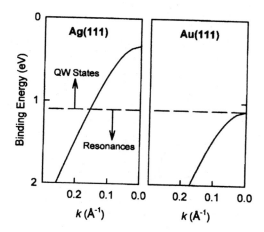

Fig. 7.9. Occupied state-band dispersion of Ag and Au along the Γ–L direction (after Chiang [5]).

potential, relative gap and symmetry gap are suggested. A simple example of the relative gap was reported for the Ag–Au(111) system [5]. Au has a band dispersion similar to that of Ag (Fig. 7.2) along the Γ–L direction. However, as illustrated in Fig. 7.9, the top of the occupied state band of Ag exists at −0.3 eV while that of Au is at −1.1 eV below E_F at L-point. Because of this discrepancy, the sp-band electrons of energy between −0.3 and −1.1 eV can feel the confinement potential at the interface in Ag films grown on Au(111) surfaces. However, in Au films on Ag(111) surfaces, all the sp-band electrons in the Au film resonate to the Ag band of the substrate, and feel no confinement potential at the interface. Actually, QWS peaks were observed in the Ag–Au(111) system in the energy range of −0.3 to −1.1 eV, while no QWS peaks appeared in the Au–Ag(111) system, as shown in Fig. 7.10 [5]. This type of interface confinement potential (due to the discrepancy between the film and substrate band energies) is called the relative gap. The other type of interface potential is the symmetry gap. This is caused by the mismatch of the symmetry of the wave functions between the film and the substrate. In the case of the Ag–Si(111) system, the interface potential is regarded to originate from both the relative and symmetry gap factors.

In the Ag films on the Si(111) surfaces, the QWSs appeared only in the occupied state, because of the characteristic band dispersions of Ag along the Γ–L direction. However, QWSs appear above E_F for the materials with no band-gap in the empty state. As an example, Fig. 7.11 shows the scanning tunneling spectroscopic results of the Yb(111) films grown on W(110) surfaces [18]. In the figure, the sample bias on the horizontal axis corresponds to the electron energy referred to E_F. The differential conductance is roughly proportional to the density of states of the electrons [19]. The peaks represent the QWSs in the empty state due to the confinement of the electrons in the (111)-oriented Yb films. As expected for the QWSs of the electrons with the conventional free-electron-type dispersion, the peaks shift downwards and the peak–peak intervals become shorter with increasing Yb film thickness.

274 Nanomaterials: From Research to Applications

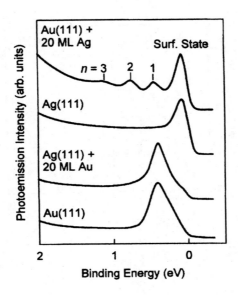

Fig. 7.10. UPS spectrum from Ag–Au(111), Ag(111), Au–Ag(111), and Au(111) surfaces (after Chiang [5]).

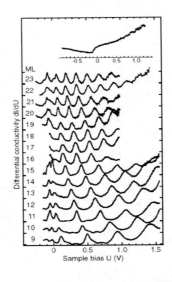

Fig. 7.11. STS (Scanning tunneling spectroscopy) spectrum of Yb(111) films grown on W(110) surfaces (after Wegner et al. [18]).

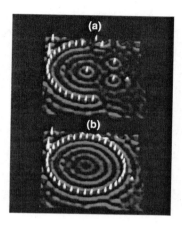

Fig. 7.12. Quantum corral construction (after Hla et al. [20]).

7.2.2 Quantum confinement of surface-state electrons along surface-parallel directions

Since the invention of STM, the study of confinement of the surface-state electrons has been advancing tremendously. STM allows us direct imaging of the standing waves of the confined electrons at surfaces. Furthermore, the STM tip enables us to manipulate individual atoms at surfaces. These advantages are utilized to construct man-designed nanostructures and observe the standing-wave patterns of the surface electrons confined along the surface-parallel direction in the nanostructures at the surfaces.

A typical example is shown in Fig. 7.12 [20]. The figure shows a three-dimensional STM image during construction (a) and after completion of a quantum corral with 36 Ag atoms on the Ag(111) surface (b). Most normal metals such as Ag have a Shockley-type surface state with free-electron-like dispersion [21]. At the surfaces, the surface-state-electrons move freely with an effective mass determined by the surface dispersion. However, the adatoms on the surface scatter these electrons. Near each adatom, the incoming and scattered surface electron waves interfere to reveal a concentric standing-wave pattern, as shown on the right-hand side of Fig. 7.12(a). By arranging these adatoms densely, we can construct a continuous potential barrier wall for the surface-state electrons. In Fig. 7.12(b), the concentric arrangement of 36 Ag atoms makes a cylindrical potential well with a diameter of 31.2 nm. Under the confinement potential $V(r)$ of circular symmetry, the quantum confinement of the surface-state electrons is given as a solution of the two-dimensional Schrodinger equation

$$\left\{ -\frac{\hbar^2}{2m^*} \left(\frac{\partial^2}{\partial r^2} + \frac{1}{r}\frac{\partial}{\partial r} + \frac{1}{r^2}\frac{\partial^2}{\partial \theta^2} \right) + V(r) \right\} \Psi(r,\theta) = E\Psi(r,\theta)$$

where m^* is the effective mass of the surface electrons, r is the distance from the center of the potential well, and θ is the angular coordinate [7]. Although the confinement by $V(r)$ is usually not so perfect in a real system, the dominant feature of the quantum confinement is obtained by approximating $V(r)$ as a hard-wall confinement potential. In this approximation, the Schrodinger equation is easily solved, to give the eigenstate and function of

$$E_{n,l} = \frac{\hbar^2 j_{l,n}^2}{2m^* a^2}$$

and

$$\Psi_{n,l}(r,\theta) \propto J_l\left(\frac{j_{l,n}r}{a}\right) \exp(il\theta),$$

where n and l are the radial and azimuthal quantum numbers, respectively, a the radius of the potential wall, and J_l the lth Bessel function. Since the Bessel function has a radial distribution as

$$J_l(kr) \sim \sqrt{\frac{2}{\pi kr}} \cos\left(kr - \frac{1}{2}l\pi - \frac{1}{4}\pi\right),$$

the quantized surface state electrons show concentric standing-wave patterns as shown in Fig. 7.12(b). More detailed analysis of the quantized surface electrons in the cylindrical potential well has been reported in [22].

As well as the adatoms, step edges work as a potential wall to confine the two-dimensional surface-state electrons. Figure 7.13 shows an STM image of nanosize Ag islands formed

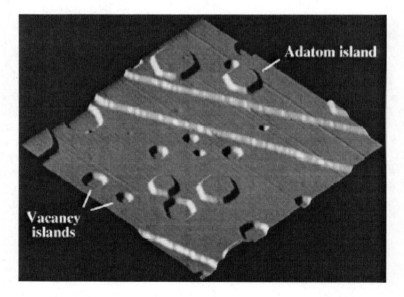

Fig. 7.13. STM image of Ag adatom islands on Ag(111). $T = 50$ K, $V = 0.4$ V, $I = 1$ nA. 160×160 nm^2 (after Li et al. [23]).

Size Control of Nanostructures by Quantum Confinement 277

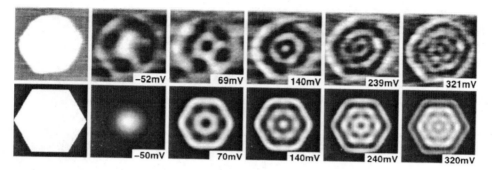

Fig. 7.14. Upper row: STM image of an hexagonal Ag island on Ag(111) (area 94 nm^2), and a series of dI/dV maps recorded at various bias voltages (at $T = 50$ K). Lower row: geometry of a hexagonal 2D confinement box, and the resulting local density of states (after Li et al. [23]).

on an Ag(111) surface [23]. In the study, the Ag surface was bombarded initially by energetic Ar$^+$ ions. This created atomic defects that served as nucleation centers of the Ag nanoislands in the subsequent deposition of Ag atoms on the surface. The Ag adatom islands (of monolayer height step) have the hexagonal shape characteristic of the crystal habit of the {111}-oriented epitaxial film. The electrons on the surfaces of the island feel the potential barriers at the peripheral step edges, and are confined into the islands. As shown in the upper row of Fig. 7.14, the standing-wave patterns of the confined surface electrons were clearly observed by STM-based dI/dV mapping (I and V are the tunneling current and bias voltage in the STM measurement. dI/dV mapping is achieved by detecting the variation in I with a periodic small modulation of V via conventional lock-in technique during STM observation), which represents the spatial distribution of the surface electron density of states (DOS) of the electrons at an energy equal to the bias voltage V more clearly than conventional STM [24]. The Shockley surface state has its bottom at -67 meV with reference to the Fermi level, and has a free-electron-like band dispersion with an effective mass of 0.42. Thus, the standing wave appears at bias voltages above approximately -50 meV. Since the wavelength of the surface electrons decreases with energy, the standing-wave pattern becomes more complicated with increasing STM bias voltage. As shown in the lower row of Fig. 7.14, the dI/dV map agrees well with the theoretically calculated distribution of the quantized surface electron wave function.

In the studies mentioned above, the sizes of the confinement structures are 30 nm in Fig. 7.12 and 10 nm in Fig. 7.14. Thus, as roughly estimated in the Introductory Remarks section, the low temperatures are not so critical from the viewpoint of suppressing the thermal disturbance of the quantized electrons, though the STM images were obtained at 6 K and 50 K in Figs. 7.12 and 7.14. However, with respect to stopping the thermal diffusion of adatoms to stabilize the quantum corral, a low temperature is inevitably necessary, as shown in Fig. 7.12. However, the islands, as shown in Fig. 7.14, are much more rigid and tolerant against thermal diffusion. Thus, the standing-wave pattern is expected to be observed even at room temperature on top of the nanoscale islands. A previous study has shown this to be true [25]. Figure 7.15 shows STM images of Cu

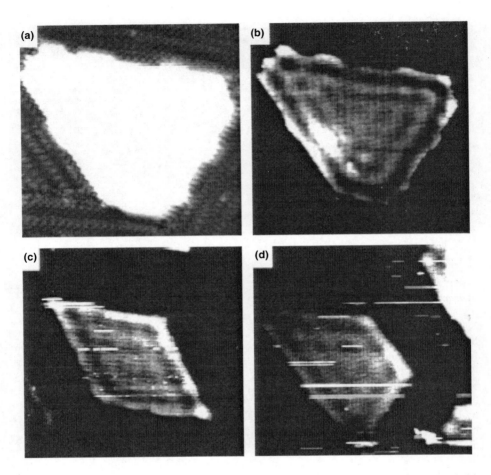

Fig. 7.15. Four STM images of a Cu(111) surface showing regions of clean surface bounded by N-islands. The size is 15 × 15 nm^2. $V = -0.01$ V, $I = 1.0$ nA (after Silva and Leibsle [25]).

islands formed on Cu(111) surfaces. The Cu islands were introduced to the surface by N ion bombardment which was followed by annealing at 500 K. The sample was then cooled to room temperature, and STM observation was carried out. The bombardment and subsequent annealing process produced Cu(111) islands with a size of several tens of nanometers at the surfaces. The islands showed the standing-wave patterns of confined and quantized surface electrons clearly at room temperature.

7.3 Size Control of Nanostructures via Quantum Confinement

7.3.1 Theoretical background

The quantization of the electrons causes the size-dependent change in the electronic energy of nanostructures. Thus, the nanostructure would have a specific *magic* size,

Size Control of Nanostructures by Quantum Confinement

at which the electronic energy could have a minimum. This is the basic idea behind the size control of the nanostructures via the electron quantum confinement. In this subsection, we take the surface-normal quantization of the conduction electrons in the thin films on substrates as an example, and explain the theoretical background in more detail. For simplicity, we regard the film on the substrate as a free-standing film of thickness L in the following discussion. The conduction electron has a free-electron-like dispersion. The effective mass of the electron (m) is assumed to be isotropic. The energy (E) of the electron is referred to as the bottom of the conduction band.

In conventional bulk materials, E relates to the wave vector of the electron (k) as,

$$E = \frac{\hbar^2}{2m} k^2.$$

electron, k occupies the volume under the Fermi sphere of radius k_F uniformly in the reciprocal space. However, in the thin films, the electrons are quantized along the surface-normal direction (z). This makes k discrete along the k_z direction as

$$k_z = \frac{n\pi}{L}.$$

Here, n denotes the quantum number. As a result, k occupies not all the volume in the Fermi sphere uniformly, but occupies only the disks corresponding to the quantum states below the Fermi level as illustrated in Fig. 7.16 [26]. Thus, in the thin films, the total energy of the conduction electrons (E_{el}) is given by

$$E_{el} = 2 \cdot \frac{A}{(2\pi)^2} \sum_{n=1}^{n_0} \left\{ \int_0^{\sqrt{k_F^2 - k_z^2}} \frac{\hbar^2}{2m} (k_{//}^2 + k_z^2) \cdot (2\pi k_{//} dk_{//}) \right\} = \frac{A\hbar^2}{8\pi m} \sum_{n=1}^{n_0} (k_F^4 - k_z^4),$$

where 2 originates from the spin, A is the surface area of the thin film, n_0 is the quantum number of the highly occupied quantum state, and $k_{//}$ is the wave vector in the k_x–k_y

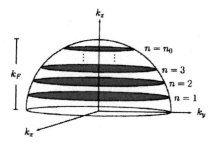

Fig. 7.16. Subband structures associated with the confinement of electrons to a quantum well along k_z direction (after Czoschke et al. [26]).

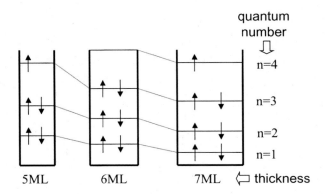

Fig. 7.17. Oscillatory behavior of the highest occupied QWS with the film thickness.

plane. Meanwhile, n_0 is determined by a constraint that the number of electrons in the quantum states below the Fermi level ($1 \leq n \leq n_0$),

$$N_{\text{el}} = 2 \cdot \frac{A \cdot L}{(2\pi)^2} \sum_{n=1}^{n_0} \pi(k_F^2 - k_z^2)$$

should be equal to the product of the bulk electron density and the volume of the thin films. As a consequence, the electronic energy of the thin film oscillates with film thickness. The reason for this is shown intuitively in Fig. 7.17. In the figure, the conduction electrons in the thin film are distributed in subbands associated with the vertical quantum states n ($1 \leq n \leq n_0$). The total number of electrons N_{el} increases with the film thickness L, and is allotted to the QWSs below the highest occupied state. On the other hand, the energy separation between QWSs decreases. This causes the number of QWSs below the highest occupied state to increase. In these trends, the energy level of the highest occupied state increases and decreases with film thickness. In a crude approximation, we can recognize that electronic energy is governed dominantly by the energy level of the highest occupied state. In this respect, the figure suggests that the electronic energy of the thin film oscillates, and takes minimums at specific magic thicknesses. As a consequence, the thin film prefers to take the magic heights strongly.

Czoschke et al. estimated the change in the electronic energy of the thin film numerically [26]. In Fig. 7.18, their calculation of the electron energy of Pb(111) film is shown as a function of the film thickness L. As illustrated by the blue line, the electronic energy oscillated with film thickness. Here, note that the thin dotted line was the calculation for the electrons confined between infinite potential barriers at the free-standing film surfaces. The steep increase in energy with the thickness decrease is due to the marked increase in the quantized state energy of the electrons confined perfectly in the film. In more realistic films with finite potential barriers, the increase in the electronic energy at the small-thickness region is not so large, due to the spilling of electrons outside the surfaces, as indicated by the thick dotted and thick solid lines in the figure. However, the oscillation of electronic energy still persists. This means that the films with finite

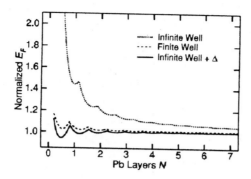

Fig. 7.18. The Fermi energy of a Pb(111) film as a function of thickness, normalized to its bulk free-energy (after Czoschke *et al.* [26]).

potential barriers still favor to take specific thicknesses, at which the electronic energy has a local minimum due to the quantum confinement of the conduction electrons.

More realistically, the thin film is not standing freely, but is bound to a substrate surface. The potential barrier at the film surface and film–substrate interface is finite and is not equivalent. Anyway, the confinement of the electron in the thin film is not perfect. The spilling out of the electrons at the interface causes an interface dipole layer, and smears out, to some extent, the quantum effect of the confined electrons. Zhang *et al.* modeled the electron spilling at the interface as a capacitor of the size of electron screen length (several Å), and examined the stability of the thin film thickness via quantum confinement [27]. They showed theoretically that the electronic energy of the thin film shows several characteristic types of film thickness dependence on the competition between the quantum confinement, electron spilling at the interface, and interface-induced Friedel oscillations (which control the energetic gain by the electron spilling at the outer surface of the thin film). The competition is delicate, and strongly depends on the combination of films and substrates. However, their calculation showed that electronic energy surely reveals oscillatory thickness dependence in several systems.

7.3.2 Experimental evidence

A. Quantum effect on very flat thin films with magic heights. The formation of flat-top thin films of unusual thickness uniformity was reported for various combinations of metal films and substrate materials. On the other hand, QWS formation of the confined conduction electrons was observed in most of these systems. This strongly suggests that the quantum confinement is the origin of the growth of the flat-top films of uniform thickness. In the following, we focus on the systems in which the quantum effect on the flat film formation was elucidated.

(a) Ag/GaAs(110) In the conventional growth of Ag films on GaAs(110) surfaces at room temperature, three-dimensional Ag islands of various heights are formed on the surfaces, as shown in the STM image of Fig. 7.19(a). However, atomically flat, epitaxial

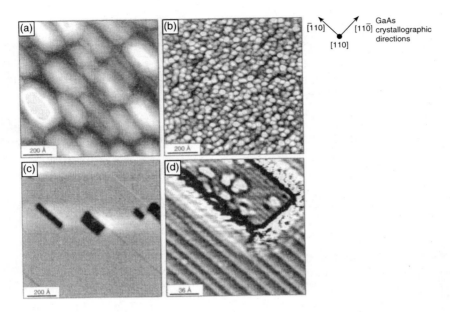

Fig. 7.19. (a) Silver three-dimensional islands created by depositing 15 Å of Ag at room temperature on a clean GaAs(110) surface. (b) Low-temperature STM image (145 K) showing the nanocluster morphology. (c) Silver two-dimensional film created by depositing 15 Å of Ag on a clean Gas(110) surface at LT (135 K) and annealing the film to RT. (d) Curvature image of one of the pits in the film (after Smith *et al.* [28]).

Ag films can be obtained on the GaAs(110) substrate surface by the two-step growth; deposition of Ag at a low temperature (approximately 100 K) (Fig. 7.19(b)) followed by slow annealing to room temperature, Fig. 7.19(c) [28–31]. As can be seen in Fig. 7.19, Ag of 15 Å was deposited on the cleaved GaAs(110) surfaces. This resulted in the formation of 3D Ag islands of typically 20–30 Å height by room-temperature deposition (Fig. 7.19(a)). In contrast, much smaller nanoislands nucleated on the substrate due to the reduced surface Ag atom diffusion at 145 K (Fig. 7.19(b)). In the subsequent annealing to room temperature, the small islands changed to a very flat thin film with a few pits (small dark rectangles, Fig. 7.19(c)).

A photoemission spectroscopic (PES) study showed that QWSs evolved during the annealing of the Ag islands deposited at a low temperature [30]. In Fig. 7.20, the Ag film of 53 Å was deposited at 100 K on the GaAs(110) surface and was annealed to room temperature. This figure shows the change in PES during the annealing. With the increase in annealing temperature, QWS peaks become distinct as the Ag film obtains a very flat morphology.

In the Ag/GaAs(110) system, the initial Ag coverage is crucial for the morphology of the thin films obtained after the annealing in the two-step growth [28,29]. An example is shown in Fig. 7.21 [29]. In the case of a small initial coverage, the Ag film has voids of bare GaAs, although other areas are covered by interconnected flat-top Ag

Size Control of Nanostructures by Quantum Confinement 283

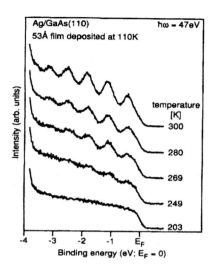

Fig. 7.20. Valence photoelectron spectra of a 53 Å silver layer on GaAs(110), deposited at 100 K, for several annealing temperatures (after Evans *et al.* [30]).

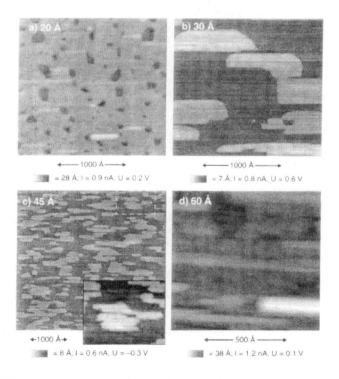

Fig. 7.21. STM images for 20, 30, 45, and 60 Å Ag deposition onto GaAs(110) (after Neuhold *et al.* [29]).

islands of uniform height (Fig. 7.21(a)). In contrast, all the surface is perfectly covered by a flat film of unique thickness if the initial coverage is above the critical thickness (Fig. 7.21(b),(c)). Excess Ag atoms form two-dimensional islands on the flat film of unique height. The existence of the critical initial thickness is a direct consequence of the quantum effect. The theory, which considered the competition between quantum confinement, electron spilling at the interface, and the interface-induced Friedel oscillation in the thin film [27], predicted that the second derivative of the electronic energy $(\partial^2 E(L)/\partial L^2) < 0$ (i.e., the film is unstable) for the height of the thin film $L < L_c$, while $(\partial^2 E(L)/\partial L^2) \geq 0$ (film is stable) for $L \geq L_c$ in the Ag/GaAs(110) system. The critical thickness L_c was calculated to be 5 ML in the theory, which agrees very well with the L_c observed in the experiment.

(b) Pb/Si(111)   Without using the two-step growth method, we can obtain flat-top islands of unique height by low temperature growth in the Pb/Si(111) system [32–35]. Figure 7.22(a) [33] shows the STM image of the Pb islands grown on the Si(111) 7 × 7 surfaces at 200 K. The Pb coverage was 3 ML. The surface was covered completely by

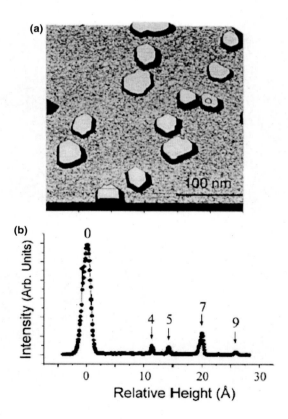

Fig. 7.22. (a) STM image of Pb islands grown on Si(111) substrate at 200 K and coverage of 3.2 ML, $V = 2$ V, 300 × 300 nm^2. (b) The distribution intensity of these islands as a function of their thicknesses. (after Su et al. [33]).

a wetting layer on which flat-top Pb islands of unique height are nucleated. As shown in Fig. 7.22(b), the Pb islands strongly preferred a specific height of 7 ML.

By increasing the temperature slightly to 207 K, the preferred height of the flat-top Pb islands changed from 7 to 9 ML [34]. In general, the flat-top Pb islands on the Si(111) surfaces were found to favor an odd layer height. A typical example is shown in Fig. 7.23. Figure 7.23 [35] shows the STM images depicting the time-dependent morphological change of the Pb thin films at 195 K. In the images, the local heights of the Pb islands are indicated in ML units. On the As-deposited surface (Fig. 7.23(a)), the Pb islands had

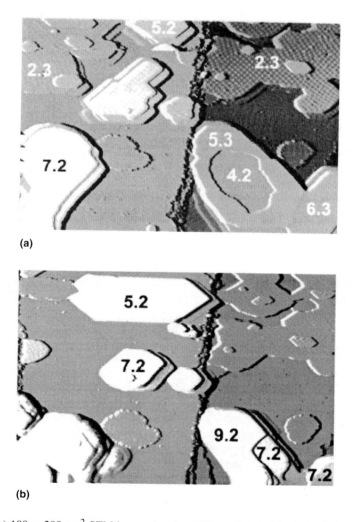

Fig. 7.23. (a) 100×200 nm^2 STM image showing different height islands (stable and unstable). (b) The same area 40 min later showing the evolution to more stable heights (after Hupalo and Tringides [35]).

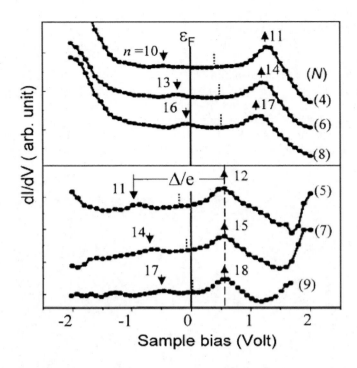

Fig. 7.24. A series of dI/dV spectra taken on individual islands of varying thickness numbered in parentheses from $N = 4$ to 9 (after Su et al. [33]).

heights of 2, 4, 5, 6, and 7 ML. However, after 40 min. (Fig. 7.23(b)), the islands of 4 and 6 ML disappeared, and only the islands of 5, 7, and 9 ML survived. This means that Pb islands of even layer height are unstable, but islands of odd layer height are sufficiently stable.

The stability of the odd-layer-height Pd films originates from the quantized nature of the conduction electrons. Figure 7.24 shows STS spectra, which are roughly proportional to the electron density of states (DOS) [19], of Pd islands of even ($N = 4,6,8$) and odd ($N = 5,7,9$) layer heights [33]. The spectra exhibit peaks of QWS-associated subbands. In the even-layer-height islands, the highest occupied state (HOS) and lowest unoccupied state (LUS) appeared at approximately -0.5 and $+1.5$ eV, respectively. The HOS and LUS shifted to the lower energy side in the odd-layer-height islands. Figure 7.25 [33] shows the island-thickness-dependent change of the central energy (δ) between HOS and LUS, referred to as the Fermi level. As shown in the figure, δ oscillates with a bilayer period. A local minimum appears at every odd layer thickness. Therefore, the odd layer height is more energetically preferable than the odd layer height for the Pd films. The bilayer period is a direct manifestation of the quantum effect. The Fermi wavelength of the Pd(111) islands is $\lambda = 0.366$ nm. Meanwhile, the Pd layer spacing is $d = 0.286$ nm. Therefore, the relation of $d = 3/4\lambda$ holds as a good approximation. On the other hand, the quantized condition of the electrons along the surface-normal direction

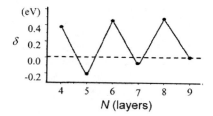

Fig. 7.25. The energy separation s(δ) from the middle positions of the Highest Occupied State and the Lowest Unoccupied State to the Fermi level as a function of island thickness (after Su et al. [33]).

is given by

$$L = N \cdot d = \left(n + \frac{1}{2}\right) \cdot \left(\frac{\lambda}{2}\right),$$

where L denotes the film thickness, N denotes the number of the layers, and n denotes the quantum number. Because of the relation of $d = 3/4\lambda$, this quantized condition is satisfied only at every odd layer height N. It is in good agreement with the experimental results.

Besides the formation of unique-height, flat-top Pd films on the Si(111) substrates, the formation of such films by low-temperature growth and annealing on the Ge(100) surfaces was also reported [37]. The quantum effect on the specific height selectivity was also elucidated in this system.

(c) Ag/Si(111) Flat-top Ag islands of a unique height can be grown on Si(111) 7 × 7 surfaces by the two-step growth process. Figure 7.26(a) [38] shows the STM images of the Ag islands. The nominal coverage was 1 ML. The islands were obtained by the deposition at 150 K and subsequent annealing to room temperature. In the STM image, the wetting layer was imaged as a dark area. 0.5 ML of Ag atoms was expelled for the formation of the wetting layer. The remaining Ag atoms were nucleated into the flat islands on the wetting layer. Figure 7.26(b) indicates that the films have two preferred heights, marked by A and C in the figure. The heights A and C correspond to the heights of the wetting layer and the islands, respectively. All the islands have a unique height of 2 ML (i.e., 5 Å).

The preference of the 2 ML height is still persistent in further deposition. Figure 7.27 shows the STM image of the surface with a nominal Ag coverage of 2.2 ML. Comparing Figs. 7.26 and 7.27, it is clear that the flat islands maintained the 2 ML height, and percolated with the increase in Ag coverage. At a larger coverage, the two-step growth still results in the formation of flat films of a unique height. The STM image of the nominally 6 ML coverage film is shown in Fig. 7.28 [39]. This Ag film was deposited at 50 K and then annealed to room temperature. Unfortunately, the height was not described, the cross section of the STM image showed the formation of a flat film of unique height (probably thicker than 2 ML) on the surface.

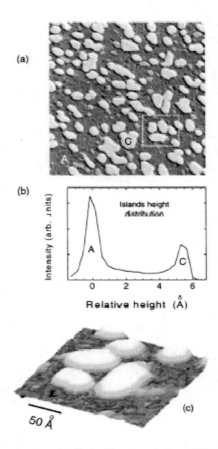

Fig. 7.26. (a) STM topography for a 1-ML Ag film deposited on Si(111) at 150 K and annealed to room temperature, 100 × 100 nm², (b) Island height distribution obtained from (a), showing the strong preferred height C of the islands. The intensity A represents the Ag wetting Layer. (c) Three-dimensional view of the area marked in (a), showing the flatness of the top of the plateau islands (after Gavioli et al. [38])

In the above studies, the relationship between QWSs and the unique island heights was not elucidated experimentally. However, a UPS study revealed the QWS affects the preference of the specific heights for flat-top films in the Ag/Fe(100) system [40]. Luh et al. deposited Ag films of thicknesses $N = 1$ to 15 ML on an Fe(100) surface at 110 K, and observed the change in the intensity of the QWS peaks in UPS in the annealing of the films. Their results are shown in Fig. 7.29. Figure 7.29 shows that the film of 6 ML height dissolved into 5- and 7-ML films (left), and the 3-ML-height film (right) dissolved into 2- and 4-ML films by the annealing. They found that the films of $N = 1$-, 2-, and 5-ML heights are extremely stable against annealing up to 800 K. Meanwhile, films of other thicknesses are unstable and bifurcate into a film of $N - 1$ and $N + 1$ ML at around 400 K. A theoretical consideration supported that their result

Size Control of Nanostructures by Quantum Confinement   289

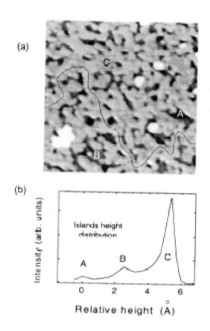

Fig. 7.27. (a) STM topography for a 2.2-ML Ag film deposited on Si(111) at 150 K and annealed to room temperature (100 × 100 nm^2), showing the presence of a percolated network, as indicated By the dotted line. (b) Islands height distribution obtained from (a), indicating that the percolated network has the same height C as that of the plateaus shown in Fig. 7.26(a) (after Gavioli *et al.* [38]).

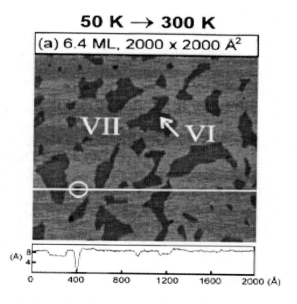

Fig. 7.28. STM image of Ag deposition above 6 ML reveals planar films with single height variations and very few multilayer pits (after Huang *et al.* [39]).

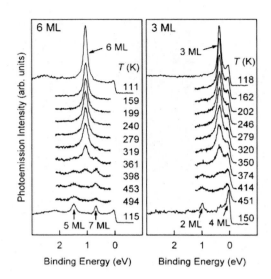

Fig. 7.29. Normal emission spectra of Ag films on Fe(100) with initial thickness of 6 ML (left) and 3 ML (right) taken with a photon energy of 15 eV. The spectra, from top to bottom, were taken at increasingly higher temperatures. Quantum-well peaks corresponding to various thicknesses are indicated (after Luh et al. [40]).

is due to the thickness-dependent change in quantized conduction electron energy in the Ag films [40].

B. Quantum effect on lateral size control of nanoislands. As well as the height of the flat-top islands, the lateral size of nanoislands could be controlled by quantum confinement. In 2005, Morgenstern et al. used a fast-scanning tunneling microscope and followed the decay of hexagonal Ag adatom islands on top of larger Ag adatom islands on Ag(111) surfaces [41]. Figure 7.30 shows typical STM images of the Ag adatom islands (insets) and their time-dependent decay of area. The initial island sizes were (a) 300, (b) 200, and (c) 1000 nm^2. The temperature was 330 K for (a) and room temperature for (b) and (c). As time passed, the adatoms of the top island diffused downward, and the area of the top island decreased gradually. However, at some specific sizes, the decay rate decreased significantly to cause a plateau in the decay curves of the top island area. In particular, the decay rate slowed down for islands with diameters of 6, 9.3, 12.6, and 15.6 nm. This is the result of the lateral (i.e., surface parallel direction) quantum confinement of the surface electrons in the adatom island. As stated in the theoretical background, electronic energy becomes unstable if the highest occupied QWS is close to the Fermi level. In contrast, the structure becomes energetically stable if the highest occupied QWS is far below the Fermi level. This condition is satisfied for islands with a size equal to $\lambda/4 + n(\lambda/2)$, where λ is the Fermi wavelength. For the surface electrons at Ag(111), $\lambda/2$ is 3.7 nm. Thus, islands are theoretically expected to be stabilized by the quantum effect at lateral sizes of 5.55, 9.25, 12.95, and 16.65 nm. This agrees very well with the experimentally observed specific sizes. This suggests that QWSs could be utilized to control the lateral size of nanoscale islands.

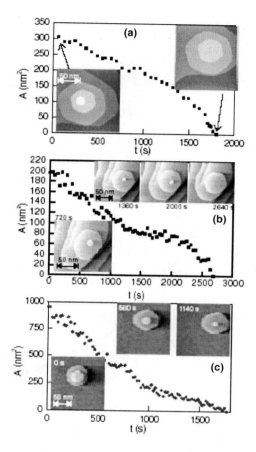

Fig. 7.30. Area development in time for decay of the top layer island of different adatom island stacks; STM images are snapshots of the STM movie at the indicated times. Parameters are (a) the initial size of bottom island $A = 250$ nm^2, $I = 40$ pA, $V = 2$ V, $T = 330$ K. (b) $A = 3500$ nm^2, $I = 5$ nA, $V = 0.3$ V, RT. (c) $A = 2800$ nm^2, $I = 1.5$ nA, $V = -1$V, RT (after Morgenstern et al. [41]).

7.4 Practical and Future Applications

For nanoscale structures, the energy levels of QWSs are determined strictly on the basis of structural size. Thus, we can tailor the electronic states to give nanostructures a new, novel function via the control of the size of nanostructures. As an example, we survey the tuning of surface reactivity of nanoscale islands by electron quantum confinement in this subsection.

Figure 7.31(a) [42] shows the low-energy electron microscopy (LEEM) image of Mg films on a W(110) substrate. After the deposition of a small amount of Mg atoms, the W(110) surfaces were covered by a Mg film of local thickness of 5–15 atomic layers. (In LEEM, the contrast reflects the height of the Mg film. The local area of larger

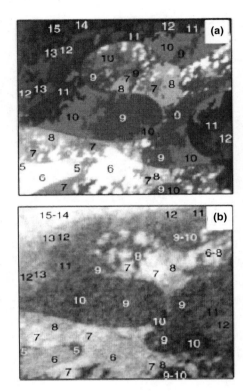

Fig. 7.31. 6×5 μm images of a Mg film in an advanced growth stage. (a) 1.3 eV LEEM image; the indicated number of atomic layers corresponding to the microregions following the reflectivity changes during the film growth at 120°C with deposition rate of 0.1 atomic layer/min. (b) XPEEM image of the same Mg film after exposure to 9 L of O_2 at 50°C, where the contrast corresponds to the Extent of local Mg oxidation (after Aballe et al. [42]).

thickness had a darker contrast. The local layer thickness is indicated by figures in the LEEM image.) Then, the Mg film was exposed to oxygen molecules. Figure 7.31(b) is the X-ray photoelectron emission microscopy (XPEEM) image of the exposed surface. The oxygen exposure was 9 Langmuir (L). In the XPEEM, the brightness of the contrast is correlated to the oxidation of the Mg film surface. The figure indicates that the oxidation was markedly enhanced on the film with specific heights of 6–8 and 13–15 atomic layers.

The specifically enhanced surface reactivity for the oxidation is due to the film thickness-dependent shift of QWSs. Figure 7.32 shows the valence band spectra of Mg films with thicknesses of 5–12 atomic layers [42]. Along with the 1.6 eV surface-state peak, two sets of QWSs are visible. One is below the surface state. The other is above the surface state, and crosses the Fermi level with increasing film thickness. When the upper QWS crosses the Fermi level, the electron density of states at the Fermi level increase. The surface electron just below the Fermi level is efficiently transferred to the unoccupied $1\pi_g$ molecular orbital of the adsorbed oxygen. This weakens the O–O bond, and triggers

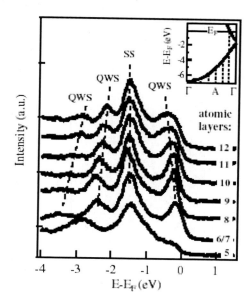

Fig. 7.32. Valence band spectra obtained from microregions with the indicated thickness. The inset shows the Mg(0001) band structure in the ΓA direction (normal to the film). The allowed wave-vector-energy (k, E) values for 7 atomic layers are calculated from the simple Bohr-Sommerfeld rule, assuming perfect reflectivity at the surface and interface and are indicated by dotted vertical lines (after Aballe et al. [42]).

the dissociation of the adsorbed oxygen molecule and subsequent oxidation of the Mg film surface. As shown in Fig. 7.32, the crossing of the Fermi level occurs at a height of approximately 6–7 atomic layers. A strong correlation of the Fermi level crossing of QWS and the enhancement of the oxidation reaction was observed clearly in Fig. 7.33 [42]. Thus, the specific enhancement of the oxidation on the film with specific heights is reasonably attributed to the Fermi level crossing of the QWS.

A correlation between the Fermi level crossing of QWS and the enhancement of the adsorption bond of the CO molecule was also reported for the Cu thin films grown on fccFe/Cu(100) surfaces [43].

In the above examples, the key strategy for enhancing the surface chemical reactivity is the increase in surface electron state density just below the Fermi level by making the QWS cross the Fermi level via thickness control. However, as described in the previous section, the QWSs are favorable to be apart from the Fermi level with respect to the structural stability. Thus, it is very important to induce attractive physical properties for the structurally stable, size-controllable thin films via the QWSs apart from the Fermi level.

One method is to utilize the QWSs as an energy filter of hot carriers generated by ultraviolet light irradiation. The irradiation of UV light generates hot carriers over a broad energy range near the surface region. Some of these hot carriers diffuse to the surface,

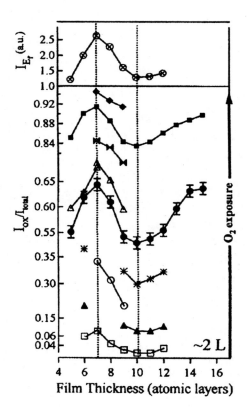

Fig. 7.33. (Lower panel) Plots of the relative weight of the Mg 2p oxide component, obtained in several experimental runs. Data indicated with the same symbols correspond to the same O_2 exposure. (Upper panel) Photoemission intensity at the Fermi level measured for different microregion thickness before oxygen exposure with energy resolution of 0.25 eV (after Aballe et al. [40]).

and occasionally attach to the HOMO/LUMO of the adsorbed molecules resonantly. This results in a remarkable enhancement of nonthermal chemical reactions at the surfaces [44,45]. However, in the conventional UV-assisted surface reaction, we have no option to select a special reaction of a specific molecule, because the energy distribution of hot carriers is too wide to cover many HOMOs/LUMOs of various molecules adsorbed onto the surfaces. However, the QWSs in the nanometer-height thin films can filter the energy of hot carriers to select only one HOMO or LUMO of a specific molecule among various adsorbed molecular species. Even if the highest occupied QWS is apart from the Fermi level to stabilize the nanostructures, we can tailor the energy levels of the QWSs to resonate to the HOMO or LUMO of the selected molecule by suitable choices of film thickness, band structure, and the film/interface potential barrier of the thin films. In this case, the film has a unique height all over the surface, via the quantum effect with the highest occupied QWS apart from the Fermi level. This reduces the broadening of the filtering energy associated with the energy level of the QWS. As a consequence, the filtering with the QWS resonates to the specific HOMO or LUMO of the selected molecule with an incredibly high Q-value (i.e., a perfect selectivity is realized). We expect that this will open a new route of artificial design of new, high-performance catalytic nanomaterials.

7.5 Summary

The quantum effect on electronic energy becomes marked and persists even at high temperatures in nanostructures. The energy levels of the QWSs are determined by the band dispersion and size of the nanostructures. Meanwhile, nanostructures favor specific sizes at which QWSs are sufficiently apart from the Fermi level. This can be utilized to control the size of nanostructures in both surface-normal and surface-parallel directions.

The size-dependent control of the energy level of the QWSs can be utilized to realize new and unique physical properties of nanostructures. As an example, we showed the enhancement of the surface chemical reactivity via the Fermi level crossing of QWSs in the nanometer-thickness thin films reported in this Chapter.

References

[1] N.W. Ashcroft and N.D. Mermin, Solid State Physics, Saunders College: Philadelphia, 1976.

[2] C.W.J. Beenakker and H. van Houten, Solid State Physics. **44** (1991) 1.

[3] S. Gasiorowicz, Quantum Physics, Wiley: New York, 1974.

[4] G. Ertl and J. Kuppers, Low Energy Electrons and Surface Chemistry, VCH Verlagsgesellschaft mbH: Weinheim, 1985.

[5] T.-C. Chiang, Surf. Sci. Rept. **39** (200) 181.

[6] K. Sumitomo, T. Kobayashi, F. Shoji and K. Oura, Phys. Rev. Lett. **66** (1991) 1193.

[7] J.H. Davis, The Physics of Low-Dimensional Semiconductors, Cambridge University Press: Cambridge, 1998.

[8] M. Watai, Masters thesis, Tokyo Institute of Technology, 2005.

[9] M. Watai and H. Hirayama, Phys. Rev. B **72** (2005) 085435.

[10] A.L. Wachs, A.P. Shapiro, T.C. Hsieh and T.-C. Chiang, Phys. Rev. B **33** (1986) 1460.

[11] G. Neuhold and K. Horn, Phys. Rev. Lett. **78** (1997) 1327.

[12] I. Matsuda, H.W. Yoem, T. Tanikawa, T. Tono, T. Nagao, S. Hasegawa and T. Ohta, Phys. Rev. B **63** (2001) 125325.

[13] A. Arranz, J.F. Sancez-Royo, J. Avila, V. Perez-Dieste, P. Dumas and M.C. Asensio, Phys. Rev. B **65** (2002) 195410.

[14] K. Pedersen, T.B. Kristensen, T.G. Pedersen, P. Morgen, Z. Li, S.V. Hoffmann, Surf. Sci. **482–485** (2001) 735.

[15] H. Hirayama, T. Kawata and K. Takayanagi, Phys. Rev. B **64** (2001) 195415.

[16] Y.R. Shen, The Principles of Nonlinear Optics, John Wiley and Sons: New York, 1984.

[17] H. Hirayama, T. Komizo, T. Kawata and K. Takayanagi, Phys. Rev. B **63** (2001) 155413.

[18] D. Wegner, A. Bauer and G. Kaindl, Phys. Rev. Lett. **94** (2005) 126804.

[19] D. Bonnell, Ed., Scanning Probe Microscopy and Spectroscopy, John Wiley and Sons: New York, 2000.

[20] S.W. Hla, K.F. Rraun and K.H. Rieder, Phy. Rev. B **67** (2003) 201402.

[21] A. Zangwill, Physics at Surfaces, Cambridge University Press: Cambridge, 1988.

[22] M. Crommie, C. Lutz and D. Eigler, Science **262** (1993) 218.

[23] J. Li, W.-D. Schneider and R. Brendt, Phys. Rev. Lett. **80** (1998) 3332.

[24] D. Bonnel, Ed., Scanning Probe Microscopy and Spectroscopy, John Wiley & Sons: New York, 2001.

[25] S.L. Silva and F.M. Leibsle, Surf. Sci. **441** (1999) L904.

[26] P. Czoschke, H. Hong, L. Basile and T.-C. Chiang, Phys. Rev. B **72** (2005) 075402.

[27] Z. Zhang, Q. Niu and C.-K. Shih, Phys. Rev. Lett. **80** (1998) 5381.

[28] A. Smith, K. Chao, Q. Niu and C.-K. Shih, Science **273** (1996) 2026.

[29] G. Neuhold, L. Bartels, J. Paggel and K. Horn, Surf. Sci. **376** (1997) 1.

[30] D.A. Evans, M. Alonso, R. Cimino and K. Horn, Phy. Rev. Lett. **70** (1993) 3483.

[31] H. Yu, C. Jiang, Ph. Ebert, X.D. Wang, J.M. White, Q. Niu, Z. Zhang and C.K. Shih, Phys. Rev. Lett. **88** (2002) 016102.

[32] Ph. Ebert, K.-J. Chao, Q. Ninn and C. K. Shih, Phys. Rev. Lett. **83** (1999) 3222.

[33] W.B. Su, S.H. Chang, W.B. Jian, C.S. Chang, L.J. Chen and T.T. Tsong, Phys. Rev. Lett. **86** (2001) 5116.

[34] M. Hupalo, S. Kremmer, V. Yeh, L. Brebil-Bautista, E. Abram and M.C. Tringides, Surf. Sci. **493** (2001) 526.

[35] M. Hupalo and M.C. Tringides, Phys. Rev. B **65** (2002) 115406.

[36] V. Yeh, L. Berbil-Bautista, C.Z. Wang, K.M. Ho and M.C. Tringides, Phys. Rev. Lett. **85** (2000) 5158.

[37] L. Floreano, D. Cvetko, F. Bruno, G. Bavdek, A. Cossaro, R. Gotter, A. Verdini and A. Morgante, Prog. Surf. Sci. **72** (2003) 135 and references therein.

[38] L. Gavioli, K.R. Kimberlin, M.C. Tringides, J.F. Wendelken and Z. Zhang, Phys. Rev. Lett. **82** (1999) 129.

[39] L. Huang, S.J. Chey and J.H. Weaver, Surf. Sci. Lett. **416** (1998) L1110.

[40] D.-A. Luh, T. Miller, J.J. Paggel, M.Y. Chou and T.-C. Chiang, Science **292** (2001) 1131.

[41] K. Morgenstern, E. Lagesgaard and F. Besebacher, Phys. Rev. Lett. **94** (2005) 166104.

[42] L. Aballe, A. Brinov, A. Locatelli, S. Heun and M. Kiskinova, Phys. Rev. Lett. **93** (2004) 196103.

[43] A. Danesse, F.G. Curti and R. Bartynski, Phys. Rev. B **70** (2004) 165420.

[44] A. de Meijere, H. Hirayama and E. Hasselbrink, Phy. Rev. Lett. **70** (1993) 1147.

[45] E. Hasselbrink, H. Hirayama, A. de Meijere, F. Weik, M. Wolf and G. Ertl, Surf. Sci. **269/270** (1992) 235.

CHAPTER 8

Grain Boundary Dynamics in Ceramics Superplasticity

Fumihiro Wakai and Arturo Domínguez-Rodríguez

Abstract

Ceramic materials can be deformed extensively in tension above approximately half the absolute melting point provided they have a fine (<1 μm diameter) equiaxed grain size that is stable during deformation. Studies of ceramics superplasticity have evolved during the past decades, because of the potential applications of superplastic forming. Micrograins move past one another by grain boundary sliding during deformation. The analysis of grain boundary dynamics provides an insight into the role of grain boundaries in superplasticity and grain growth. The progress in microstructural control leads to the development of high-strain rate superplasticity in ceramics. Here, we report the recent progress and the current status of the understanding of ceramics superplasticity focusing on how grain boundaries affect deformation behavior.

Keywords: **Superplasticity, grain boundary sliding, grain growth, ceramics.**

8.1 Introduction

Superplasticity is phenomenologically defined as an ability of a polycrystalline material to exhibit extraordinarily large elongation at elevated temperatures and at relatively low stresses [1]. It is a property commonly found in many metals, alloys, intermetallics, and ceramics when the grain size is very small: less than several micrometers for metals and less than one micrometer for ceramics [2–5]. Such behavior is of interest, firstly, because the ability to achieve large strains makes superplastic forming an attractive option for the manufacture of complex shaped components in the metal industry. Ceramics, e.g., oxides, nitrides, and carbides, are hard, strong and stiff materials. They are brittle and lack the ductility of metals at ambient temperatures. The application of superplasticity makes it possible to fabricate ceramic components just like superplastic metals. Secondly, in spite of decades of investigation, there is only limited understanding of the contribution

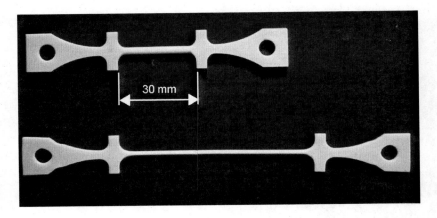

Fig. 8.1. First demonstration of superplastic elongation of Y_2O_3-stabilized tetragonal ZrO_2 polycrystals (Y-TZP) [7]. Specimens before and after deformation are shown.

of various possible mechanisms to superplasticity. The exact mechanism of ceramics superplasticity still remains controversial.

Since Garvie [6] proposed a concept of transformation toughening in zirconia, ZrO_2-based ceramics have been used for tough and wear-resistant components. The discoveries of ceramics superplasticity in Y_2O_3-stabilized tetragonal ZrO_2 polycrystals (Y-TZP) [7] (Fig. 8.1) and its composite [8] triggered considerable research to clarify differences and similarities in the deformation mechanisms between metals and ceramics [4,9]. Many efforts have been devoted to improve the superplasticity of ZrO_2-based ceramics over the last decades. For example, a markedly high elongation of 2510% has been achieved at a strain rate of 0.085 s^{-1} in a ZrO_2-based composite [10] recently.

The research on ceramics superplasticity has further expanded from ZrO_2-based ceramics to silicon nitride (Si_3N_4) [3,11], silicon carbide (SiC), bioceramics, and oxide superconductors [12]. Probably superplasticity is the common nature of fine-grained polycrystalline materials and also nanocrystalline materials at intermediate temperatures.

8.2 Motion and Topological Evolution of Grains

Polycrystals deform under an applied load at elevated temperatures. The creep deformation mechanisms include the diffusional transport of atoms and the movement of lattice dislocations by glide and climb. The grains become elongated along the stress axis during creep. On the other hand, fine grains retain an approximately equiaxed shape even after the extremely large elongations (>1000%) in superplasticity.

Micrograins apparently move past one another during superplastic flow. The relative motion of grains was directly observed in the superplasticity of Y-TZP [13]. Any motion of a rigid grain consists of the combination of both translation and rotation. The relative motion of two adjoining grains has components parallel and vertical to their common grain boundary. The grain boundary sliding is the component parallel to the grain boundary. The contribution of grain boundary sliding to the total strain is determined by the relative motion of neighboring grains [14,15], and, as well as in metals, it is approximately 70% in the superplasticity of fine-grained ceramics [16,17]. The motion of two adjoining grains vertical to the grain boundary is equivalent to the 'non-conservative' motion of the grain boundary to the crystalline lattice. 'Non-conservative' motion occurs when the interface motion is coupled to long-range diffusional fluxes of the components to/from the interface [18]. An example of non-conservative motion is diffusional creep, in which the grain boundary acts as a source/sink of vacancies. The lattice sites at the boundary are created or destroyed, which in turn, cause the relative motion of crystalline grains. The relative motion, or Lifshitz sliding, is illustrated as a shift of the marker line in Fig. 8.2(a). In classical models of diffusional creep, the volume of the grain is unchanged, and the grains are elongated. While the topological change is not considered in diffusional creep, the motion of grains involves grain switching, the rearrangement of grains by grain boundary sliding, in superplasticity. The sliding of rigid grains generates cavities and cracks inevitably. Then, the essential mechanism of superplasticity is the accommodation process of grain boundary sliding so that the polycrystalline materials can be stretched extensively without fracture. Grain boundary sliding, grain rotation, diffusion, dislocation motion, and grain boundary migration are the main processes associated with superplasticity [1,2].

The structure of the grain boundary network in polycrystalline solids has characteristics similar to those of soap froth and emulsions. Ashby and Verral [23] constructed topological models of superplasticity from the observation of the flow of soap froth and emulsions. The soap-froth model (Fig. 8.2(b)) illustrates the variation in structure with strain for the periodic array of hexagonal bubbles. The soap froth rapidly restores a hexagonal structure after grain switching at $\gamma = 1.15$. Gifkins [26] considered that the boundary 'mantle' of the grain behaved differently from the central 'core' of the grain. Grains change their shape by deformation of the 'mantle' in his core-mantle model (Fig. 8.2(c)). Since the

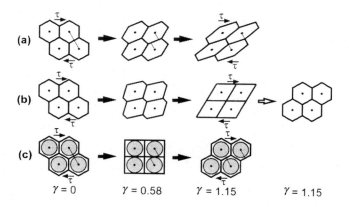

Fig. 8.2. Shear deformation in regular array of grains [25]. (a) Diffusional creep; (b) Soap-froth model; (c) Core-mantle model. (The gray circle shows the 'core'.)

grains recover their initial hexagonal shape at $\gamma = 1.15$, shear deformation can proceed beyond this strain. In both the soap-froth model and the core-mantle model for superplasticity, one grain loses two interfaces and gains two interfaces in the strain period $\gamma = 1.15$, to retain the equiaxed shape. This topological law describes the relationship between grain switching and the shape of regular grains.

The topology of grain boundaries is modified not only by the movement of grains but also by the 'conservative' motion of grain boundaries in grain growth [19–22]. The 'conservative' motion is defined as the motion of an interface, which occurs in the absence of a diffusion flux of any component in the system to/from the interface [18]. The positions of the two grains adjoining the interface remain fixed during conservative motion. Figure 8.3 is a snapshot taken from a movie of the grain growth. The elemental processes of topological change in grain growth are grain switching (T1 process) and the disappearance of grains (T2 process). Small grains shrink and disappear, and the mean size of the remaining grains increases. Figure 8.4 shows an example of how one grain changes its size and shape with time. The state of the grain is classified according to its number of faces f, or coordination number. Grain switching changes f, i.e., grain switching creates or eliminates a face of the grain.

Here, we illustrate grain switching in three dimensions using an aggregate of grains as a model (Fig. 8.5) [22]. As the grains A and B shrink, a small triangular interface between A and B vanishes ((2) → (4)); this is face-elimination switching. When the volumes of grain A and grain B are larger than those of grains C, D, and E, the reverse process proceeds ((4) → (2)); this is face-creation switching.

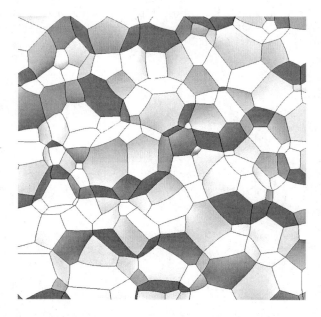

Fig. 8.3. Snapshot of microstructural evolution during grain growth [20].

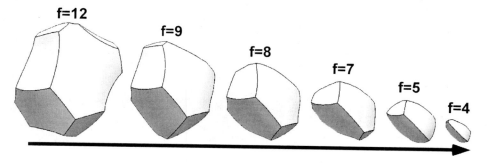

Fig. 8.4. Topological evolution of a grain in grain growth [22]. f is the number of facets of the particle.

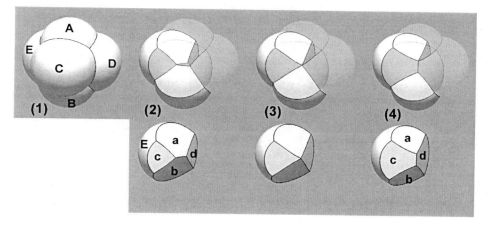

Fig. 8.5. Grain switching in three dimensions [22]: (1) surface of five grains; (2) initial stage; (3) intermediate stage; (4) final stage. The upper figures show the internal interface. The interfaces among the encircling grains (C, D, and E) are shown in light green. The lower figures show the interfacial patterns of encircling grain E. The faces a, b, c, and d show the interfaces to grains A, B, C, and D, respectively.

An example of grain switching by movement of grains is shown in Fig. 8.6. The aggregate consists of four free particles (C, D, E, and F), which are located between two sintered grains (A and B). The front particle F is omitted from Fig. 8.5 to show the structure of the internal interfaces of the aggregate. In Fig. 8.6(a), grains A and B contact; this is the On-state. When an external tension force is applied on grains A and B, two particles move as a result of non-conservative motion of grain boundaries. If atom diffusion takes place minimizing the sum of surface energy and grain boundary energy, the A-B interface will disappear (Fig. 8.6(b)), and two grains will be separated (Off-state) (Fig. 8.6(c)). This is a minimal model for superplasticity. A new rectangular interface between grains D and F emerges (Fig. 8.6(c)). Although the elimination and creation of rectangular interfaces occur simultaneously in this six-grain model of grain switching, the disappearance of

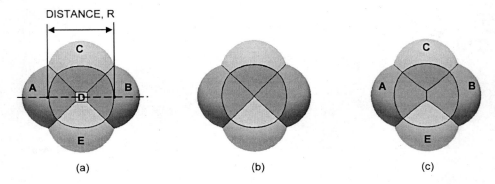

Fig. 8.6. Geometry of six-grain cluster in 3D. (a) On-state; (b) critical state; (c) Off-state. The cluster consists of four free particles (light gray particles C, D, E, and F) which are located between two constrained particles (dark gray particles A and B). The structure of internal interface is shown by omitting the front grain F.

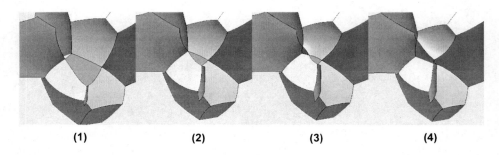

Fig. 8.7. Process of disappearance of rectangular face during three-dimensional simulation of grain growth [22].

a four-sided interface can be decomposed into face-creation switching and face-elimination switching of triangular interfaces as shown in Fig. 8.7.

Ashby and Verral [23] identified *grain switching* as a geometrical requirement to maintain the equiaxed shape of grains during superplasticity. In their two-dimensional model, grains exchange places with their neighbors. The model was modified to consider symmetrical diffusion path by Spingarn and Nix [24] afterward. Grain neighbor switching was directly observed during superplastic deformation of Y-TZP in the cross section of a specimen [13]. In three dimensions, face-creation switching and face-elimination switching occur independently during superplastic deformation as well as that in grain growth. Figure 8.8 shows the difference between creep and superplasticity in three dimensions schematically [25]. While grains are elongated in diffusional creep and dislocation creep without grain switching, the grains maintain an approximately equiaxed shape during deformation due to grain switching. One grain is caged in a cluster of neighboring

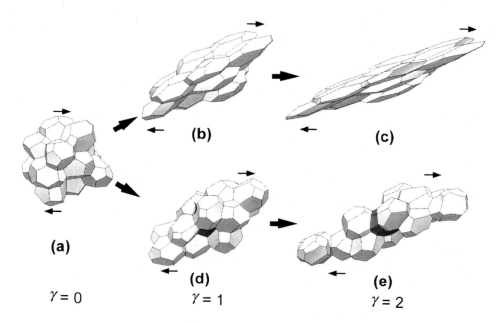

Fig. 8.8. Difference between creep and superplasticity in shear deformation [25]. (a) One grain is caged in a cluster of neighbor grains. (b) and (c) The grains of the cluster elongate in diffusional creep or dislocation creep. The caged grain remains inside after the deformation. (d) and (e) The grains retain an approximately equiaxed shape when grain switching occurs. Some neighbor grains separate from the center grain and the red center grain emerges from the inside.

grains (Fig. 8.8(a)). As a result of face-creation switching, a red grain emerges from inside of the cluster of grains (Fig. 8.8(e)). Gifkins [26] proposed *grain emergence* as a model of the rearrangement of grains in three dimensions. His *grain emergence* is equivalent to this *cage break-up* of the cluster.

The rate of grain growth during superplastic deformation is often higher than that caused by annealing alone (static grain growth). The enhancement of grain growth is termed *strain-enhanced grain growth*, or *dynamic grain growth*. The mechanistic origin of dynamic grain growth is probably related to grain boundary migration, which is enhanced by grain boundary sliding [27], grain switching [28], or grain boundary diffusion [29]. The grain switching during grain growth also contributes to the maintenance of the equiaxed shape of grains [30], if the grain boundary mobility and the grain boundary energy are isotropic.

Recently, the cooperative manner of grain boundary sliding, which reveals itself through sliding of grain groups as an entity, has been demonstrated in some superplastic alloys [31]. Such cooperative grain boundary sliding has been suggested to play an important role in fine-grained ceramics also [32]. However, the direct observation of grain rearrangement during the superplasticity of Y-TZP revealed that the influence of cooperative grain boundary sliding was very limited [13].

8.3 Physical Characteristics of Ceramics Superplasticity

The grain compatibility during grain boundary sliding is maintained by a concurrent accommodation process, which involves diffusion, dislocation motion, and grain boundary migration. Most of the physical models proposed for superplasticity consider that the accommodation mechanism is the rate-controlling process [1,2]. The importance of each accommodation process varies with materials, temperature, and stress.

The strain rate in superplasticity of many metals and ceramics is often expressed by the following semi-empirical equation [5,33],

$$\dot{\varepsilon} = \frac{AGb}{kT} \left(\frac{b}{d}\right)^p \left(\frac{\sigma - \sigma_0}{G}\right)^n D \qquad (8.1)$$

where $\dot{\varepsilon}$ is the strain rate, b is Burgers vector, G is the shear modulus, σ is the stress, σ_0 is the threshold stress, n is the stress exponent, D is the diffusion coefficient [$= D_0 \exp(-Q/kT)$, where D_0 is the frequency factor, Q is the activation energy and k is Boltzmann's constant], d is the grain size, and p is the grain-size exponent. p is two when the rate controlling process is lattice diffusion, and three when the rate controlling process is grain boundary diffusion. The threshold stress σ_0 is zero in many cases. The threshold stress depends on the nature of the grain boundaries [5,34].

Crystalline grains in Y-TZP are, as well as that in metals, directly bonded atomically at grain boundaries [35]. To understand the intrinsic mechanism of superplasticity, the deformation of Y-TZP with a minimum amount of impurities has been extensively analyzed. The deformation of high-purity Y-TZP is characterized by a stress exponent of $n \approx 2$ at comparatively higher stresses, whereas the deformation at lower stresses is associated with $n \geq 3$ as shown schematically by the solid line in Fig. 8.9 [36–39]. The values differ from that of grain boundary sliding accommodated by diffusion, which predicts a nearly Newtonian flow ($n = 1$) [23]. A hypothesis of interface-controlled Coble creep [40] was proposed to explain this discrepancy. Although the mechanical data can be consistent with the interface-controlled Coble creep [41], the retention of the equiaxed grain shape after superplasticity is apparently in conflict with the expectation of grain elongation accompanying diffusional creep [42]. This conflict is partly resolved if significant grain growth occurs during deformation [30], because grain switching also accompanies grain growth. However, a computer simulation [43] showed that grain growth must occur more extensively than that can be observed during the superplasticity of Y-TZP. Alternatively the grain boundary sliding with threshold stress ($n = 2$, $p = 2$ in Eq. (8.1)) can predict the mechanical data of Y-TZP [5,9,44]. Originally the grain boundary sliding model with $n = 2$ was developed for superplastic alloys, in which intragranular dislocation plays an important role in the accommodation process [45]. While the phenomenological constitutive equation agrees with the experimental results of Y-TZP, the role of intragranular dislocation in grain boundary sliding is not yet clearly elucidated, because flow stress in superplasticity is much lower than the stress required for dislocation activity [46].

While no or very limited dislocation activity was reported by Primdahl et al. [35], dislocation pileup has been observed by Morita and Hiraga [47]. The dislocation pileup

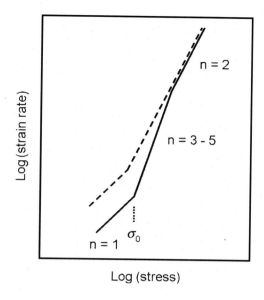

Fig. 8.9. Relationship between stress and strain rate of Y-TZP. (High-purity Y-TZP, solid line. Impurity-doped Y-TZP, dashed line.) Stress exponent n changes at the threshold stress σ_0.

suggests that stress concentration is present at multiple grain junctions to increase the applied stress by factors of $\approx 14-25$ [47–49]. The stress concentration at a triple junction can be very high transiently during grain boundary sliding, but decreases with time by diffusion [50]. The stress concentration will not be so high in the steady state of grain boundary sliding, thus the intragranular movement of dislocations can play little or no significant role in deformation [48]. However, the origin of non-Newtonian flow in the superplasticity of Y-TZP is not yet clear.

8.4 Grain Refinement and Suppression of Grain Growth

The semi-empirical equation, Eq. (8.1), provides strategies for achieving superplasticity at higher strain rates and at lower temperatures, i.e., (i) grain refinement, and (ii) diffusion enhancement.

The essential requirement for superplasticity is a small grain size. The strain rate increases with decreasing grain size. While, various thermomechanical processes, for example, severe plastic deformation [51], are used to develop a fine microstructure for metals, fine-grained ceramics are usually produced by sintering ultrafine powder. To avoid grain growth during sintering, a low sintering temperature is used in stress-assisted sintering [52], for example, hot-isostatic pressing, hot pressing, and sinter forging. A method to sinter dense nanocrystalline ceramics without final-stage grain growth was developed by exploiting the difference in kinetics between grain boundary diffusion in sintering and grain boundary migration in grain growth [53].

Nanocrystalline materials with grain sizes less than 100 nm have many interesting material properties [54,55], which are particularly dependent on the nature of the grain boundaries. For example, the *low-temperature superplasticity* [56] of nanocrystalline materials is favorable for making the use of superplastic forming even more wide spread.

The thermal stability of fine microstructures at elevated temperatures is required for the achievement of superplasticity. Grain boundary mobility depends strongly on the presence of solute or impurity segregation at grain boundaries. The rate at which solute and impurity atoms can diffuse along the boundary is slow relative to the rate at which the boundary can otherwise move [18]. The sluggish grain growth in Y-TZP is controlled primarily by the solute-drag effect of Y^{3+} ions segregating along grain boundaries [57].

The motion of grain boundaries is impeded also by second-phase particles, which are dispersed throughout the material. In two-phase composites, grain growth through grain boundary migration is coupled with Ostwald ripening via long-range diffusion, if there are limited mutual solubilities among the phases [58]. This coupling results in inhibited grain growth for both phases. Many fine-grained two-phase and multi-phase composites exhibit superplasticity, e.g., ZrO_2-Al_2O_3 [8,16,59], ZrO_2-mullite [60], ZrO_2-spinel [61], and ZrO_2-Al_2O_3-spinel [10,62].

Superplasticity has often been found at relatively low strain rates, typically approximately 10^{-5} to 10^{-3} s^{-1}. When superplasticity occurs at strain rates considerably higher than 10^{-2} s^{-1}, it is conveniently called *high-strain-rate superplasticity* [2]. High-strain-rate superplasticity was observed in the ZrO_2-Al_2O_3-spinel composite [62] at 0.4 s^{-1}. The ceramic specimen could be elongated twice within 2 s. High strain rate superplasticity of the ZrO_2-Al_2O_3-spinel composite [10,62] was achieved at a temperature (1650°C) two hundred degrees higher than the temperature for usual superplasticity (1450°C), mainly because grain growth was suppressed by dispersion of second- and third-phase particles.

The addition of the second phase affects the superplasticity and creep of composites in two ways. First, it modifies the continuum deformation mechanics. The effect of volume fraction of the second phase on strain rate can be predicted by a rheology model [60] or by a composite theory [63]. Second, the constituent atoms of the second phase affect interface related deformation characteristics, e.g., grain boundary diffusion, and will be discussed in the following section in detail.

8.5 Diffusion Enhancement

The addition of impurity atoms and solute atoms has a wide variety of effects on the grain boundary diffusivities and the lattice diffusivities of both solvent and solute species including combinations of enhancing or retarding effects. The solute atoms can change the vacancy concentration and the jump probability. From the atomistic view of point defects, differences in ionic radii and in valence contribute to the differences in diffusion coefficient. It is well known that the doping of small amounts of impurities significantly affects sintering and diffusional creep [64].

The superplasticity of Y-TZP is affected by the addition of small amounts of impurities, particularly transition metals [65], and Al and Si [66–69]. These impurities segregate at grain boundaries, and increase the strain rate of Y-TZP at lower stresses. The deformation behavior of impurity-doped Y-TZP is characterized by $n = 2$ and $p = 2$ over a wide stress range. The transition of the stress exponent in high purity Y-TZP disappears by doping of impurities, or the apparent threshold stress becomes zero.

In an early review [3], it was illustrated that larger cations of lower valence decreased grain-boundary mobility by solute drag. The effect of impurity segregation on superplasticity has been systematically studied recently. Cations with a small ionic size decrease flow stress, while large cations increase flow stress [70]. These results suggest that small dopant cations enhance grain boundary diffusion. On the other hand, the solute atom Y^{3+} segregates at grain boundaries of high purity Y-TZP [3,35], and has been considered as a possible origin of threshold stress [9,34]. The threshold stress becomes apparently negligible by co-segregation of impurity atoms, e.g., Al^{3+} or Si^{4+}, with Y^{3+}, but its physical mechanism has not been fully elucidated.

It is believed that the solution of TiO_2 and MgO in a ZrO_2 crystal enhances lattice diffusion [71]. The *low-temperature and high-strain rate superplasticity* of ZrO_2, was developed by the co-doping of TiO_2 and CaO [71], and also by the co-doping of TiO_2 and MgO [72]. The degree of enhancement of deformation is dependent on the ionic radius of the dopant cation [73].

The diffusion coefficient of ceramics depends on their chemical composition (stoichiometric or nonstoichiometric) and their crystal structure. Silicon carbide (SiC) is a covalent material that adopts network structures featuring vertex sharing of $[SiC_4]$. Although the diffusion coefficient in SiC is very slow, the superplasticity can be achieved by doping with a small amount of boron [74]. The doped boron segregates at grain boundaries, and takes the place of silicon, forming bonds in a local environment that is similar to that in the B_4C structure [75]. It is supposed that the segregated boron promotes deformation by grain boundary diffusion. On the other hand, hydroxyapatite ($Ca_{10}(PO_4)_6(OH)_2$) can be superplastically deformed at a relatively low temperature of 1000°C due to its fast diffusivity [76]. Hydroxyapatite is biocompatible with bone, and can be used for orthopedic and dental implants.

The addition of a liquid phase can enhance the deformation of polycrystalline solids significantly. The distribution of the liquid phase amongst the crystalline grains is dependent on its amount and the ratio of γ_{gb}/γ_{sl}, where γ_{gb} and γ_{sl} are the grain boundary energy and the interface energy between the solid and liquid phases, respectively. The grain boundary is only partially covered with liquid when $\gamma_{gb}/\gamma_{sl} < 2$, and most of the grain boundaries do not have intergranular glass film. It had been believed that glass-doped Y-TZP had intergranular amorphous film, but high resolution microscopy revealed that there was no amorphous film at grain boundaries as shown in Fig. 8.10(a) [67,77].

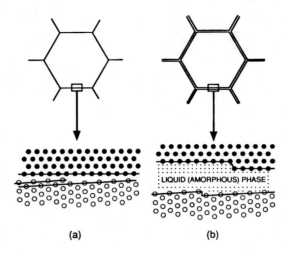

Fig. 8.10. Atomic structures of grain boundary. (a) The line defect at clean grain boundary is modeled by a boundary dislocation. (b) The line defect at grain boundary with thin liquid film becomes a step [77].

On the other hand, when $\gamma_{gb}/\gamma_{sl} \geq 2$, the liquid phase covers the faces of all grains as shown in Fig. 8.10(b). In creep and superplasticity of some advanced ceramics, the intergranular liquid phase acts as a lubricant for grain boundary sliding, and also as a path for material transport by solution-precipitation process [78,79]. Silicon nitride (Si_3N_4) is a promising material for engine components. A glass phase pocket and a thin glass film [80] with a thickness approximately 1 nm often remain at the grain boundaries of liquid-phase sintered Si_3N_4. The glass behaves as a liquid at elevated temperatures. The superplasticity and creep of Si_3N_4 is affected by the viscosity of intergranular liquid and the solubility of the solid to the liquid, and have been reviewed recently [81,82]. The mechanism of the superplasticity of glass-containing Si_3N_4 is different from the intrinsic superplasticity of metals and ceramics that do not contain glass film. The 'classical' microstructural requirement for superplasticity is a material consisting of equiaxed fine grains. The liquid phase promotes the grain boundary sliding, so that even the materials which contain rod-shaped β-Si_3N_4 can be superplastically deformed. In this 'non-classical' superplasticity the anisotropic grains tend to align with strain, producing a fiber-strengthening effect [83]. This phenomenon leads to strengthening and toughening of Si_3N_4 by compressive deformation [84].

SiAlON is a solid solution of Si_3N_4, in its structure, some Si and N atoms are replaced by Al and O atoms. The strain rate is enhanced by reducing the viscosity of glass, for example, Li-doped SiAlON [85] deforms 10 times faster than Si_3N_4. The liquid-enhanced creep of SiAlON undergoes a transition from Newtonian behavior ($n = 1$) to *shear thickening* behavior ($n = 0.5$) at a characteristic stress [86]. The cause of shear thickening was attributed to the formation of the direct contact of grains through intergranular liquid film. The shear thickening behavior has not been observed in some Si_3N_4, which also have a similar intergranular film [87–89].

8.6 Superplastic Forming

The process of superplastic forming (SPF) of ceramics can be an advanced manufacturing method for producing complex thin-sheet components [90]. The superplastic forming concurrent with the diffusion bonding (SPF/DB) process, which is used in the production of titanium alloys in the aerospace industry, is also applicable to ceramics. The superplasticity has been used for bonding of dissimilar ceramic material sheets [91]. In the ceramic industry, complex-shaped components with accurate dimensions have been usually fabricated by sintering the shaped powder compact. Therefore, in comparison to the sintering, two factors are important for the practical application of superplastic forming: 1) reliability, and 2) efficiency.

First, it is necessary to assure the reliability of the products. The source of variability in strength is related to flaws, particularly for ceramics due to its inherent brittleness. Many voids, or cavities are formed in the extremely deformed specimens in tension [92–95]. The cavity growth is regarded as an inverse process of sintering. The suppression of cavitation is of utmost importance for the superplastic forming of ceramics. A fast grain boundary diffusion, a small grain size, and a small ratio of grain boundary energy to surface energy are necessary for suppressing the nucleation of cavities [96]. The cavities grow much larger than the grain size of the matrix. The cavity growth is controlled by macroscopic plastic deformation of the matrix, and the cavity size increases exponentially with strain [93,94]. The spherical cavities are elongated parallel to the tensile direction together with the flow of the matrix as shown in Fig. 8.11. The cavitation can be suppressed by selecting appropriate forming conditions, so that ZrO_2-toughened ceramics can maintain an excellent strength even after an elongation of more than 100%.

From the viewpoint of efficiency, the combination of sintering and forming, for example, sinter forging [97], is most promising, because both densification and net-shaping are achieved simultaneously. The cavities and voids are further decreased by sinter forging in

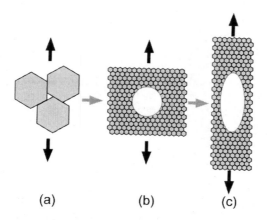

Fig. 8.11. Schematic illustration of cavitation. (a) Cavity nucleation at grain boundary, (b) cavity growth, (c) cavity elongation by plastic mechanism.

compression. A very high strength and a high fracture toughness of Si_3N_4 were achieved by superplastic sinter forging due to the reduced flaw size and the grain alignment [98]. Spark plasma sintering (SPS) is a recently developed technique that enables ceramics to be fully densified at comparatively low temperatures and in very short times. The process is similar to conventional hot pressing, and the powder compact (green ceramics) is heated by a direct-pulsed DC current in the graphite die. Very fast densification has been reported in oxides and Si_3N_4-based ceramics. Furthermore, the compressive deformation of SiAlON exceeded 10^{-2} s^{-1} at 1500°C in the SPS apparatus [99]. This result suggests a favorable relationship between high-strain rate superplasticity and ultrafast sintering.

High-strain rate superplasticity in ceramics made the SPF an attractive technology for shaping components efficiently. However, the microstructural optimization for superplasticity is not necessarily a demand of the market. For practical applications, superplasticity should be compatible to other properties, hardness, toughness, wear resistance, and strength. The high-strain rate superplasticity of useful materials with excellent properties, such as Si_3N_4, will be a target for future development.

8.7 Future Prospects

The grain boundary dynamics is a key to understand and control the microstructural evolution in materials science. The structure of a grain boundary network evolves dynamically in superplasticity, as well as that in grain growth and sintering. The research on ceramics superplasticity has elucidated some of the basic principles in the topological transformation of grains, and the role of grain boundary chemistry in deformation. Although significant advances have been made in developing and improving the superplasticity of various ceramics, our understanding of the deformation mechanism on the atomic level is still limited.

At present, there still remains a wide gap between high-temperature deformation, where atoms and vacancies move and jump, and the first principles calculation of the electron density of grain boundaries at absolute zero temperature [100,101]. The use of large scale molecular dynamics (MD) demonstrates the grain boundary sliding and emission of dislocations in the deformation of nanocrystalline material [102]. However, to study the long-range diffusion of atoms in superplasticity, a breakthrough in simulation technique is necessary to overcome the inherent time restriction of MD simulation. In the coming decades, new experimental techniques and computer simulations will give new insights on grain boundary dynamics. The fundamental knowledge obtained in ceramics superplasticity will be useful for many research fields in materials science.

References

[1] J.W. Edington, K.N. Melton and C.P. Cutler, Prog. Mater. Sci. **21** (1976) 61.

[2] T.G. Nieh, J. Wadsworth and O.D. Sherby, Superplasticity in Metals and Ceramics, Cambridge Univ. Press: Cambridge, 1997.

[3] I.W. Che and L.A. Xue, J. Am. Ceram. Soc. **73** (1990) 2585.

[4] A.H. Chokshi, A.K. Mukherjee and T.G. Langdon, Mater. Sci. Eng. R **10** (1993) 237.

[5] M. Jiménez-Melendo, A. Domínguez-Rodríguez and A. Bravo-León, J. Am. Ceram. Soc. **81** (1998) 2761.

[6] R.C. Garvie, R.H. Hannink and R.T. Pascoe, Nature **258** (1975) 703.

[7] F. Wakai, S. Sakaguchi and Y. Matsuno, Adv. Ceram. Mater. **1** (1986) 259.

[8] F. Wakai and H. Kato, Adv. Ceram. Mater. **3** (1988) 71.

[9] M. Jiménez-Melendo and A. Domínguez-Rodríguez, Phil. Mag. A **79** (1999) 1591.

[10] B.N. Kim et al., Scripta Mater. **47** (2002) 775.

[11] F. Wakai et al., Nature **344** (1990) 421.

[12] J.M. Albuquerque, M.P. Harmer and Y.T. Chou, Acta Mater. **49** (2001) 2277.

[13] R. Duclos, J. Euro. Ceram. Soc. **24** (2004) 3103.

[14] T.G. Langdon, Acta Metall. Mater. **42** (1994) 2437.

[15] T.G. Langdon, Mater. Sci. Eng. **A174** (1994) 225.

[16] S. Ishihara et al., Mater. Trans. JIM. **40** (1999) 1158.

[17] R. Duclos, J. Crampon and C. Carry, Phil. Mag. Letter. **82** (2002) 529.

[18] A.P. Sutton and R.W. Balluffi, Interfaces in Crystalline Materials, Oxford University Press: Oxford, 1995.

[19] H.V. Atkinson, Acta Metal. **36** (1988) 469.

[20] F. Wakai, N. Enomoto and H. Ogawa, Acta Mater. **48** (2000) 1297.

[21] F. Wakai, Y. Shinoda, S. Ishihara and A. Domínguez-Rodríguez, Phil. Mag. B **81** (2001) 517.

[22] F. Wakai, Y. Shinoda, S. Ishihara and A. Domínguez-Rodríguez, J. Mater. Res. **16** (2001) 2136.

[23] M.F. Ashby and R.A. Verrall, Acta Metall. **21** (1973) 149.

[24] J.R. Spingarn and W.D. Nix, Acta Metall. **26** (1978) 1389.

[25] F. Wakai, Y. Shinoda, S. Ishihara and A. Domínguez-Rodríguez, Acta Mater. **50** (2002) 1177.

[26] R.C. Gifkins, Metall. Trans. **7A** (1976) 1225.

[27] D.S. Wilkinson and C.H. Càceres, Acta Metall. **32** (1984) 1335.

[28] E. Sato and K. Kuribayashi, ISIJ International **33** (1993) 825.

[29] B.N. Kim, K. Hiraga, Y. Sakka and B.W. Ahn, Acta Mater. **47** (1999) 3433.

[30] A.H. Chokshi, J. Euro. Ceram. Soc. **22**, (2002) 2469–2478.

[31] M.G. Zelin and A.K. Mukherjee, Mater. Sci. Eng. A **208** (1996) 210.

[32] H. Muto and M. Sakai, Acta Mater. **48** (2000) 4161.

[33] A.K. Mukherjee, Mater. Sci. Eng. A **322** (2002) 1.

[34] D. Gómez-García, C. Lorenzo-Martín, A. Muñoz-Bernabé and A. Domínguez-Rodríguez, Phil. Mag. **83** (2003) 93.

[35] S. Primdahl, A. Thölen and T.G. Langdon, Acta Metall. Mater. **43** (1995) 1211.

[36] T.G. Nieh and J. Wadsworth, Acta Metall. Mater. **38** (1990) 1121.

[37] D.M. Owen and A.H. Chokshi, Acta Mater. **46** (1998) 667.

[38] A. Bravo-Leon, M. Jiménez-Melendo, A. Domínguez-Rodríguez and A.H. Chokshi, Scripta Met. et Mater. **34** (1996) 1155.

[39] A. Domínguez-Rodríguez, A. Bravo-León, J. Ye and M. Jiménez-Melendo, Mat. Sc. Eng. A **247** (1998) 97.

[40] M.Z. Berbon and T.G. Langdon, Acta Mater. **47** (1999) 2485.

[41] I. Charit and A.H. Chokshi, Acta Mater. **49** (2001) 2239.

[42] C. Herring, J. Appl. Phys. **21** (1950) 437.

[43] B.N. Kim and K. Hiraga, Acta Mater. **48** (2000) 4151.

[44] M. Jiménez-Melendo and A. Domínguez-Rodríguez, Acta Mater. **48** (2000) 3201.

[45] A. Ball and M.M. Hutchison, Met. Sci. J. **3** (1969) 1.

[46] A. Muños, D. Gómez-García, A. Domínguez-Rodríguez and F. Wakai, J. Euro. Ceram. Soc. **22** (2002) 2609.

[47] K. Morita and K. Hiraga, Acta Mater. **50** (2002) 1075.

[48] N. Balasubramanian and T.G. Langdon, Scripta Mater. **48** (2003) 599.

[49] K. Morita and K. Hiraga, Scripta Mater. **48** (2003) 1403.

[50] R. Raj and M.F. Ashby, Metall. Trans. **2** (1971) 1113.

[51] R.Z. Valiev, R.K. Islamgaliev and I.V. Alexandrov, Prog. Mater. Sci. **45** (2000) 103.

[52] M.J. Mayo, Int. Mater. Rev. **41** (1996) 85.

[53] I.W. Chen and X.H. Wang, Nature **404** (2000) 168.

[54] J. Karch, R. Birringer and H. Gleiter, Nature **330** (1987) 556.

[55] F.A. Mohamed and Y. Li, Mater. Sci. Eng. A **298** (2001) 1.

[56] X. McFadden *et al.*, Nature **398** (1999) 684.

[57] K. Matsui *et al.*, J. Am. Ceram. Soc. **86** (2003) 1401.

[58] D. Fan and L.Q. Chen, Acta Mater. **45** (1997) 4145.

[59] L. Clarisse *et al.*, Acta Mater. **45** (1997) 3843.

[60] C.K. Yoon and I.W. Chen, J. Am. Ceram. Soc. **73** (1990) 1555.

[61] K. Morita, K. Hiraga and Y. Sakka, J. Am. Ceram. Soc. **85** (2002) 1900.

[62] B.N. Kim, K. Hiraga, K. Morita and Y. Sakka, Nature **413** (2001) 288.

[63] J.D. French *et al.*, J. Am. Ceram. Soc. **77** (1994) 2857.

[64] W.D. Kingery, H.K. Bowen and D.R. Uhlmann, Introduction to Ceramics, John Wiley & Sons: New York, 1976.

[65] C.M.J. Hwang and I.W. Chen, J. Am. Ceram. Soc. **73** (1990) 1626.

[66] M. Nauer and C. Carry, Scripta Metall. Mater. **24** (1990) 1459.

[67] Y. Ikuhara, P. Thavorniti and T. Sakuma, Acta Mater. **45** (1997) 5275.

[68] E. Sato, H. Morioka, K. Kuribayashi and D. Sundararaman, J. Mater. Sci. **34** (1999) 4511.

[69] L. Donzel, E. Conforto and R. Schaller, Acta Mater. **48** (2000) 777.

[70] K. Nakatani et al., Scripta Mater. **49** (2003) 791.

[71] M. Oka, T. Tabuchi and T. Takashi, Mater. Sci. Forum **304–306** (1999) 451.

[72] Y. Sakka et al., Adv. Eng. Mater. **5** (2003) 130.

[73] J. Mimurada et al., J. Am. Ceram. Soc. **84** (2001) 1817.

[74] Y. Shinoda, T. Nagano, H. Gu and F. Wakai, J. Am. Ceram. Soc. **82** (1999) 2916.

[75] H. Gu, Y. Shinoda and F. Wakai, J. Am. Ceram. Soc. **82** (1999) 469.

[76] F. Wakai, Y. Kodama, S. Sakaguchi and T. Nonami, J. Am. Ceram. Soc. **73** (1990) 457.

[77] P.H. Imamura, N.D. Evans, T. Sakuma and M.L. Mecartney, J. Am. Ceram. Soc. **83** (2000) 3095.

[78] R. Raj and C.K. Chyung, Acta Metall. **29** (1981) 159.

[79] F. Wakai, Acta Metall. Mater. **42** (1994) 1163.

[80] D.R. Clarke, J. Am. Ceram. Soc. **70** (1987) 15.

[81] D.S. Wilkinson, J. Am. Ceram. Soc. **81** (1998) 275.

[82] J.J. Meléndez-Martínez and A. Domínguez-Rodríguez, Prog. Mater. Sci. **49** (2004).

[83] X. Wu and I.W. Chen, *J. Am. Ceram. Soc.* **75** (1992) 2733.

[84] N. Kondo, T. Ohji and F. Wakai, J. Am. Ceram. Soc. **81** (1998) 713.

[85] A. Rosenflanz and I.W. Chen, J. Am. Ceram. Soc. **80** (1997) 1341.

[86] I.W. Chen and S.L. Hwang, J. Am. Ceram. Soc. **75** (1992) 1073.

[87] R.J. Xie, M. Mitomo and G.D. Zhan, Acta Mater. **48** (2000) 2049.

[88] G.D. Zhan, M. Mitomo, R.J. Xie and K. Kurashima, Acta Mater. **48** (2000) 2373.

[89] J.J. Melendez-Martinez, D. Gómez-Garcia, M. Jiménez-Melendo and A. Domínguez-Rodriguez, Phil. Mag. **84** (2004) 3375.

[90] J. Wittenauer, T.G. Nieh and J. Wadsworth, J. Am. Ceram. Soc. **76** (1993) 1665.

[91] A. Domínguez-Rodríguez, F. Guiberteau and M. Jimenez-Melendo, J. Mater. Res. **13** (1998) 1631.

[92] D.M. Owen, A.H. Chokshi and S.R. Nutt, J. Am. Ceram. Soc. **80** (1997) 2433.

[93] Z.C. Wang, N. Ridley and T.J. Davies, J. Mater. Sci. **34** (1999) 2695.

[94] K. Hiraga, K. Nakano, T.S. Suzuki and Y. Sakka, J. Am. Ceram. Soc. **85** (2002) 2763.

[95] S. Harjo *et al.*, Mater. Trans. **43** (2002) 2480.

[96] A.G. Evans, J.R. Rice and J.P. Hirth, J. Am. Ceram. Soc. **63** (1980) 368.

[97] K.R. Venkatachari and R. Raj, J. Am. Ceram. Soc. **70** (1987) 514.

[98] N. Kondo, Y. Suzuki and T. Ohji, J. Am. Ceram. Soc. **82** (1999) 1067.

[99] Z. Shen, H. Peng and M. Nygren, Adv. Mater. **15** (2003) 1006.

[100] C. Molteni, G.P. Francis, M.C. Payne and V. Heine, Phys. Rev. Lett. **76**, (1996) 1284.

[101] H. Yoshida, Y. Ikuhara and T. Sakuma, Acta Mater. **50** (2002) 2955.

[102] H. Van Swygenhoven, Science **296** (2002) 66.

CHAPTER 9

Nanostructure Control for High-Strength and High-Ductility Aluminum Alloys

Tatsuo Sato

Abstract

High-performance aluminum alloys with high strength and high ductility have become markedly attractive for applications to light-weight automobiles, aircrafts, and various components. To achieve such a high strength and high ductility for aluminum alloys, the nanostructure control of precipitates and grain boundary morphologies is essential. The initial phase decomposition and nano-composite structures have been realized to be effective in producing high-performance aluminum alloys. Recently, it has been clarified that nanoclusters are formed in the initial stage of phase decomposition and play very important roles in producing nanoprecipitates. Microalloying elements are also expected to modify the nanostructures of aluminum alloys. Nanoclusters act as useful nucleation sites for precipitates in grains and near grain boundaries. Several techniques have been applied to characterize nanostructures, including precise electrical resistivity measurements, calorimetry, and high-resolution transmission electron microscopy (HRTEM). One of the most advanced techniques is the three-dimensional atom probe (3DAP) technique. The author's group has successfully detected several types of nanoclusters by the 3DAP technique and established a fundamental method of controlling nanostructures using initial nanoclusters. It has also been clarified that precipitate-free zones are important in improving the ductility of alloys.

Keywords: **aluminum alloys, nanoclusters, precipitation, age hardening, microalloying.**

9.1 Introduction

Aluminum alloys with a high mechanical strength and ductility have become extremely attractive as structural materials for reducing weight in various applications such as automobiles, high-speed trains, and aircrafts. A number of fabrication techniques have been developed to increase the mechanical strength and ductility of alloys. Among them, precipitation strengthening is one of the most essential techniques of increasing the mechanical strength of aluminum alloys. Small additions of various alloying elements are of prime importance in modifying microstructures and improving mechanical properties [1,2]. Such microalloying effects have been widely applied to various alloy systems. Microalloying elements in aluminum alloys alter properties by changing the morphology, chemistry, structure, spatial distribution, and size of precipitates. In Al–Cu alloys, solute atom clusters are formed in the very early stage of phase decomposition due to the attractive interaction between microalloying elements and Cu atom clusters. In the case of Al–Cu alloys with the Mg addition, rapid hardening occurs in a very short aging time at low temperatures. In a recent work, the origin of such rapid hardening is proposed to be the interaction between dislocations and Cu/Mg/vacancy complex clusters [3]. The analyses of small clusters in the early stage of phase decomposition have recently been performed using high-resolution transmission electron microscopy (HRTEM) and three-dimensional atom probe with field ion microscopy (3DAP-FIM) [3,4]. It has been realized that nanoscale clusters (i.e., nanoclusters), which are generally formed in the initial stage of phase decomposition, are extremely important in controlling precipitate microstructures and resultant alloy properties. Microalloying elements are considered to affect directly the formation kinetics of nanoclusters; however, the details have not been sufficiently clarified.

The formation behavior of nanoclusters has been examined by electrical resistivity measurements, calorimetry, and HRTEM. Recently, a 3DAP method has been developed and becomes an extremely powerful technique of identifying materials on an atomic scale [5]. A computer simulation using a Monte Carlo method is a powerful technique of determining the kinetics of phase decomposition associated with the diffusion via vacancies, and may also provide atomic-scale information of microstructural evolution [6].

In this chapter, the development of high-strength and high-ductility aluminum alloys is briefly introduced, and the importance of nanocluster control in achieving refined precipitate microstructures and resultant mechanical properties are discussed. The identification techniques of nanoclusters are also introduced together with a theoretical computer calculation for predicting nanocluster formation. The fundamental behaviors of microalloying elements in the formation of nanoclusters are also introduced.

9.2 History of High-Strength and High-Ductility Aluminum Alloys

The mechanical properties of commercial wrought aluminum alloys are shown in Fig. 9.1 to demonstrate the currently achieved status. Figure 9.1 shows the relationship between the tensile strength and elongation of nominated commercial 1000 to 7000 series alloys under various fabrication conditions. The alloys denoted by solid symbols are age-hardenable alloys. The 2000 and 7000 series alloys exhibit the highest strength, which makes them useful materials for aircrafts and high-speed trains. The simultaneous increase in both strength and ductility is highly required.

The high-strength aluminum alloys have achieved their superior strength by producing the effect of precipitation strengthening. The age-hardening phenomenon was first

Nanostructure Control for High-Strength and High-Ductility Aluminum Alloys 317

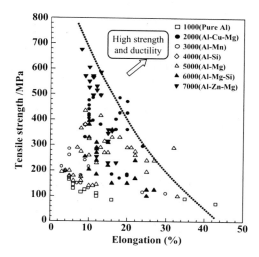

Fig. 9.1. Plots of mechanical properties, tensile strength, and elongation, for various wrought aluminum alloys. The alloys denoted by solid symbols are age-hardenable alloys. The required properties are high strength and high ductility.

discovered by Wilm [7] in Al–Cu–Mg alloys in 1906 and has been extensively applied to the development of high-strength aluminum alloys such as the 2017 aluminum alloy (Al–Cu–Mg-based alloy) in 1917, 2024 aluminum alloy (Al–Cu–Mg-based alloy) in 1939, and 7075 aluminum alloy (Al–Zn–Mg–Cu-based alloy) in 1943. In particular, these alloys have been developed to achieve a high mechanical strength for application as aircraft materials. In the 1970s, however, the requirement for alloys with a high fracture toughness and a high corrosion resistance has become the main issue rather than that for high-strength alloys because of the strong demand for highly reliable alloys. In the 1980s and 1990s, new powder metallurgy (PM) alloys have been developed using fine powders fabricated by rapid solidification methods such as atomization. The extremely fine microstructures of PM alloys composed of fine intermetallic compounds are effective in increasing the elevated temperature resistance as well as mechanical strength. A high cooling rate and well-designed alloy compositions produce amorphous alloys without crystal structures. Nano-composite alloys consisting of very fine crystalline phases in the amorphous phase have also been fabricated. These nano-composite alloys have an extremely high strength of more than 1500 MPa, almost twice the strength of conventional high-strength aluminum alloys [8]. It is clear that the nanoscale-controlled microstructures are effective in increasing mechanical strength. Figure 9.2 shows the history of the development of high-strength aluminum alloys. An extremely high strength has been achieved using the controlled nanostructures [9].

Precipitation strengthening is one of the essential techniques of increasing the mechanical strength of alloys. The phenomena occurring at the early stage of phase decomposition are known to be very important in controlling precipitation microstructures and

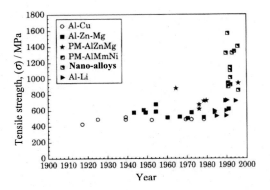

Fig. 9.2. The history of high-strength aluminum alloys showing the increase in tensile strength with year.

the resultant properties. Age-hardening curves of the alloys basically contain the initial hardness increase, maximum hardness, and subsequently decreased hardness with aging time, which are called the under-aging, peak-aging, and over-aging stages, respectively. The characteristic change in the strength or hardness of the alloys is associated with the interaction between precipitate particles and dislocations, as shown in Fig. 9.3.

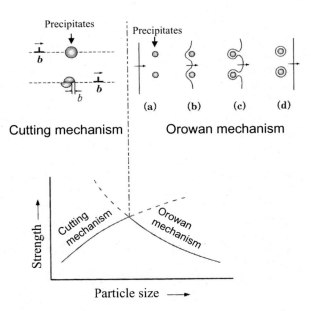

Fig. 9.3. Strength as a function of precipitate particle size, correlated to the interactions between precipitate particles and dislocations.

In the under-aging stage, particles are cut by moving dislocations (the cutting mechanism), whereas in the over-aging stage, particles are not cut and dislocations move away from particles leaving dislocation loops (the Orowan mechanism). This simply indicates that the mechanical strength of alloys can be increased by controlling precipitate microstructures.

Recently, the early-stage phenomena have become the focus again. Al–Mg–Si alloys, which are attractive for automobile applications, exhibit complicated precipitation phenomena and microstructure changes depending on aging conditions and alloy compositions [10,11]. In particular, the early stage of phase decomposition and two-step aging behavior are complicated. The phenomena are considered to be closely associated with the formation of nanoscale clusters in the early stage. Al–Cu–Mg alloys show very rapid early-stage hardening, which is different from that exhibited by Al–Cu binary alloys. Ringer et al. [12] proposed that the early-stage rapid hardening is due to the formation of solute clusters different from Guinier-Preston (GP) zones or Guinier-Preston Bagaryatsky (GPB) zones (See Section 9.3). They called such a phenomenon 'cluster hardening.' It has become very important to clarify the initial structural evolution in order to increase strength and ductility.

9.3 Discovery of GP Zones

The age-hardening phenomenon was first discovered in the Al–Cu–Mg alloys quenched from high temperatures and aged at room temperature (RT) in 1906 [7]. In aluminum alloys, the age-hardening phenomenon is very attractive for use in increasing the mechanical strength and has been extensively applied to the development of new alloys. The mechanism of increasing the strength by aging had not been clarified until Guinier [13] and Preston [14] discovered Cu-rich solute clusters, which are coherent with the aluminum matrix. Interesting research reports were published in a special issue [15] after the memorial symposium on GP zones held in 1988 to celebrate 50 years of the GP zone discovery. Although a number of studies on GP zones have been carried out since Guinier and Preston discovered their existence in 1938, many research studies have still been performed. It has been clarified that GP zones are formed in many alloys such as Al–Zn, Al–Ag, Al–Mg–Si, Al–Zn–Mg, Cu–Be, Cu–Fe, Fe–Mo, and Fe–Cu. However, details of the structure, compositions, and formation kinetics are not yet sufficiently established.

By HRTEM and high-angle annular dark-field STEM (HAADF-STEM), the stacking of Cu-rich layers and lattice distortion have been investigated [16–24]. Typical high-resolution images of GP zones in an Al–Cu alloy are shown in Fig. 9.4. GP zones are formed in a disk- or plate-shaped morphology on {100} matrix planes with typically single Cu-rich layer zones (GP(1) zone) and two or more than two Cu-rich layers separated by three Al layers (GP(2) zone). Recently, a 3DAP technique has been applied to investigate the composition of GP zones [25]. Theoretical investigations have also been carried out to explain the stable morphology, composition, and thermodynamics [26,27]. In Table 9.1, the morphology and crystallographic orientation of GP zones in various aluminum alloys are listed together with intermediate and stable phases. The GP zones formed in Al–Cu–Mg alloys are generally called GPB zones, as described in Table 9.1.

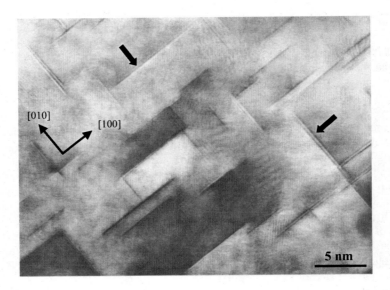

Fig. 9.4. Transmission electron micrograph of typical GP zones as indicated by arrows formed in an Al–4 mass% Cu alloy aged at 423 K for 86.4 ks.

9.4 Clusters in the Early Stage of Phase Decomposition

Most of the precipitation-strengthening aluminum alloys produce metastable phases, such as GP zones and intermediate phases, which play important roles in increasing the mechanical strength of alloys. The formation of solute clusters and GP zones is generally detected by electrical resistivity measurement, X-ray diffuse scattering, and calorimetry in various alloys.

The electrical resistivity changes during aging at 273 K for Al–4 mass% Cu alloys containing various microalloying elements are shown in Fig. 9.5 [10]. The resistivity of the Al–Cu binary alloy increases in the early stage of aging and then decreases after reaching a maximum value. The increased resistivity indicates that a number of small clusters or GP zones are formed in the initial stage of aging. The microalloying element of Mg suppresses the initial increase and subsequently accelerates the resistivity increase to give a greater maximum, suggesting that the Mg addition affects the formation of clusters or GP zones. Similarly, the electrical resistivity changes in Al–Zn alloys containing some microalloying elements during aging at room temperature (RT), as shown in Fig. 9.6 [10]. The resistivity increases in the initial stage and then after a maximum decreases, to reach a plateau. The plateau stage corresponds to the metastable state of GP zones; therefore, solute clusters with various compositions are considered to be formed before the plateau stage. Microalloying elements also affect the initial clustering and GP zone formation.

It is not easy to give a general definition to solute clusters and GP zones. Here, however, the following definitions are given to clusters and GP zones in aluminum alloys for convenience [11,28].

Table 9.1. Morphology and crystallography of GP zones in the matrix of typical age-hardenable aluminum alloys. Intermediate and stable phases are also described.

Alloys	GP zones	Intermediate phases	Stable phases
Al–Cu	GP(1) zones: Disk-shaped with a Cu monolayer structure GP(2) zones(θ''): Plate-shaped with 2 Cu layers separated by 3 Al layers GP(1) and GP(2) zones are formed on the {100} planes of the Al matrix	θ': Tetragonal with Al$_2$Cu Semi-coherent, formed on the {100} planes of the Al matrix	θ: Tetragonal with Al$_2$Cu Incoherent and irregular shaped
Al–Cu–Mg	GPB zones: Needle-shaped elongated in the <100> directions of the Al matrix (GPB zones are possibly classified into GPB(1) and GPB(2) zones)	S$'$: Orthorhombic with Al$_2$CuMg Semi-coherent, formed on the {210} planes in the <100> directions of the Al matrix	S: Orthorhombic with Al$_2$CuMg Incoherent
Al–Mg–Si	GP zones: Formed on the {100} planes of the Al matrix, containing Mg and Si Needle GP zones(β''): Needle-shaped elongated in the <100> directions of the Al matrix, containing Mg and Si	β': Hexagonal, with Mg$_2$Si Rod-shaped elongated to the <100> directions of the Al matrix, semi-coherent Another type of the intermediate phase is formed in excess Si alloys	β: Cubic with Mg$_2$Si Plate-shaped on the {100} planes of the Al matrix Incoherent
Al–Zn–Mg	GP zones: Spherical-shaped, containing Mg and Zn	η': Hexagonal with MgZn$_2$, spherical or elongated shape T$'$: Hexagonal with Mg$_{32}$(Al,Zn)$_{49}$ (formed in the high Mg/Zn ratio alloys)	η: Hexagonal with MgZn$_2$, Incoherent T: Hexagonal with Mg$_{32}$(Al,Zn)$_{49}$
Al–Li	Ordered structure with the L1$_2$-type	δ': Spherical shaped with Al$_3$Li, having the L1$_2$-type ordered structure	δ: Cubic with AlLi, Preferentially formed at grain boundaries Incoherent

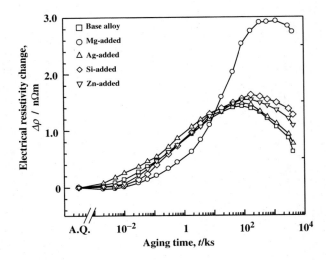

Fig. 9.5. Electrical resistivity changes during aging at 273 K for Al–4 mass% Cu-based alloys containing microalloying elements, Mg, Ag, Si, and Zn.

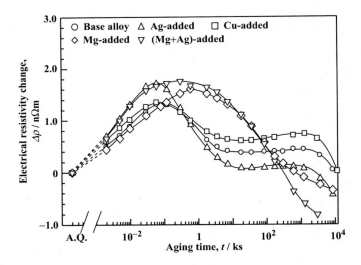

Fig. 9.6. Electrical resistivity changes during aging at room temperature (RT) for Al–4.5% Zn-based alloys containing 0.1% Ag, Cu, Mg, and (0.1% Mg + 0.1% Ag), (mol%).

Clusters: The crystal structure is FCC, which is coherent with the aluminum matrix. The composition is lower than that of the metastable state. The size is smaller than about 2 nm, and the shape is irregular or spherical.

GP zones: The crystal structure is FCC, which is coherent with the aluminum matrix. The composition is the same as that of the metastable state. The size is larger than about 2 nm, and the shape is distinct (e.g., spherical, needle, and disk).

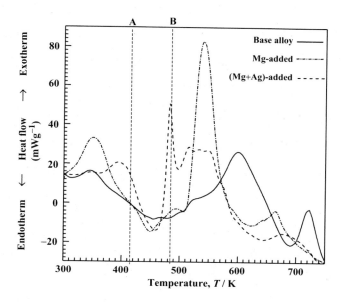

Fig. 9.7. Differential scanning calorimetry (DSC) curves for the as-quenched Al–4 mass% Cu, Mg-added, and (Mg + Ag)-added alloys. The points indicated by A and B show the temperatures at which TEM observation was performed.

The formation of clusters at low temperatures is also detected by calorimetric techniques such as the differential scanning calorimetry (DSC) and adiabatic specific heat measurements. Figure 9.7 shows the DSC curves for as-quenched Al–Cu alloys with microalloying elements of Mg and (Mg + Ag) [10]. The exothermic peaks from 350 to 400 K are due to the formation of clusters or GP zones. The clear exothermic peaks observed indicate the formation of solute clusters or GP zones at low temperatures. The DSC measurement is a useful technique of detecting the formation of small clusters.

9.4.1 Cluster hardening

In Al–Cu–Mg alloys, rapid early hardening is observed during aging in the temperature range of 320–470 K. Ringer *et al.* [12] proposed that such rapid hardening is associated with the formation of clusters of Cu and Mg atoms, and called it 'cluster hardening.' As an example, the rapid hardening behavior of the Al–Cu–Mg alloy aged at 323 K after several quenching processes is shown in Fig. 9.8 [29]. The hardness markedly increases within a few minutes even at low temperatures such as RT and 323 K. Reich and coworkers [3,4] reported the existence of co-clusters of Mg and Ag atoms in the Al–1.9 at% Cu–0.3 at% Mg–0.2 at% Ag alloy aged at 453 K for 5 s using the 3DAP technique. Cu atoms are not included at this stage. In the alloy aged for 120 s, Cu atoms aggregate into Mg–Ag co-clusters. This indicates that the Mg–Ag co-clusters act as the formation sites for solute Cu clusters.

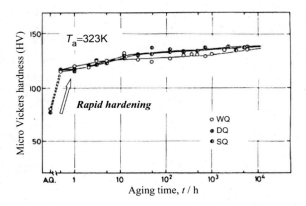

Fig. 9.8. Hardness changes with aging time at 323 K for an Al–4% Cu–1.5% Mg alloy (mass%) showing extremely rapid hardening in the initial stage (WQ: water-quenched, DQ: directly quenched, SQ: step-quenched).

9.4.2 Microalloying elements for cluster formation

Understanding the role of trace elements or microalloying elements in age-hardenable aluminum alloys is also very important in controlling microstructures and properties [10,30–32]. The typical phenomena related with microalloying elements are described below for the Al–Cu and Al–Mg–Cu alloys.

Al–Cu alloys. The Al–Cu system is one of the most fundamental alloys that show age hardening and has been extensively investigated in terms of phase decomposition, kinetics, and microstructural changes. The generally accepted precipitation sequence of this system is as follows:

$$\text{SSSS} \rightarrow \text{GP(1) zones} \rightarrow \text{GP(2) zones (or } \theta'') \rightarrow \theta' \rightarrow \theta - \text{Al}_2\text{Cu} \qquad (9.1)$$

Here, SSSS means a super saturated solid solution. The age-hardening curves for Al–Cu alloys containing microalloying elements of Mg and Ag are shown in Fig. 9.9 [10]. The characteristic increase in hardness is observed in the Al–Cu binary and Mg or (Mg + Ag)-added alloys in Fig. 9.9(a). In the Mg-added alloy, the age hardening is suppressed in the initial stage of aging at RT and then accelerated in the subsequent stage compared with that in the Al–Cu binary alloy. At 373 K (Fig. 9.9(b)), the age-hardening of the Mg-added alloy is greatly accelerated from the beginning of aging. On the other hand, the age-hardening of the (Mg + Ag)-added alloy is markedly suppressed in the initial stage of aging and subsequently greatly accelerated compared with that of the two other alloys. This tendency is more pronounced at 373 K. The Ag-added alloy (single addition) exhibits age-hardening curves almost similar to those of the Al–Cu alloy, suggesting that the single addition of Ag negligibly affects the age-hardening kinetics. TEM micrographs and corresponding diffraction patterns for the Al–Cu and Mg-added alloys aged at RT for 18 ks are shown in Fig. 9.10 [11]. An imaging plate is used to record the weak intensity of diffraction patterns precisely for both alloys. The diffraction patterns of both the alloys

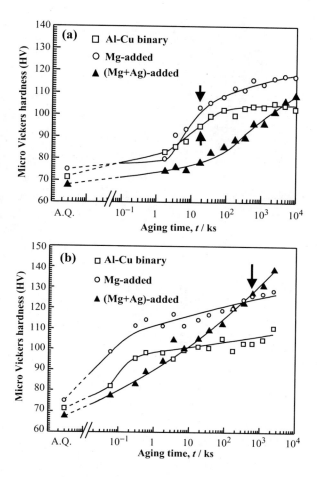

Fig. 9.9. Changes of micro Vickers hardness in Al–4 mass% Cu binary, Mg-added, and (Mg + Ag)-added alloys during aging at (a) RT and (b) 373 K.

exhibit streaks running toward the <100> matrix directions, showing that GP zones are formed at this aging stage. The quantitative analysis of the streaks of both the alloys reveals that the streaks of the Mg-added alloy are broader than those of the binary alloy, demonstrating that the GP zones of the Mg-added alloy are much finer than those of the binary alloy. Mg produces finer GP zones and possibly affects the nucleation of GP zones. To determine the reason for the greatly increased hardness of the (Mg + Ag)-added alloy aged at 373 K, TEM observation is performed. The microstructure observed exhibits fine platelets coexisting with the conventional GP(1) zones aligned in the <100> matrix directions. The fine platelets seem to be aligned on the {111} matrix planes. To confirm this point, HRTEM images are taken from the alloy aged at 373 K for 604.8 ks, and an example is shown in Fig. 9.11 [11]. Small platelets with a single layer contrast are observed to be aligned on the {111} matrix planes. The precipitates on the {111} planes are fully coherent with the matrix and exhibit a strain field contrast.

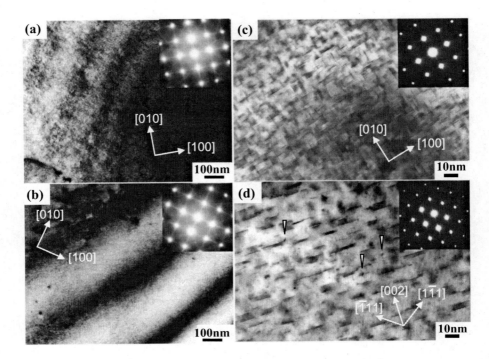

Fig. 9.10. Electron micrographs and diffraction patterns of Al–4 mass% Cu, Mg-added and (Mg + Ag)-added alloys. (a) Al–4 mass% Cu: RT, 18 ks, (b) Mg-added: RT, 18 ks, (c) Al–4 mass% Cu: 373 K, 604.8 ks, and (d) (Mg + Ag)-added: 373 K, 604.8 ks.

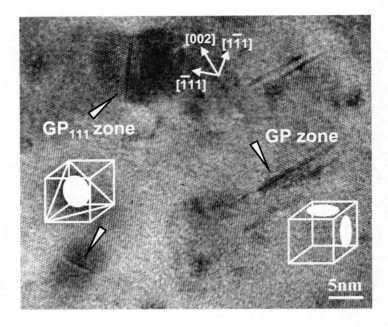

Fig. 9.11. High-resolution TEM image showing new type GP_{111} zones formed on the {111} planes of the matrix together with GP zones formed on the {100} planes in the (Mg + Ag)-added alloy aged at 373 K for 604.8 ks.

These platelets are called the GP$_{111}$ zones in this work, which are different from the conventional GP(1) zones. From the above observation, the highest hardness is attributed to the characteristic microstructures of the GP$_{111}$ zones on the {111} matrix planes, which are the slip planes of the FCC structure. The combined addition of (Mg + Ag) is observed to produce fine GP$_{111}$ zones. It is also suggested that the combined addition of Mg and Ag affects the nucleation of the fine {111} platelets. A similar suggestion is also reported by Mukhopadhyay [33].

(a) Three-dimensional atom probe (3DAP)

As discussed in our previous articles [1,10,34], small clusters of nanoscale size (nanoclusters) are formed in the initial stage of phase decomposition of age-hardenable aluminum alloys. The nanoclusters are expected to affect the formation of GP zones and metastable phases. In this work, we have applied the 3DAP technique to detect directly the nanoclusters formed in the alloys containing microalloying elements. The results obtained using this technique can reveal each atom in the specimen with its three-dimensional position and identification, and that this technique is a powerful tool for analyzing microstructures on an atomic scale. Figure 9.12 shows the 3DAP atom maps of the Al–Cu and Mg-added alloys aged at RT [11]. To avoid confusion, only the Cu and Mg atoms are displayed in the maps. In the binary alloy, the Cu atoms aggregate to form small clusters at RT.

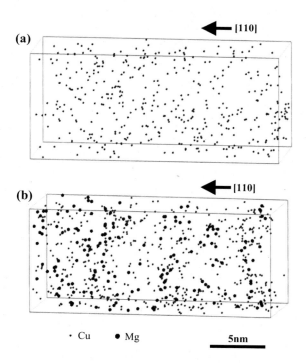

Fig. 9.12. 3DAP atom maps showing nanoclusters in Al–4 mass% Cu binary and Mg-added alloys. (a) Al–4 mass% Cu: RT, 21.6 ks, (b) Mg-added: RT, 12.6 ks.

The results of the statistical analysis of atom positions reveal that the Cu atoms are no longer distributed homogeneously even in the very initial stage of aging. In the Mg-added alloy, the Cu and Mg atoms aggregate to form small clusters that are closely correlated. Therefore, the Cu and Mg atoms are incorporated inside the nanoclusters. It is also reported that the quenched-in excess vacancies are incorporated inside the nanoclusters [11]. The nanoclusters containing solute atoms (Cu), microalloying atoms (Mg), and vacancies are called complex clusters. On the basis of the results of our computer simulation, the complex clusters are found to act as effective nucleation sites for GP zones. Figure 9.13 shows the 3DAP atom maps of the binary and (Mg + Ag)-added alloys aged at 373 K [11]. In Fig. 9.13(b), some GP zones are observed to be aligned in the <100> matrix directions. On the contrary, a number of small clusters containing Cu, Mg, and Ag atoms are found in Figs. 9.13(c) and (d). In Fig. 9.13(d), some of the clusters are elongated on the {111} matrix planes. This indicates that the clusters tend to aggregate on the {111} planes and resultantly, accelerate the formation of the GP_{111} zones.

(b) Computer simulation

The nanoclusters containing solute atoms, microalloying atoms, and vacancies are expected to act as nucleation sites for the subsequent precipitates. Therefore, identifying the characteristic features of nanoclusters is extremely important. However, identifying nanoclusters experimentally is difficult. Thus, we performed a computer simulation to analyze the very early stage of phase decomposition. Figure 9.14 shows the results of the computer simulation for the Al–Cu and Al–Cu–Mg alloys showing the process of cluster formation [35]. With an increasing number of Monte Carlo steps (MCs) or time, a number of clusters are formed. In the simulation, the addition of Mg clearly indicates that the size of nanoclusters decreases and the number density increases. Figure 9.15 shows one of the typical examples demonstrating the formation sequence of the complex clusters and nanoclusters. A Mg/vacancy cluster is initially formed (Fig. 9.15(b)), indicating that vacancies are trapped by Mg atoms. In the next step, Cu atoms aggregate on the Mg/vacancy cluster to form a Cu/Mg/vacancy complex cluster (Figs. 9.15(c) and (d)). This is an important sequence that represents the heterogeneous nucleation behavior of GP zones. This process is termed 'Nanocluster Assist Processing (NCAP)' for the GP zone formation [10]. Mg atoms trap vacancies to retard phase decomposition (vacancy trapping [36]) and subsequently aggregate to form Cu/Mg/vacancy complex clusters that accelerate the GP(1) zone formation. Similarly, the (Mg + Ag) combined addition also produces strong vacancy trap sites in the form of Mg/Ag/vacancy clusters, and then contributes to the formation of Cu/Mg/Ag/vacancy nanoclusters, which effectively act as the nucleation sites for the GP_{111} zones formed on the {111} matrix planes, as schematically illustrated in Fig. 9.16. This is again the heterogeneous nucleation of GP zones on nanoclusters. Microalloying elements are one of the important factors in NCAP available to improve alloy properties.

The effects of microalloying elements on the cluster or GP zone formation are interesting and important in controlling the precipitate microstructures. The effects of microalloying elements are basically understood by considering the interaction parameters between

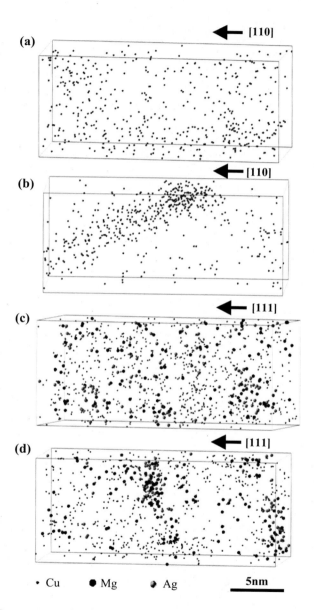

• Cu   ● Mg   ● Ag   5nm

Fig. 9.13. 3DAP atom maps showing nanoclusters in Al–4 mass% Cu binary and (Mg + Ag)-added alloys aged at 373 K. (a) Al–4 mass% Cu: 373 K, 0.6 ks, (b) Al–4 mass% Cu: 373 K, 86.4 ks, (c) (Mg + Ag)-added alloy: 373 K, 0.6 ks, (d) (Mg + Ag)-added alloy: 373 K, 86.4 ks.

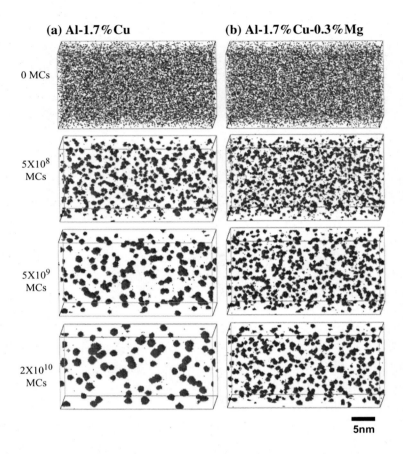

Fig. 9.14. Monte Carlo simulation images showing nanocluster formation in (a) Al–4 mass% Cu and (b) Al–4 mass% Cu–0.3 mass% Mg alloys at RT. The microalloying of Mg causes refined nanoclusters.

elements or vacancies and microalloying elements. Figure 9.17 shows a pair interaction map representing the interaction energies calculated using the first principal theory between Mg and X (elements), and between a vacancy and X. The positive and negative values indicate the repulsive and attractive interactions. The map clearly shows that the interactions of Mg–Ag, Cu–Mg, and Mg-vacancy are negative, indicating that the interactions are attractive. According to this map, the clusters of Mg/Cu/vacancy and Mg/Ag/vacancy are preferentially formed in solid solutions. Recently, Zhu et al. [37] have calculated short-range order parameters to predict possible cluster formation in Al–Cu–Mg, Al–Cu–Mg–Ag, and Al–Cu–Mg–Si alloys using a quasi-chemical model for higher order systems. They reported that Mg-rich Mg–Cu co-clusters are most likely formed rather than Mg clusters in an Al–Cu–Mg alloy. In an Al–Cu–Mg–Si alloy, Si tends to form co-clusters with Mg and Cu, consuming Cu and Mg, which jeopardizes the formation of Cu–Mg clusters and Ω precipitates. In an Al–Cu–Mg–Ag alloy, Ag has an overwhelming tendency to form co-clusters with Mg, which leads to the formation of

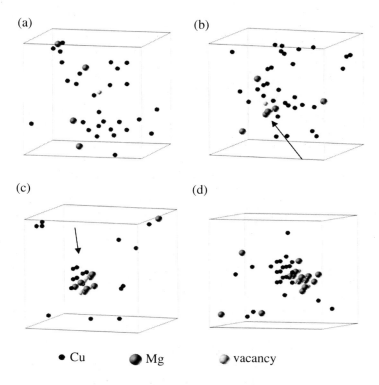

Fig. 9.15. Formation sequence of complex clusters and nanoclusters obtained in the computer simulation for Mg-added alloy. (a) Start, (b) 5×10^5 MCs, (c) 5×10^6 MCs, and (d) 5×10^8 MCs.

Ag–Mg–Cu co-clusters. They concluded that Ag–Mg–Cu co-clusters most likely act as the precursors for Ω precipitates rather than Mg–Ag co-clusters.

Al–Mg–Cu(–Ag) alloys. The effect of nanoclusters on the control of precipitate microstructures and alloy properties is also demonstrated in Al–3% Mg–1% Cu(–0.3% Ag) alloys (in mass%). Figure 9.18 shows age-hardening curves for the Al–Mg–Cu and Al–Mg–Cu–Ag alloys aged at 443 K [38]. Both alloys show a very rapid initial hardening process within a few minutes at 443 K. The binary alloys of Al–3% Mg and Al–1% Cu with the same composition as Al–3% Mg–1% Cu do not show age hardening, suggesting that the combination of Mg and Cu is effective in inducing age hardening [39]. Furthermore, the microalloying element of Ag is extremely effective in increasing age hardening. In the TEM micrographs of both alloys aged at 443 K for 1210 ks, the almost peak hardness stage as indicated by arrows, are shown in Fig. 9.19. In the Al–Mg–Cu alloy, large rod-shaped precipitates (S′-Al$_2$CuMg) are observed together with small amounts of fine precipitates. In the Ag-added alloy, extremely fine spherical precipitates are observed to be distributed homogeneously. The fine spherical precipitates are believed to be the T′ phase; however, they are not completely identified yet. It is clear that the small addition of Ag causes marked changes in the precipitate morphologies and microstructures, which are useful in increasing the

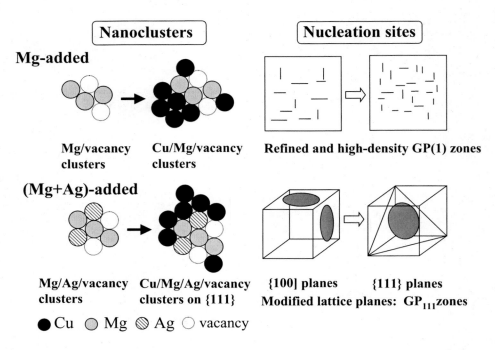

Fig. 9.16. Nanoclusters formed in the Mg-added and (Mg + Ag)-added alloys. The nanoclusters in the initial stage trap vacancies, causing retarded aging (vacancy trapping phenomenon), whereas complex nanoclusters formed in the subsequent stage accelerate the nucleation of GP zones by performing as nucleation sites. In the (Mg + Ag)-added alloy, Cu/Mg/Ag/vacancy clusters are formed on the {111} matrix planes and enhance the formation of new type GP_{111} zones on the {111} matrix planes.

hardness. The TEM observation, however, cannot reveal the microstructure changes in the initial stage of rapid hardening. The 3DAP atom maps for the Al–Mg–Cu–Ag alloy aged at 443 K for 3.6 ks, indicated by an arrow in Fig. 9.18, are shown in Fig. 9.20. The distribution of each element, Mg, Cu, or Ag, shows its contribution to the cluster formation. It is clearly identified that very small solute clusters containing Ag are formed, indicating that Ag accelerates the formation of clusters. As described in this section, Al–Cu alloys, Ag has an attractive interaction with Mg; therefore, Mg/Al co-clusters are preferentially formed in the initial stage and act as effective nucleation sites for the refined precipitates.

9.4.3 Two-step aging behavior

After quenching from solid solution temperature, pre-aging is carried out at temperature T_1 for time t_1, then the second aging is performed at temperature T_2 ($T_2 > T_1$) for time t_2. This aging sequence is called the two-step aging. In general, the pre-aging affects the precipitation behavior in the second aging, kinetics, microstructures, and properties

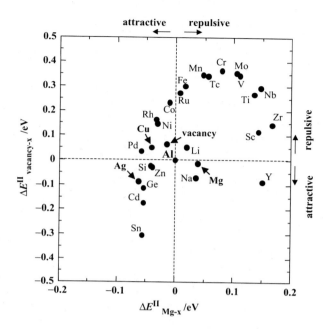

Fig. 9.17. Pair interaction map for various elements, X, with Mg and vacancy. Typical elements in Al–Cu–Mg–Ag alloys are represented by arrows, showing that these elements are effective to form nanoclusters of Mg/Cu/vacancy and Mg/Ag/vacancy.

[40,41]. The typical two-step aging behavior is described below for the Al–Zn–Mg and Al–Mg–Si alloys.

Al–Zn–Mg alloys. Among commercial aluminum alloys, the Al–Zn–Mg-based system is a typical age-hardening alloy with the highest strength and has been mainly used in the aerospace industry. The precipitation sequence is as follows:

$$\text{SSSS} \rightarrow \text{GP zones} \rightarrow \text{metastable } \eta' \rightarrow \text{stable } \eta \text{ (MgZn}_2\text{)} \tag{9.2}$$

The formation mechanism and composition of the GP zones and η' phase have been widely investigated. Recently, Sha and Cerezo [42] have reported the cluster formation based on their experimental results obtained using a 3DAP technique. They found small Mg-rich clusters (or GP-I zones) (<30 solute atoms with Zn/Mg = ~0.9) in the Al–Zn–Mg–Cu alloy after short aging times (within 30 min) at 394 K, together with large GP-I zones (Zn/Mg = 1.0). They also pointed out that the dominant mechanism for the η' formation is the transformation of small GP-I zones via the elongated clusters and not the nucleation on large zones.

In the high-strength age-hardenable alloy, Al–Zn–Mg, the pre-aging treatment is effective in producing refined precipitates and the resultant increased age-hardening [43]. Figure 9.21 shows age-hardening curves at 423 K for various heat treatments together

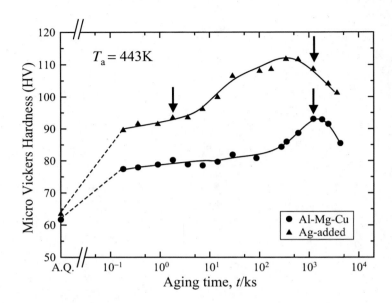

Fig. 9.18. Age-hardening curves for Al–3% Mg–1% Cu and Ag-added (0.4%) alloys (mass%) aged at 443 K, showing the rapid initial hardening. TEM observation and 3DAP analysis were performed at the stages indicated by arrows.

with TEM micrographs in an Al–Zn–Mg alloy. With increasing aging time at RT (pre-aging), the peak hardness increases compared with that in the single aging at 423 K after water quenching (WQ). The pre-aging treatment causes a positive effect, i.e., two-step aging with a positive effect, in an Al–Zn–Mg alloy. The direct quenching (DQ) causes negligible age hardening at 423 K. Such quenching results in coarse precipitates (η' phase), whereas the two-step aging results in refined precipitates (η' phase), as shown in the TEM micrographs. It is clear that small clusters and GP zones are effective in forming refined η' precipitates and increasing age hardening.

Al–Mg–Si alloys. The Al–Mg–Si system is also a typical age-hardening alloy with high formability, high corrosion resistance, and high weldability, and is very useful as an automobile material. These alloys are strengthened by paint baking due to the precipitation of the β'' phase. However, the phase decomposition in the initial stage before the formation of the β'' phase is not fully established [44–47]. Two-step aging phenomena are also more complicated in the Al–Mg–Si alloys than in the Al–Zn–Mg alloys. The generally accepted precipitation sequence is as follows:

$$\text{SSSS} \rightarrow \text{atomic clusters} \rightarrow \text{GP zones} \rightarrow \beta'' \rightarrow \beta' \rightarrow \beta\text{-Mg}_2\text{Si} \tag{9.3}$$

Marioara et al. [48] reported that the pre-β'' phase can be derived from an Al lattice by replacing the Al atoms with the Mg and Si atoms of $(\text{Al} + \text{Mg})_5\text{Si}_6$ composition and by relaxing the lattice to accommodate different atomic radii. They also reported that the transition from an SSSS with a large amount of quenched-in vacancies to the β'' phase

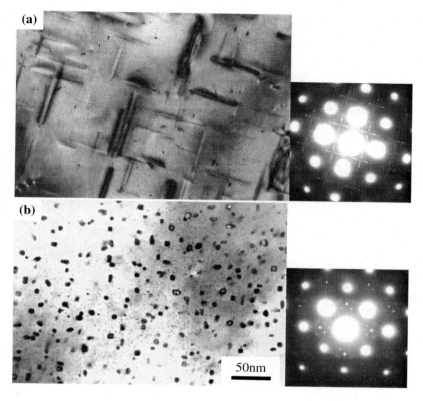

Fig. 9.19. TEM micrographs and diffraction patterns for (a) Al–3% Mg–1% Cu and (b) Ag-added (0.4%) alloys (mass%) aged at 443 K for 1210 ks, as indicated by arrows in Fig. 9.18.

is a process involving the formation of several intermediate phases. Marioara et al. [49] also pointed out that atomic clusters are created during storage at RT. Then, completely coherent pre-β'' particles (or GP-I zones) are formed from these atomic clusters. If the annealing temperature is higher than 398 K, the β'' phase will form from the pre-β'' phase precipitates. However, it has recently become evident that the β'' phase contains Mg and Si at a ratio close to 1 rather than 2. Andersen et al. [50] investigated the structure of the β'' phase by HRTEM and electron diffraction analyses. They determined that the β'' phase has a smaller Mg/Si ratio with the composition Mg_5Si_6 than the equilibrium phase β-Mg_2Si. More detailed analyses were performed by Edwards et al. [51] by atom probe field ion microscopy (APFIM) with the combined methods of DSC and TEM. They proposed the following precipitation process:

Mg- and Si- independent clusters → Mg/Si co-clusters → small unknown

precipitates → β'' needle-shaped precipitates → β' rod-shaped precipitates,

B' lath-shaped precipitates → β-Mg_2Si (9.4)

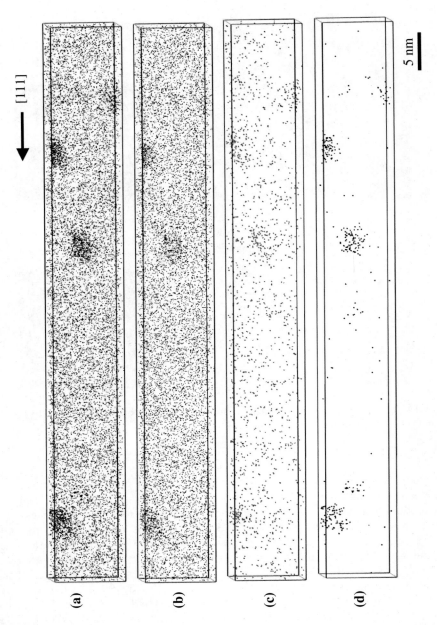

Fig. 9.20. 3DAP atom maps showing nanoclusters in an Al–3% Mg–1% Cu–0.4% Ag alloy (mass%) aged at 443 K for 3.6 ks. (a) Mg, Cu, and Ag atoms, (b) Mg atoms, (c) Cu atoms, and (d) Ag atoms. The nanoclusters contain Mg, Cu, and Ag atoms.

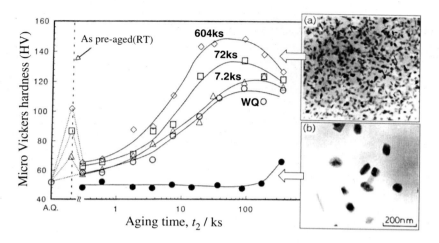

Fig. 9.21. Age-hardening curves for an Al–5% Zn–2% Mg alloy (mass%) aged at 423 K after direct quenching (DQ), water quenching (WQ), and pre-aged at RT for various aging times. The TEM micrographs indicate that the DQ produces coarse precipitates and two-step aging at RT causes refined precipitates (a) and DQ produces coarse precipitates (b).

They reported that the Mg/Si ratio in the intermediate precipitates and co-clusters is close to 1. Dutta and Allen [52] also proposed that the observed effects in Al–Mg–Si alloys result from the clustering of Si atoms. As described above, the importance of clusters formed in the initial stage of precipitation has recently been realized.

The two-step aging behavior is rather complicated in the Al–Mg–Si alloys. Figure 9.22 shows age-hardening curves for an Al–Mg–Si alloy single aged after water

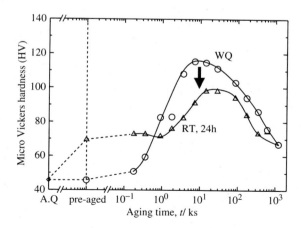

Fig. 9.22. Age-hardening curves for an Al–1.4 mass% Mg_2Si alloy aged at 453 K after water-quenched (WQ) and pre-aged at RT for 86.4 ks (two-step aging).

quenching (WQ) and two-step aging (pre-aging at RT for 24 h) at 453 K. Such pre-aging causes decreased peak hardness, indicating that the pre-aging at RT results in a negative effect, different from the case of an Al–Zn–Mg alloy. In detail, it has been clarified that the two-step aging behavior changes complicatedly depending on the pre-aging temperature [53,54]. Figure 9.23 shows the various aging treatment procedures for clarifying the two-step aging behavior of Al–Mg–Si alloys. Figure 9.24 shows hardness

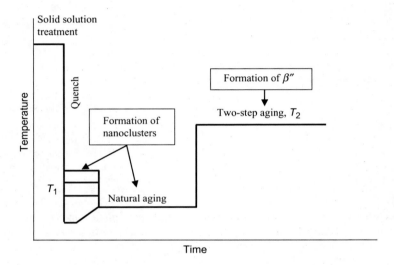

Fig. 9.23. Schematic illustration showing the two-step aging treatment procedure for Al–Mg–Si alloys.

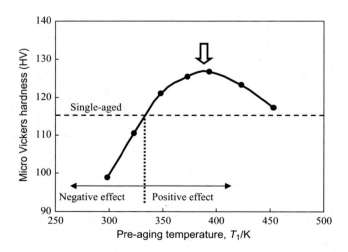

Fig. 9.24. Hardness of an Al–1.4 mass% Mg_2Si alloy two-step aged at 453 K for 36 ks after pre-aging at various temperatures for 3.6 ks.

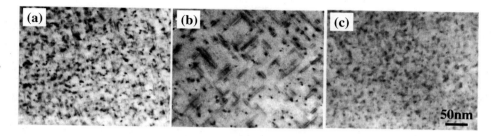

Fig. 9.25. TEM micrographs of the Al–1.4 mass% Mg_2Si alloy aged with various conditions. (a) single aging (453 K, 14.4 ks), (b) conventional two-step aging (RT, 86.4 ks → 453 K, 14.4 ks), and (c) pre-aging and two-step aging (373 K, 0.6 ks → RT, 86.4 ks → 453 K, 14.4 ks).

after two-step aging at 453 K for 36 ks (almost a peak hardness condition) as a function of pre-aging temperature [54]. Compared with the hardness of the single-aged condition the pre-aging at temperatures lower than 340 K causes a negative effect, whereas the pre-aging at temperatures higher than 340 K produces a positive effect. The maximum hardness is obtained at the pre-aging temperature of about 380 K. The hardness change depending on the pre-aging temperature is well associated with the TEM micrographs, as shown in Fig. 9.25. The conventional two-step aging (pre-aged at RT) results in the microstructure of coarse precipitates, as shown in Fig. 9.25(b), compared with the microstructure of the single-aged alloy. The short aging at 373 K before pre-aging at RT produces refined precipitates, as shown in Fig. 9.25(c). These complicated behaviors are understood by considering the nanoclusters formed during pre-aging in Al–Mg–Si alloys. There are many TEM observations on the precipitates formed by conventional aging. However, a few TEM observations have been carried out for GP zones or clusters. Particularly, it is difficult to detect clusters formed during pre-aging. Figure 9.26 shows HRTEM images showing nanoclusters and GP zones in an Al–Mg–Si alloy aged at 295 K for 6000 ks [55]. A very small contrast region of 1 nm observed in the enlarged image (b) is considered to be a solute cluster or a GP zone. The dark contrast of the region suggests that some lattice strain is introduced inside the cluster. In the image (c), a single-layer zone is observed. It is difficult to obtain further information from the TEM image.

(a) Three-dimensional atom probe (3DAP)

The 3DAP technique has been applied to identify solute clusters in Al–Mg–Si alloys. Figure 9.27 shows 3DAP atom maps in the Al–Mg–Si alloy aged at RT and 373 K, showing clearly the existence of nanoscale solute clusters. The Mg and Si atoms are shown, but the Al atoms are not represented to avoid complicated images. In Fig. 9.27(a), Si-rich, Mg-rich, and Mg–Si clusters are observed. In Fig. 9.27(b), clusters composed of Mg and Si, which are larger than those in Fig. 9.27(a), are observed to be formed even in the alloy aged for a short time at 373 K. Although the difference between the two clusters observed in the alloy aged at RT and 373 K is not clear, it is confirmed by the 3DAP technique that solute clusters are actually formed. One of the clusters in Fig. 9.27(b) is analyzed

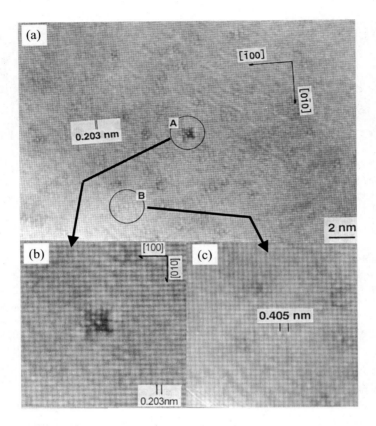

Fig. 9.26. High-resolution TEM images showing nanoclusters and GP zones in an Al–1.6 mass% Mg_2Si alloy aged at 295 K for 6000 ks. (a) Low magnification, (b) an enlarged image of an initial stage GP zone in A, (c) an enlarged image of a single layer GP zone.

to determine the Mg/Si ratio, as shown in Fig. 9.28. The cluster contains 40 Mg atoms and 24 Si atoms, indicating that the Mg/Si ratio is about 1.7.

(b) Mechanism of two-step aging behavior

On the basis of the specific heat measurement obtained in our group, we assume the formation of two types of nanocluster, Cluster(1) and Cluster(2), and propose the mechanism to explain the complicated two-step aging phenomenon of Al–Mg–Si alloys, as shown in Fig. 9.29 [53,54]. Cluster(1) is formed at temperatures below 343 K and is harmful for the formation of the β'' phase, whereas Cluster(2) is formed at temperatures above 343 K and is useful for the formation of the β'' phase. Two clusters are formed competitively depending on the pre-aging conditions. One should be careful in controlling the cluster formation in an industrial process. Vacancies are also assumed to be incorporated into the clusters. More detailed atomic scale analyses are required in the Al–Mg–Si alloys.

Nanostructure Control for High-Strength and High-Ductility Aluminum Alloys   341

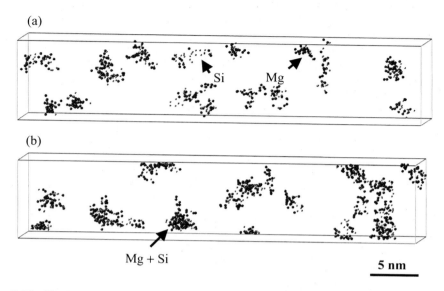

Fig. 9.27. 3DAP atom maps showing nanoclusters in an Al–1.4 mass% Mg_2Si alloy (Mg and Si atoms are represented). (a) RT, 1210 ks, (b) 373 K, 3.6 ks.

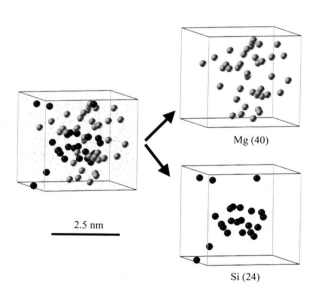

Fig. 9.28. Number of Mg and Si atoms in a nanocluster formed in an Al–1.4 mass% Mg_2Si alloy aged at 373 K for 3.6 ks.

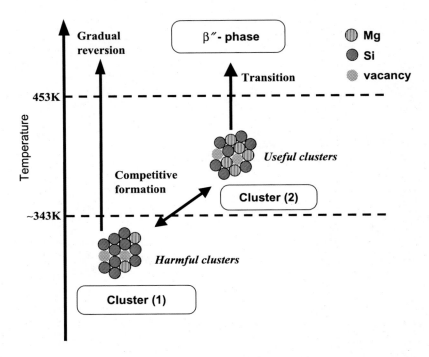

Fig. 9.29. Schematic illustration showing the two-step aging mechanism in Al–Mg–Si alloys. Two types of clusters, Cluster(1) and Cluster(2), are formed competitively during pre-aging and affect the nucleation of the β'' phase. Cluster(1) is harmful for the nucleation of the β'' phase, whereas Cluster(2) is useful for the nucleation of the β'' phase.

9.5 Ductility and PFZ Control by Nanoclusters

The ductility of alloys is also affected by precipitate-free zones (PFZs) near grain boundaries. Figure 9.30 shows the relationship between proof stress and elongation for Al–Zn–Mg-based alloys aged at different temperatures for several aging times [56]. Ag is also added as a microalloying element. The addition of Ag and low-temperature aging are effective in increasing strength and elongation. The increased strength and ductility are basically due to the decreased width of PFZs. Figure 9.31 shows the TEM micrographs of the Al–Zn–Mg and Al–Zn–Mg–Ag alloys aged at 433 K for 86.4 ks. It is clear that the width of PFZ markedly decreases by the addition of Ag. The size of grain boundary precipitates also decreases by the addition of Ag. The low-temperature aging is also effective in decreasing the width of PFZs. Decreasing the width of PFZs is effective in increasing ductility. The detailed mechanism for the decreased width of PFZs or disappearance of PFZs is not clear. The distribution profiles of solute atoms and quenched-in vacancies are assumed to influence the nucleation and growth of precipitates. The microalloying element of Ag is expected to enhance the formation of nanoclusters that accelerate the nucleation of precipitates near grain boundaries. Nanoclusters are presumably useful in controlling PFZs and grain boundary precipitates.

Nanostructure Control for High-Strength and High-Ductility Aluminum Alloys 343

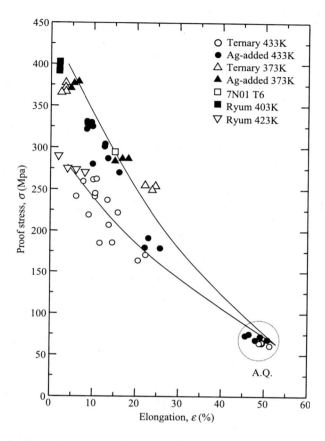

Fig. 9.30. Relationship between proof stress and elongation for Al–4.9% Zn–1.8% Mg ternary, Ag-added (0.3%), 7N01 alloys aged at various temperatures (mass%).

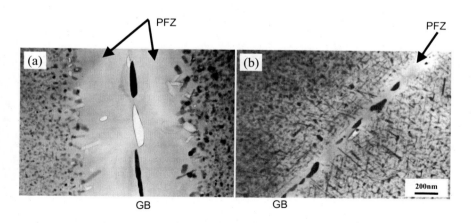

Fig. 9.31. TEM micrographs showing precipitate-free zones (PFZs) in Al–4.9% Zn–1.8% Mg and Ag-added (0.3%) alloys aged at 433 K for 86.4 ks (mass%). The microalloying element of Ag is effective to decrease the width of PFZ.

9.6 Summary

The history of the development of high-strength aluminum alloys is briefly introduced. It is described that the precipitate microstructures are very important in increasing the mechanical strength and ductility of aluminum alloys. Most of the high-strength aluminum alloys achieve their strength by the metastable precipitates, such as GP zones, coherent with the aluminum matrix. Controlling GP zones is one of the most important key factors in alloy development. Therefore, a number of investigations have been carried out to understand the formation mechanism, kinetics, and structures using many techniques, such as electrical resistivity measurements, hardness measurements, calorimetric analyses, X-ray diffraction techniques, and HRTEM. Recently, it has been realized that very small nanoclusters are formed before the formation of GP zones and affect the formation of GP zones and metastable phases. Microalloying elements are effective in controlling the formation kinetics and structures of nanoclusters. Nanoclusters act as nucleation sites for the subsequent precipitates, enhancing precipitation strengthening. In some cases, however, nanoclusters suppress the formation of the subsequent precipitates, decreasing strength. Therefore, it is essentially important to determine the characteristic features of nanoclusters. The 3DAP technique recently developed is extremely useful in identifying nanoclusters on an atomic scale. It is also required to clarify the role of quenched-in excess vacancies on the formation of nanoclusters. The positron annihilation measurement is one of the useful methods of identifying vacancies. The theoretical analyses, such as the first principle calculation and simulation, are also important in understanding the phenomena occurring in the very early stage of precipitation. Aluminum alloys are superior lightweight materials and are expected to be used for various structural materials. The increased strength and ductility of aluminum alloys are expected to further extend the commercial applications. The newly proposed concept of the nanocluster assist processing (NCAP) is one of the useful methods of achieving markedly improved properties.

Acknowledgments

The present study was partly conducted under the 'Nanotechnology Metal Project' supported by the New Energy and Industrial Technology Development Organization (NEDO) and The Japan Research and Development Center for Metals (JRCM).

References

[1] T. Sato, Mater. Sci. Forum **331–337** (2000) 85.

[2] Y. Baba, J. Jpn. Inst. Metals **30** (1966) 679.

[3] L. Reich, S.P. Ringer and K. Hono, Phil. Mag. Lett. **79** (1999) 639.

[4] L. Reich, M. Murayama and K. Hono, Acta Mater. **46** (1998) 6053.

[5] K. Hono, Acta Mater. **47** (1999) 3127.

[6] S. Hirosawa and T. Sato, Modell. Simul. Mater. Sci. Eng. **9** (2001) 129.

[7] A. Wilm, Metallurgie **8** (1911) 225.

[8] A. Inoue and T. Masumoto, J. Jpn. Inst. Light Metals **42** (1992) 299.

[9] T. Sato, Mater. Jpn. **36** (1997) 685.

[10] T. Sato, S. Hirosawa, K. Hirose and T. Maeguchi, Metall. Mater. Trans. A **34A** (2003) 2745.

[11] T. Sato, K. Hirose and S. Hirosawa, Proc. 9th Inter. Conf. Aluminum Alloys (ICAA9) (Brisbane, Australia), (2004) 956.

[12] S.P. Ringer, K. Hono, T. Sakurai and I.J. Polmear, Scr. Mater. **36** (1997) 517.

[13] A. Guinier, Nature **142** (1938) 569.

[14] G. D. Preston, Nature **142** (1938) 570.

[15] T. Sato and T. Takahashi, Scripta Metall. **22** (1988) 941.

[16] V. Gerold, Z. Metallkd. **45** (1954) 593.

[17] V. Gerold, Acta Crystallogr. **11** (1958) 230.

[18] Y. Ando, K. Mihama, T. Takahashi and Y. Kojima, J. Cryst. Growth **24–25** (1974) 581.

[19] T. Sato and A. Kamio, Mater. Sci. Eng. A **146** (1991) 161.

[20] H. Fujita and C. Lu, Mater. Trans. JIM. **33** (1992) 892.

[21] H. Fujita and C. Lu, Mater. Trans. JIM. **33** (1992) 897.

[22] M. Karlik and B. Jouffrey, Acta Mater. **45** (1997) 3251.

[23] T.J. Konno, K. Hiraga and M. Kawasaki, Scripta Mater. **44** (2001) 2303.

[24] T.J. Konno, M. Kawasaki and K. Hiraga, J. Electron Microsc. **50** (2001) 105.

[25] K. Hono, T. Sakurai and H.W. Pickering, Metall., Trans. A **20A** (1989) 1585.

[26] C. Wolverton, Phil. Mag. Lett. **79** (1999) 683.

[27] M. Takeda, H. Oka and I. Onaka, Phys. Status. Solidi. (a) **132** (1992) 305.

[28] J.F. Nie, B.C. Muddle, H.I. Aaronson, S.P. Ringer and J.P. Hirth, Metall. Mater. Trans. A **33A** (2002) 1649.

[29] T. Takahashi and T. Sato, J. Jpn. Inst. Light Metals **35** (1985) 41.

[30] S. Hirosawa, T. Sato, J. Yokota and A. Kamio, Mater. Trans. JIM. **39** (1998) 139.

[31] S. Hirosawa, T. Sato, A. Kamio and H.M. Flower, Acta Mater. **48** (2000) 1797.

[32] S. Hirosawa and T. Sato, Mater. Sci. Forum **396–402** (2002) 649.

[33] A.K. Mukhopadhyay, Mater. Trans. **38** (1997) 478.

[34] K. Hirose, S. Hirosawa and T. Sato, Mater. Sci. Forum **396–402** (2002) 795.

[35] T. Sato and K. Matsuda, J. Jpn. Inst. Light Metals **53** (2003) 449.

[36] H. Kimura and R. Hasiguti, Acta Metall. **9** (1961) 1076.

[37] A. Zhu, B.M. Gable, G.J. Shiflet and E.A. Starke Jr., Acta Mater. **52** (2004) 3671.

[38] Y. Suzuki, A. Hibino, T. Muramatsu, H. Horosawa and T. Sato, Proc. 9th Inter. Conf. Aluminum Alloys (ICAA9) (Brisbane, Australia), (2004) 258.

[39] P. Ratchev, B. Verlinden, P. De Smet and P. Van Houtte, Acta Mater. **46** (1998) 3523.

[40] G.W. Lorimer and R.B. Nicholson, Acta Metal. **14** (1966) 1009.

[41] D.W. Pashley, M.H. Jacobs and J.T. Virtz, Phil. Mag. **16** (1967) 51.

[42] G. Sha and A. Cerezo, Acta Mater. **52** (2004) 4503.

[43] H. Inoue, T. Sato, Y. Kojima and T. Takahashi, Metall. Trans. A **12A** (1981) 1429.

[44] I. Kovacs, J. Lendvai and E. Nagy, Acta Metall. **20** (1972) 975.

[45] K. Matsuda, S. Ikeno, H. Matsui, T. Sato, K. Terayama and Y. Uetani, Metall. Mater. Trans. A **36A** (2005) 2007.

[46] K. Matsuda, T. Kawabata, Y. Uetani, T. Sato and S. Ikeno, J. Mater. Sci. **37** (2002) 3369.

[47] M. Murayama and K. Hono, Acta Mater. **47** (1999) 1537.

[48] C.D. Marioara, S.J. Andersen, J. Jansen and H.W. Zandbergen, Acta Mater. **49** (2003) 321.

[49] C.D. Marioara, S.J. Andersen, J. Jansen, H.W. Zandbergen, Acta Mater. **51** (2003) 789.

[50] S.J. Andersen, H.W. Zandberg, J. Jansen, C. Traholt, U. Tundal and O. Roiso, Acta Mater. **46** (1998) 3283.

[51] G.A. Edwards, K. Stiller, G.L. Dunlop and M.J. Couper, Acta Mater. **46** (1998) 3893.

[52] I. Dutta and S.M. Allen, J. Mater. Sci. Lett. **10** (1991) 323.

[53] K. Yamada, T. Sato and A. Kamio, Mater. Sci. Forum **331–337** (2000) 669.

[54] K. Yamada, T. Sato and A. Kamio, J. Jpn. Inst. Light Metals **51** (2001) 215.

[55] K. Matsuda and S. Ikeno, J. Jpn. Inst. Light Metals **50** (2000) 23.

[56] T. Ogura, S. Hirosawa and T. Sato, Proc. 9th Inter. Conf. Aluminum Alloys (ICAA9) (Brisbane, Australia), (2004) 1061.

PART IV

Nanostructure Architecture for Engineering Applications

CHAPTER 10

Nanoporous Materials from Mineral and Organic Templates

Kiyoshi Okada and Kenneth J.D. MacKenzie

Abstract

Nanoporous materials have attracted great interest because of their excellent porous properties, which are expected to lead to a variety of applications exploiting their ordered nanospace structures. In general they are prepared using chemical reagents as the inorganic sources, structurally directed by the use of organic templates. Thus, an ordered nanospace structure is formed by the polymerization of inorganic monomers assisted by organic templating molecules. By contrast, many studies have been reported using minerals as the inorganic sources to prepare unique ordered nanoporous materials which can display different properties from materials prepared using chemical reagents. In these cases, the mineral starting materials play an important role in building up ordered nanospace structures originating from their unique initial structures, i.e., these reactions are self-templating. Thus, the resulting nanoporous materials constitute a new family of ordered nanospace structures. In this chapter, the preparation and properties of nanoporous materials from mineral and organic templates are outlined and compared with nanoporous materials derived from chemical reagents by the use of organic templates.

Keywords: **nanoporous materials, mineral templates, organic templates, porous properties, sorption, solid acidity, selective leaching.**

10.1 Historical Background and Development

Porous materials are widely used as catalysts, catalyst supports, filters, membranes, ion exchangers, adsorbents, etc, as schematically shown in Fig. 10.1. For air environments, they are mainly used as catalysts and catalyst supports for remediation of air pollution by NO_x and SO_x mainly generated by burning of fossil fuel at thermal power stations, and from automobile emissions, etc. For water environments, porous materials are used as filters and sorbents for the separation of suspensoids, removal of toxic and harmful ions,

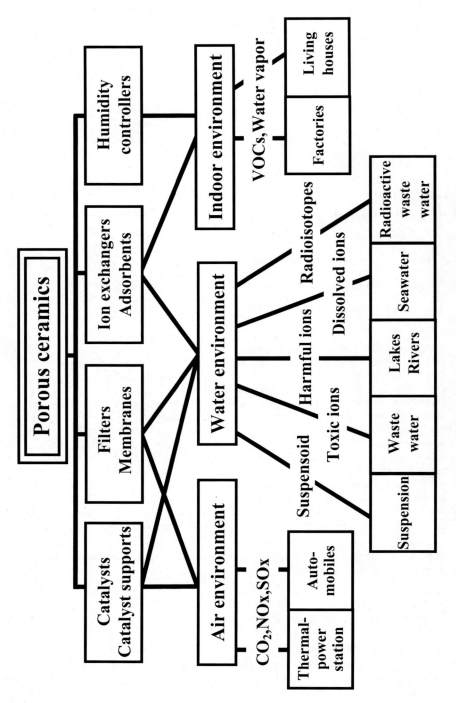

Fig. 10.1. Schematic depiction of various roles of porous materials.

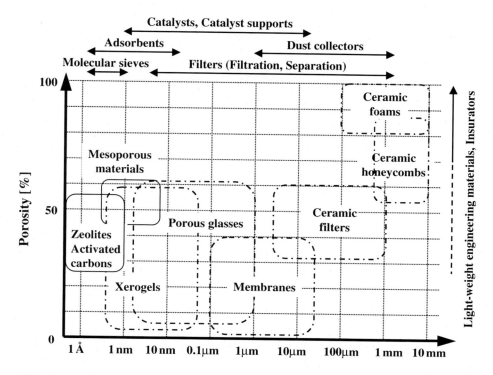

Fig. 10.2. Relationship between pore size and porosity of various typical porous materials.

salt-to-fresh water filtration, trapping of radioactive ions, etc. For indoor environments, they are used as sorbents for the removal of volatile organic compounds (VOCs), control of relative humidity, and as thermal insulators. The fundamentally important properties of these porous materials are their geometrical properties such as pore size, pore shape, surface area, pore volume (porosity), and their surface chemical properties, i.e., whether they are hydrophilic or hydrophobic. Figure 10.2 shows the pore size and porosity of various porous materials [1]. The pores are classified on the basis of their size, as micropores (< 2 nm), mesopores (2–50 nm) and macropores (> 50 nm), according to the recommendations of the International Union of Pure and Applied Chemistry (IUPAC) [2]. Ceramic membranes, filters, honeycombs, and foamed ceramics are macroporous materials. Since porous glasses prepared using spinodal phase separation have pore sizes tunable over a wide range, they can be categorized as both mesoporous and macroporous materials. Various xerogels and mesoporous materials prepared using organic templates as well as porous glasses are typical mesoporous materials. Zeolites, activated carbons, and pillared layer compounds (PLCs) are well-known microporous materials. In this chapter, we sometimes use terms such as 'nanoporous' and 'nanospace' to represent nanometer-scale pores in place of the terms 'microporous' and 'mesoporous.'

Typical preparation methods for nanoporous materials are listed in Fig. 10.3. These are divided into two groups: build-up processes and leaching processes. In the build-up

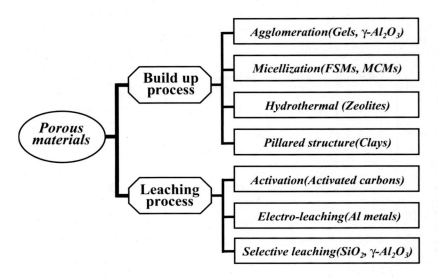

Fig. 10.3. Preparation methods for porous materials.

processes, the porous materials are formed from atoms, ions, and/or molecules, producing particles with pores which form further pores by aggregation of the particles. The most primitive preparation method is from gels, by agglomeration of fine particles formed by precipitation or sedimentation. Various xerogels and γ-Al_2O_3 are generally prepared by this method. The pore sizes formed by this process span a wide range due to the difficulty in controlling of all the important experimental parameters. However, they can be controlled to some extent by careful selection of the experimental conditions. Figure 10.4 shows the effect of mixing time of ethanol solutions of $Si(OC_2H_5)_4$ and $Al(NO_3)_3 \cdot 9H_2O$ on the pore size distributions of the resulting Al_2O_3–SiO_2 xerogels [3]. The pore sizes can be varied from 5 to 40 nm solely by decreasing the mixing time. These changes are found by TEM observation to correspond to changes in the agglomerated state of the fine particles. Since xerogels are prepared by precipitation and/or sedimentation of fine particles, the resulting porous properties tend to be influenced by the surface energy of the solvent. A solvent with high surface energy such as water produces tight agglomeration and thus smaller pore sizes and lower specific surface areas. To avoid shrinkage during drying, supercritical drying is an effective process and the resulting gels are called aerogels [4]. Aerogels have generally higher specific surface areas and pore volumes than xerogels. Mrowiec-Bialon et al. [5] prepared $CaCl_2$/silica xerogels and aerogels and compared their porous properties and water vapor sorption behavior. The different procedures used to dry xerogels and aerogels strongly influence their specific surface areas, those of the aerogels (640 and 886 m^2/g) being more than double those of the xerogels (200 and 276 m^2/g). The total pore volumes of the aerogels (1.9 and 3.9 ml/g) are also much higher than the xerogels (0.45 and 0.54 ml/g). As a result, a silica aerogel containing 29 mass% $CaCl_2$ absorbs as much as 65 mmol/g of water vapor, giving it very good humidity control properties.

Advanced preparation methods for porous materials use various organic compounds as templates, playing the role of both structure-directing agents (SDA) and pore formers.

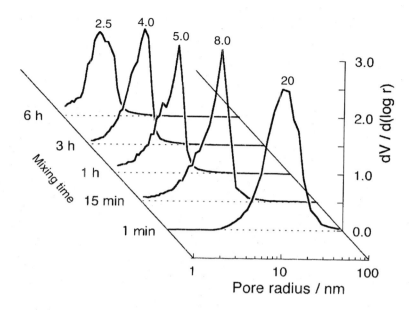

Fig. 10.4. Change of pore sizes of Al_2O_3–SiO_2 xerogels with change of mixing time for ethanol solutions of $Al(NO_3)_3 \cdot 9H_2O$ and TEOS.

Nowadays, many zeolites are synthesized using quaternary ammonium hydroxides (R_4NOH, R: CH_3-, C_2H_5-, C_3H_7-, C_4H_9-, etc) as organic templates. The relationships between the templates and the synthesized zeolites are shown schematically in Fig. 10.5 [6,7]. The organic templates dissolve in alkaline solutions as R_4N^+ ions, forming large tetrahedral structures. Aluminosilicate anions electrostatically surround the R_4N^+ ions and form crystalline three-dimensional framework structures under hydrothermal conditions. Zeolites are thus obtained after removal of the organic template by calcination at several hundred degree centigrade. Although the organic templates R_4NOH form relatively large hydrated cations in solution, the resulting sizes are limited to the micropore range (< 1 nm). By contrast, long chain alkyl ammonium ions having hydrophobic and hydrophilic groups in the head and tail are known to form micelles within certain concentration and temperature ranges [8]. The Mobil group was the first to demonstrate the formation of mesoporous silica (MCMs) with an ordered hexagonal structure, by hydrothermal reaction of a silica source (tetramethylammonium silicate) with a template of hexadodecyltrimethylammonium chloride ($C_{16}H_{33}(CH_3)_3NCl$) in alkaline solution [9]. The mesoporous silica (MPS) had a specific surface area of about 1000 m²/g, higher than previously reported for porous silicas, and a sharp pore size distribution of about 2–3 nm. Their pore size was also tunable by using single-chained quaternary ammonium surfactants of various alkyl chain lengths ($C_nH_{2n+1}(CH_3)_3N^+$, n: 8–16). The formation of cylindrical micelles and their hexagonal closest packing is the key factor in forming such an ordered porous structure. Preceding this finding, the Waseda group succeeded in forming a similar nanostructure (FSMs) using the layered mineral kanemite ($NaHSi_2O_5 \cdot 3H_2O$) [10]. The size of the nanopores capable of being synthesized increased stepwise to > 2 nm by the development of mesopores, compared

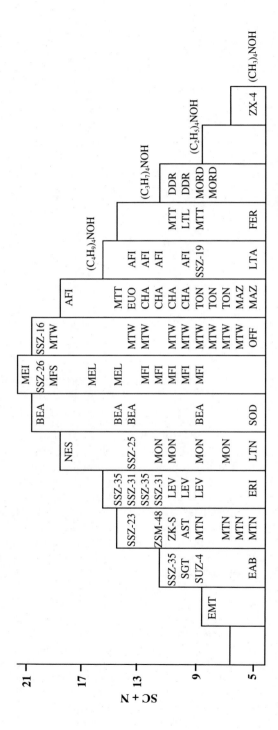

Fig. 10.5. Relationship between structure-directing agents (SDA) and the resulting zeolites. The abbreviated names of the zeolites are taken from Ref. [7].

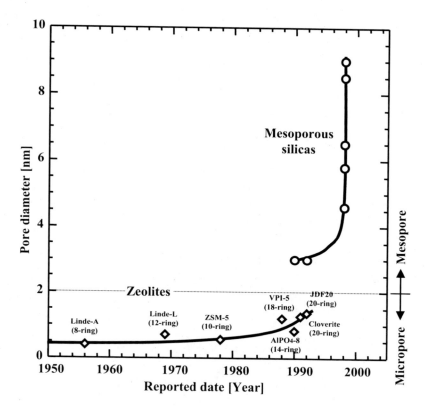

Fig. 10.6. Historical development of the maximum pore sizes of zeolites and mesoporous silicas.

with sizes of < 2 nm in zeolites. The history of changes in the maximum pore size of typical porous materials synthesized using an organic template is shown in Fig. 10.6, indicating that the pore sizes of the MPSs increased to nearly 10 nm within a very short time compared with the gradual increase in zeolite pore sizes.

In the syntheses of zeolites and MPSs, organic templates play an important role as both SDA and pore formers. Various organic compounds are used to synthesize MPSs with improved properties or characteristics. By contrast, fewer studies have been reported on inorganic sources though they offer different polymerization states ranging from monomers to framework structures. These differences may influence the porous properties of synthesized mesoporous materials even though silica is thought to dissolve and re-precipitate under hydrothermal conditions in alkaline solutions. On the other hand, the layered structure of the starting material acts as a major template in the formation of pillared layer compounds (PLCs) [11,12]. In these porous materials, the interlayer spaces of the PLC are involved in the formation of the pores by the insertion of the pillaring oxide. For this purpose, complex metal hydroxide cations are intercalated by cation exchange and converted to oxide pillars by calcining. The host layered compounds generally used are synthetic F-mica ($Na(Mg,Al)_{2-3}(Si,Al)_4O_{10}F_2$) and montmorillonite

$(Na_{1/3}(Al_{11/3}Mg_{1/3})Si_4O_{10}(OH)_2)$ with high cation exchange capacities. The most commonly used complex cation is $[Al_{13}O_4(OH)_{24}(H_2O)_{12}]^{7+}$, with a Keggin-type structure consisting of $Al(O,OH)_6$ octahedra [13] because of its high valence (providing a strong driving force for intercalation) and its stable structure in solution. The resulting PLCs show a maximum specific surface area of about 400 m^2/g, pore sizes of 1–2 nm and thermal resistance up to about 700°C [14]. Thus, the pore size generally obtained with PLCs is intermediate between zeolites and mesoporous materials.

A new group of PLCs was developed by Galarneau et al. [15] using a pillaring technique and an organic template to form micelles similar to mesoporous materials. The first step of the synthesis is cation exchange by $C_nH_{2n+1}(CH_3)_3N^+$ to expand the interlayers of smectites (e.g., hectorite, montmorillonite, and saponite). In the second step, a neutral amine $(C_nH_{2n+1}NH_2)$ and a silica source (e.g., tetraethyl orthosilicate (TEOS) $(Si(OC_2H_5)_4)$) are intercalated into the interlayers. Since the neutral amine molecules form micelles in the interlayers, TEOS hydrolyzes and polymerizes in the remaining spaces and forms pillars after calcination. Thus, the resulting porous materials show specific surface areas of about 600–950 m^2/g with cylindrical pores 1.5–2.0 nm in diameter. These materials are called porous clay heterostructures (PCHs) and have the following characteristic properties [16]:

(1) nanoporosity arising from a combination of micro- and mesoporosity,

(2) a large number of silanol groups giving hydrophilic surface properties, and

(3) relatively high acidity introduced by surface modifications of the surface silanols. It is therefore considered that PCHs are very promising candidates for catalysis and sorption.

Leaching is also useful for preparing porous materials. Activated carbon, one of the most widely used porous materials, is produced by a leaching process, called activation. Activation is performed using oxidizing gases such as steam, CO_2, CO, etc; this method is termed physical activation. Activation can also be achieved chemically and this process is called chemical activation. The chemicals generally used are alkali carbonates, alkali hydroxides, phosphoric acid, or zinc chloride. The specific surface area and pore volume of chemically activated carbons are higher than in physically activated carbons. A maximum specific surface area of about 3000 m^2/g has been achieved [17], the highest of all the porous materials. By contrast, physical activation has the advantage of producing activated carbon by a very simple and cost-effective process, widely used in industry. Activated carbon can be produced from various starting materials including carbon, i.e., coals, pitches, a variety of agricultural wastes, textiles and fabrics, papers, plastics, etc. For example, we have prepared activated carbon with a high specific surface area ($\approx$2000 m^2/g) from used newspaper [18], and honeycombs from corrugated paper [19].

A unique leaching process involves electrochemical leaching of Al metal [20] and Si [21] in acid solution. This is used for the surface oxidation of Al metal to form a protective surface layer called 'Almite.' The oxidized layer forms a nanoporous structure under certain conditions. The process characteristically forms uniform cylindrical nanopores running perpendicular to the surface of the Al plate. Such a unique nanostructure is thought to be formed by partial dissolution of the surface Al metal and precipitation of the dissolved Al as Al-hydroxide or oxide by acid and electrochemical effects. Similar nanostructures are reported on the surface of Si, giving rise to a photo-emitting phenomenon as a result of the unique nanostructure [21].

Acid treatment is known to activate clay and solid acidity can be generated by such treatment of smectites, producing activated clays [22]. Activated clay has been used as a catalyst for oil refining and a similarly prepared catalyst, K-10 is still used in industry. The treatment causes changes in the chemical composition of the clay and makes it porous by selective leaching of the constituents other than SiO_2. Thus, this process is called 'selective leaching.' Various clay minerals have been used as the starting materials and the resulting specific surface areas in the range of about 300–700 m^2/g, pore volumes in the range 0.1–0.4 ml/g pore sizes in the range 0.6–4 nm [23]. The driving force for this process is both the solubility difference under different pH conditions and the difference in dissolution rates between amorphous and crystalline phases. A good example is the preparation of mesoporous γ-Al_2O_3 by selective leaching of calcined kaolinite ($Al_2Si_2O_5(OH)_4$) using KOH solution [24]. SiO_2 and Al_2O_3 are both highly soluble in strongly alkaline solution but the dissolution rate of SiO_2 is much faster than Al_2O_3 because in this system the SiO_2 is amorphous but the Al_2O_3 is crystalline γ-Al_2O_3. We have thus prepared mesoporous γ-Al_2O_3 by the selective leaching method and have confirmed its unique properties arising from its nanostructure.

10.2 Review of the Porous Properties of Nanoporous Materials Produced Using Mineral Templates

At present, various types of nanoporous materials are synthesized using different mineral templates, mostly in combination with organic templates. Schematic models for the syntheses of these typical nanoporous materials are shown in Fig. 10.7, and their properties and characteristics are summarized in Table 10.1.

10.2.1 Mesoporous silica (MPS)

As mentioned above, MPS has been synthesized using various silica sources and organic templates. Many of the studies are more focused on the organic template, and a variety of templates has been used to form the various ordered nanostructures known at present. This may be because the micelles formed by organic templates are thought to play a major role in the formation of ordered nanoporous structures. The organic templates used for the syntheses of MPSs are listed in Table 10.2. Since the silica source is generally present as silicate anions or fine sol particles which have negative surface charges in alkaline solution (the normal synthesis conditions for MPS), cationic surfactants are commonly used for these syntheses. Anionic and neutral surfactants are, however, also used in some special cases. The effects of organic templates on the porous properties of nanoporous materials are summarized in several reviews [8,47,48]. By contrast, the silica sources used in the synthesis of MCM-41-type (hexagonal) MPSs are mostly alkoxides, especially TEOS [49]. Other sources such as

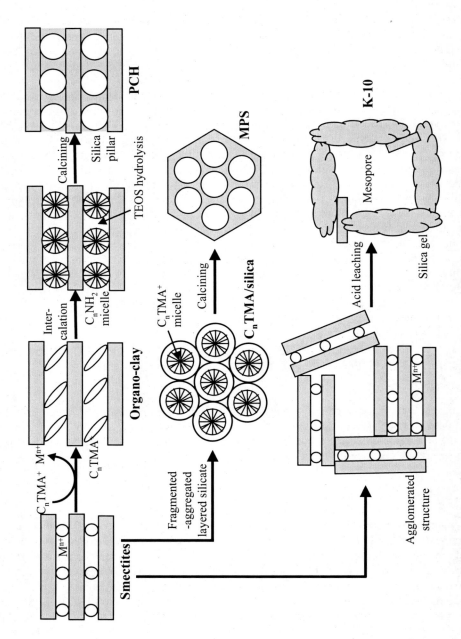

Fig. 10.7. Schematically depicted models for syntheses of various types of nanoporous materials from mineral templates.

Table 10.1. Characteristics of various nanoporous materials synthesized from various mineral templates.

Nanoporous material	Style	Mineral template Origin	Ideal formula	Pre-treatment	Si/M	S_{BET} (m²/g)	Pore size (nm)	V_p (ml/g)	Other property	References
MCM-41	Al/SiO$_2$	Mordenite	NaAlSi$_5$O$_{12}$·3H$_2$O	No	15	920	2.78	0.47	—	[25]
MCM-type	Al/SiO$_2$	Montmorillonite	Na(Al$_5$Mg)Si$_{12}$O$_{30}$(OH)$_6$	Acid	—	408	—	0.55	Catalytic	[26]
MCM-type	Al/SiO$_2$	Montmorillonite	Na(Al$_5$Mg)Si$_{12}$O$_{30}$(OH)$_6$	Acid	—	736	—	—	Catalytic	[27]
MCMs	Al/SiO$_2$	Kaolinite	Al$_2$Si$_2$O$_5$(OH)$_4$	Zeolitization	1.6–5.6	452–818	2.5–4.4	0.36–0.77	—	[28]
MCM-41	Al/SiO$_2$	Kaolinite	Al$_2$Si$_2$O$_5$(OH)$_4$	Heating, acid	5–100	–1420	2.8	–0.94	Thermal stability	[29]
FSM	Al/SiO$_2$	Kanemite	NaHSi$_2$O$_5$·3H$_2$O	No	6.1	—	2.0–3.4	—	Catalytic	[30]
FSM	SiO$_2$	Kanemite	NaHSi$_2$O$_5$·3H$_2$O	No	Infinity	1003	3.6	0.7	Surface modification	[31]
FSM	Al/SiO$_2$	Saponite	NaMg$_3$(Si$_{11}$Al)O$_{30}$(OH)$_6$	Acid	11–13	478–900	2.5–3.0	0.37–0.61	Solid acidity	[32]
PCH	Si/hectorite	F-hectorite	Li(Mg$_8$Li)Si$_{12}$O$_{30}$F$_6$	No	1.33	470–750	1.4–2.2	—	Original work	[15]
PCH	Si/hectorite	F-hectorite	Li(Mg$_8$Li)Si$_{12}$O$_{30}$F$_6$	No	1.33	175–550	2.1	—	Hg sorption	[33]
PCH	Si/saponite, montmorillonite	Saponite, montmorillonite	—	No	1.07, 2.0	734, 790	1.7	0.54–0.62	Surface modification	[34]
PCH	Si/montmorillonite	Montmorillonite	Na(Al$_5$Mg)Si$_{12}$O$_{30}$(OH)$_6$	Acid	56.1	795–951	—	0.71–0.82	Solid acidity	[35]
PCH	Si/saponite	Saponite	NaMg$_3$(Si$_{11}$Al)O$_{30}$(OH)$_6$	No	0.88–0.97	800–920	1.5–2.3	0.38–0.44	Solid acidity	[36]
PCH	Si/montmorillonite	Montmorillonite	Na(Al$_5$Mg)Si$_{12}$O$_{30}$(OH)$_6$	No	1.57	635–765	—	0.40–0.55	VOC sorption	[37]
PCH	Si/montmorillonite	Montmorillonite	Na(Al$_5$Mg)Si$_{12}$O$_{30}$(OH)$_6$	No	—	506–690	2.4–3.2	0.25–0.30	VOC sorption	[38]
PLC/PCH	Si/vermiculite	Vermiculite	—	Acid	130–200	715–1205	—	0.29–0.43	Hydrophilic	[39]
PLC	Si/montmorillonite	Montmorillonite	Na(Al$_5$Mg)Si$_{12}$O$_{30}$(OH)$_6$	Acid	2.2–3.6	315–387	—	0.20–0.48	Catalytic	[40]
PLC	Al/mica	F-mica	NaMg$_{2.5}$Si$_4$O$_{10}$F$_2$	No	—	–250	4	–0.15	Thermal stability	[41]
PLC	Ti/montmorillonite	Montmorillonite	Na(Al$_5$Mg)Si$_{12}$O$_{30}$(OH)$_6$	No	—	–400	2–20	—	Catalytic	[42]
SLPM	Al/SiO$_2$	Montmorillonite	Na(Al$_5$Mg)Si$_{12}$O$_{30}$(OH)$_6$	Acid	3.9–13.1	163–264	0.6	0.33–0.42	Gas sorption	[43]
SLPM	Al/SiO$_2$	Kaolinite	Al$_2$Si$_2$O$_5$(OH)$_4$	Heating, acid	5–100	–360	0.6–4	0.17–0.29	Gas sorption	[44]
SLPM	(Al,Mg)/SiO$_2$	Vermiculite	—	Acid	1.2–950	86–670	1–2	0.05–0.45	Solid acidity	[45]
SLPM	Mg/SiO$_2$	Antigorite	Mg$_3$Si$_2$O$_5$(OH)$_4$	Acid	2.1–330	170–410	4–6	0.08–0.23	Thermal stability,	[46]
SLPM	Si/Al$_2$O$_3$	Kaolinite	Al$_2$Si$_2$O$_5$(OH)$_4$	Heating, acid	5–100	–300	4–6	0.42–0.85	vapor sorption	[24]

Table 10.2. Organic template used for the synthesis of nanostructure materials.

Type	Chemical formula	n	m	s
Cationic	$C_nH_{2n+1}(CH_3)_3N^+$	8–22		
	$C_nH_{2n+1}(C_2H_5)_3N^+$	12, 14, 16, 18		
	$(C_nH_{2n+1})_2(CH_3)_2N^+$	10–18		
	$C_nH_{2n+1}(CH_3)_2C_{16}H_{33}N^+$	1–12		
Gemini	$C_nH_{2n+1}(CH_3)_2N^+\text{-}C_sH_{2s}\text{-}N^+(CH_3)_2C_mH_{2m+1}$	16	1–16	2–12
Anionic	$C_nH_{2n+1}COO^-$	14, 17		
	$C_nH_{2n+1}OPO_3^-$	12, 14		
	$C_nH_{2n+1}OSO_3^-$	12, 14, 16, 18		
	$C_nH_{2n+1}C_6H_4SO_3^-$	12		
Neutral	$C_nH_{2n+1}NH_2$	10–16		
	$C_nH_{2n+1}N(CH_3)_2$	10–16		
	$C_nH_{2n+1}(CH_2CH_2O)_m$	11–15	9, 12, 15, 20, 30	
	$C_nH_{2n+1}C_6H_5(CH_2CH_2O)_m$	8, 12	8, 10, 18	
	$[(C_2H_4O)_x]_{13}[(C_3H_6O)_y]_{30}[(C_2H_4O)_x]_{13}$			

a monomeric silicate (tetra-methylammonium silicate) [9], waterglass (sodium silicate hydrate) [50], silica gel [51], fumed silica [52], and even coal fly ash [53] can be used as starting materials. In the synthesis of FSM, the layered silicate mineral kanemite ($NaHSi_2O_5 \cdot 3H_2O$) was used as the silica source [10]. An attractive formation mechanism was proposed by Inagaki et al. [54], in which a kanemite silicate sheet is folded to surround the organic micelles and form a hexagonal structure. Other layered minerals such as kaolinite [28], saponite [32], montmorillonite [26], etc have been investigated as silica sources, but their composition need to be made more silica-rich by various pre-treatments. The silica sources used for the syntheses of MPSs are listed in Table 10.3.

Paulino and Schuchardt [49] investigated the effect of chain length of the alkoxyl group of the silica source on the porous structure of MPSs and found an increase in the specific surface area from 685 to 1127 m^2/g, an increased pore volume from 0.39 to 0.84 ml/g, an increased pore size from 2.3 to 3.0 nm, and an increase in the hexagonal lattice parameter a_0 from 4.5 to 4.7 nm, but a decrease in the wall thickness of the silica matrix from 1.8 to 1.2 nm as the alkoxyl chain length increases from $-OCH_3$ to $-OC_3H_7$. These data suggest that the density of silica matrix is decreased with increasing alkoxyl chain length, i.e., more spaces are formed in the framework structure during polymerization of the longer alkoxyl chain alkoxides. This trend is, however, completely opposite to the results observed in the porous properties of the silica xerogels prepared by a sol–gel method from the three Si-alkoxides, TMOS, TEOS, and TPOS [55]. Differences in the hydrolysis rates of the Si-alkoxides and the pH of the solutions are thought to be the reasons for the differences in this case.

Although the effect of the state of polymerization of the starting materials on the porous properties of the resulting MPSs has not previously been addressed, we consider that these differences should influence the amorphous framework structure of the silica matrix. In the hydrothermal synthesis of various silicates at relatively low temperatures, differences in the silica source are known to significantly influence the formation rate

Table 10.3. Characteristics of various silica sources used for the synthesis of nanostructured materials.

Source	Process	Formula	Structure	Pre-treatment	Characteristics
Alkoxide	Synthesis	$Si(OCH_3)_4$	Monomeric (Q^0)	No, partial hydrolysis	High-purity, reactive
Alkoxide	Synthesis	$Si(OC_2H_5)_4$	Monomeric (Q^0)	No, partial hydrolysis	High-purity, reactive, most common
Alkoxide	Synthesis	$Si(OC_3H_7)_4$	Monomeric (Q^0)	No, partial hydrolysis	High-purity, reactive
Alkoxide	Synthesis	$Si(OC_4H_9)_4$	Monomeric (Q^0)	No, partial hydrolysis	High-purity, reactive
Silicate	Synthesis	$((CH_3)_4N)_4SiO_4$	Monomeric (Q^0)	No	Soluble
Silicate	Synthesis	$(NH_3)_4SiO_4$	Monomeric (Q^0)	No	Soluble
Silicate	Synthesis	Na_2O–SiO_2	Chain (Q^2), layer (Q^3)	No	Soluble
Waterglass	Synthesis	Na_2O–SiO_2–H_2O	Chain (Q^2), layer (Q^3)	No	Soluble
Silica gel	Synthesis	SiO_2–H_2O	Frame (Q^4, Q^3)	No	High surface area
Fumed silica	Synthesis	SiO_2	Frame (Q^4)	No	Fine particle
Silica glass	Synthesis	SiO_2	Frame (Q^4)	No	Slow dissolution
Fly ash	Artificial	Variable	Frame (Q^4)	No, acid	Aluminosilicate glass, impurity phases
Kanemite	Natural, synthesis	$NaHSi_2O_5 \cdot 3H_2O$	Layer (Q^3)	No	Cation exchange, undulated SiO_4 layer
Kaolinite	Natural, synthesis	$Al_2Si_2O_5(OH)_4$	Layer (Q^3)	Heating, acid	High surface area, flat SiO_4 layer
Saponite	Natural, synthesis	$NaMg_3(Si_{11}Al)O_{30}(OH)_6$	Layer (Q^3, Q^2)	No, acid	Cation exchange, flat $(Si,Al)O_4$ layer
Montmorillonite	Natural, synthesis	$Na(Al_5Mg)Si_{12}O_{30}(OH)_6$	Layer (Q^3, Q^2)	No, acid	Cation exchange, flat SiO_4 layer
Vermiculite	Natural	—	Layer (Q^3, Q^2)	No, acid	Cation exchange, flat $(Si,Al)O_4$ layer

of the target crystalline phases and even form different crystalline phases in some cases. This effect is thought to be due to differences in the structures of the silicate anions in solution, indicating that the silicate anions may exist as variously sized clusters of a SiO_4 tetrahedra which are not completely hydrolyzed to monomeric SiO_4 tetrahedra [56]. Layered silicates are thought to dissolve into anion clusters by fragmentation of the SiO_4 layer structure. On the other hand, alkoxides hydrolyze to SiO_4 tetrahedra and react with micelles without passing through a highly polymerized state. In framework structures such as silica gel and fumed silica, dissolution occurs from the surface in individual units, forming monomers [57]. We therefore consider that the very high specific surface areas of the MPSs produced from heat- and acid-treated kaolinite (Table 10.1) result from these unique fragmented clusters, converting the silica matrix to a lower-density structure. The importance of the structural state of aluminosilicate anions in solution was also pointed out by Ogura et al. [58] in connection with the preparation of faujasite-type zeolite. In this way, silicate anions may take up a different structure in solution depending on the structure of the silica source and this should influence the synthesized nanostructures.

The porous properties of the MPSs prepared from various mineral and organic templates are listed in Table 10.1, which shows that the specific surface areas range from about 400 to 1400 m^2/g and the pore volumes from 0.36 to 0.94 ml/g. By contrast, the pore sizes are relatively similar, ranging only from 2 to 3 nm, because the same organic template ($C_{16}TMA^+$) is commonly used. One of the characteristics of the mineral templates is that they may contain considerable amounts of other components in addition to SiO_2. These components may affect the resulting specific surface areas. For MPSs prepared using $C_{16}TMA^+$ as the organic template, the relationship between the M/Si ratio (M = $\sum(Al + Mg + Fe)$) of the mineral templates and the specific surface area of the resulting MPSs is shown in Fig. 10.8, together with the data for MPSs prepared from non-mineral silica sources. In the mineral-templated MPSs (Fig. 10.8(a)), the specific surface

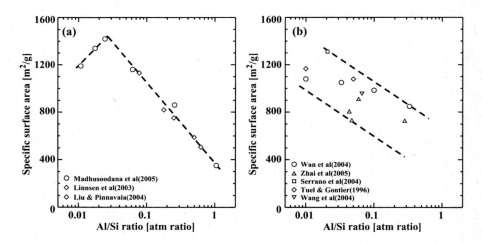

Fig. 10.8. Relationship between the M/Si (M = $\sum(Al + Mg + Fe)$) ratio of silicate sources and the resulting specific surface area for mineral (a) and non-mineral templates (b).

area clearly shows a peak at about $M/Si = 0.025$ ($Si/M = 40$). Although a maximum is not clearly apparent in MPSs prepared from non-mineral templates (Fig. 10.8(b)), a decrease of specific surface area is observed with an increasing M/Si ratio. Such an interesting relationship has not been reported previously. Since the pore sizes and hexagonal a_0 parameters are similar when the same organic template is used, even if the M/Si ratios are changed, the presence of M components, especially Al, clearly appears to exert an influence by changing the matrix density of the MPSs.

The specific surface area of MPSs prepared from heat- and acid-treated kaolinite with $Al/Si = 0.0244$ ($Si/Al = 41$) is 1420 m^2/g, the highest value of all the MPSs, even those prepared from non-mineral silica sources [29]. There are four possible reasons for this high specific surface area, namely,

(1) a high value for the external surface,

(2) micropores in the silica matrix,

(3) a low matrix density without micropores, and

(4) fine surface roughness in the matrix wall of the mesopores.

The external surface area of this MPS was calculated by the t-plot method but was negligible (< 100 m^2/g) compared with the total specific surface area. The presence of micropores was sought by careful Ar gas adsorption measurements in the very low-pressure range ($P/P_0 \geq 10^{-5}$), the micropore size distribution being calculated from this isotherm by the Horvath–Kawazoe and Saito–Foley methods. The isotherm showed no inflection point in the P/P_0 range corresponding to micropores, which suggests a lack of microporosity even though the micropore size distribution curves calculated by the above two methods clearly showed a peak at a pore size of about 0.8 nm. The geometrical surface area and pore volume of cylindrical mesopores, calculated assuming an ideally smooth surface and a silica matrix density of 2.2 g/cm^3 (that of amorphous silica) are shown in Fig. 10.9 as a function of the wall thickness. Since the pore size and hexagonal a_0 parameter of the present MPS are 2.78 and 4.61 nm, respectively, the calculated specific surface area and pore volume of the MPS is only 320 m^2/g and 0.22 ml/g, respectively. These calculated values for the ideal surface area and pore volume are significantly lower than the observed values (1420 m^2/g and 0.94 ml/g) but are found to increase steeply for thinner wall thicknesses. On the other hand, the density of silica matrix of the present MPS was calculated from the observed specific surface area and the above geometrical parameters. The calculated density is about 0.5 g/cm^3. Feuston and Higgins [59] calculated the density of the silica matrix of MPS using the molecular dynamics (MD) method; the calculated tetrahedral site densities ranged from 6.8 to 11.6 /nm^3, corresponding to matrix densities of 0.68–1.16 g/cm^3. The calculated matrix density of the present MPS is therefore a little lower than the MPS densities obtained from the MD simulation. However, the matrix density of the present MPS is increased by assuming an increase in the actual mesopore surface area due to fine surface roughness in the wall corresponding to the surface atomic arrangement of the SiO$_4$ tetrahedral framework structure. Thus, the high specific surface area of the MPS can be reasonably explained by taking into account a lowering of the silica matrix density

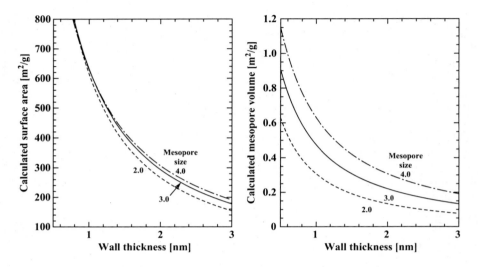

Fig. 10.9. Calculated specific surface area and pore volume of ideal cylindrical mesopores in MPS assuming the matrix density is that of amorphous silica (2.2 g/cm^3) and assuming a hexagonal lattice.

and an increase in the mesopore surface area due to the actual surface roughness of the mesopores arising from the real atomic arrangement.

In this way, mineral templates, especially kaolinite, are attractive silica sources for the synthesis of MPSs. The mineral template for microporous silica, which is obtained from heat- and acid-treated kaolinite, has a unique porous nanostructure consisting of layered SiO$_4$ tetrahedra [44]. Since the tetrahedral sheet consists of a corner shared six-membered ring structure, the fragments formed by dissolution of the microporous silica in alkali solution are expected to contain a high concentration of six-membered ring clusters in the solution. We believe that these fragment clusters enhance the formation of a bulky framework structure in the matrix of the present MPSs, producing a high specific surface area. Further investigation of clay minerals and other layer compounds as a new type of inorganic template for MPSs would be worthwhile.

10.2.2 Porous clay heterostructure (PCH)

In MPS, the nanoporous structure is more strongly directed by organic templates than mineral templates because the latter are also used as silica sources and are more or less hydrolyzed and fragmented in the alkaline solution before being re-precipitated around the organic micelles to form the MPS. By contrast, mineral sources constitute the main structural member in the formation of nanoporous PCH structures because they comprise the host layers. The starting materials for PCHs are however relatively limited because the host layer compounds must have good exchangeability and preferably swellability, to allow the intercalation of long chain alkylammonium ions and expand the interlayers to provide space for the neutral alkylamines to form micelles. In practice, only smectite group minerals such as F-hectorite, montmorillonite, and saponite have been successfully used as the starting materials of PCHs (Table 10.1).

On the basis of exchangeability, vermiculite is also a candidate starting material. Ishii *et al.* [39] prepared a nanoporous material similar to PCH using vermiculite as the host layer; the specific surface area of this material was as high as 1200 m^2/g, higher than 'normal' PCHs. However, since neutral alkylamine was not used for the formation of micelles in the interlayers, the resulting pore size showed a wider distribution than in the PCHs. Another candidate host layer compound for PCH formation is swellable F-mica [12], commonly used for PLC synthesis. No researchers, however, have yet reported the use of this as the host layer compound for PCH.

Layered double hydroxides (LDH; typically $Mg_6Al_2(OH)_{16} \cdot xH_2O$) are other candidate host materials for intercalating anionic surfactants (Table 10.2). Kopka *et al.* [60] synthesized LDHs intercalated with anion surfactants using dodecyl glycol ether sulfate ions ($C_{12}H_{25}(CH_2CH_2O)_m SO_4^-$), one of the anionic surfactants listed in Table 10.2. The organic surfactants were found to form bilayer conformations in the interlayers of LDH but formation of micelles was not reported. If neutral amines together with TEOS were introduced into the composites, it may be possible to synthesize LDH nanoporous materials similar to PCHs. However, LDHs have a disadvantage as the host layer compounds of PCHs because their thermal stability is generally lower than the smectites due to their higher hydroxyl density.

As mentioned above, the PCHs have unique characteristics in the surfaces of their porous structures due to the higher density of their silanol groups compared with MPSs, resulting in porous properties which are different from those of MPSs. It would therefore be interesting to prepare PCHs under different reaction conditions to obtain new types of PCHs with enhanced porous properties. For example, TEOS has been the only silica source used until now, but it would be worthwhile to investigate the use of different silica sources to develop new PCHs. Mineral templates are potentially useful for the synthesis of PCHs with unique properties because of their solid acidity arising from Al substitution in the mineral template. Another possible way to enhance the porous properties of PCHs is to control the aggregated structure of the porous PCH particles. If micelles are absorbed with mineral templates on the external surfaces of the PCH particles, the resulting higher-order texture could enhance the porous properties of both the interlayer mesopores and intraparticle meso- or macropores.

10.2.3 Pillared layer compound (PLC)

In the synthesis of PLCs, important factors are the choice of both the host layer compounds and pillaring ions. The fundamental necessity for the host layer compound is its ion exchangeability, as in the case of the PCHs. Thus, host layer compounds are relatively restricted by this condition to smectites (e.g., montmorillonite, saponite, and hectorite), vermiculite, and synthetic F-micas. In ion exchange, the intercalating ions are generally larger than the host ions in order to form pillars which generate the spaces in the interlayers by the ion exchange reaction. Therefore, the host layer compounds should have swellable properties in addition to ion exchangeability. Since the layer charge is related to the pillar density in the interlayers, it is preferable to control the layer charge of the host layer compound. On this basis, synthetic F-micas are more suitable host layer compounds for PLC formation than smectites and vermiculite. The higher crystallinity and thermal stability of the F-micas are also advantageous. Of the various F-micas, Na-tetrasilisic fluorine mica ($NaMg_{2.5}Si_4O_{10}F_2$), Na-taeniolite ($Na(Mg_2Li)Si_4O_{10}F_2$), and the Na-taeniolite series of fluorine micas ($Na_x(Mg_{3-x}Li_x)Si_4O_{10}F_2$: $x = 0.4$–0.8) are especially suitable as the host layer compound because they have both swellability and exchangeability [14]. By contrast, smectites have the advantage of a higher surface area than F-micas, reflecting their finer particle size. Of the smectites, montmorillonite has suitable swellability and exchangeability while saponite has suitably higher specific surface area. The choice of the host layer compound can be made according to the ultimate purpose of the product.

The host layer compounds for PLC formation generally have negative layer charge and intercalate cations into the interlayers by exchange reactions. There is however another type of layer compound with a positive layer charge, namely, intercalated anions rather than cations. A typical example is provided by the layered double hydroxides (LDH) in which variations occur by combination of two types of cation and anion, as follows; $M(II)_{1-x}M(III)_x(OH)_2X_x \cdot yH_2O$, $M(III)_{1-x}M(I)_x(OH)_2X_x \cdot yH_2O$, etc, where M(II) represents alkaline earth ions and divalent transition metal ions, M(III) is Al^{3+}, Fe^{3+}, Cr^{3+}, V^{3+}, In^{3+}, etc, M(I): Li^+, X: are halogen ions, NO_3^-, CO_3^{2-}, SO_4^{2-}, PO_4^{3-}, SiO_4^{4-}, polyoxyanions, etc [61,62]. Because of the great variation of possible combinations, LDHs constitute potential host layer compounds for PLC formation.

For the pillaring ions (guest ions), the desired properties are a bulky shape of the complex polymerized ions (preferably elongated prismatic shape) and high valence to produce a stronger driving force for intercalation and stability in the exchange solution. The following are candidate intercalating ions: $[Al_{13}O_4(OH)_{24}]^{7+}$, $[Ga_{13}O_4(OH)_{24}]^{7+}$, $[GaO_4Al_{12}(OH)_{24}]^{7+}$, $[Cr_n(OH)_m]^{(3n-m)+}$, $[Fe_n(OH)_m]^{(3n-m)+}$, $[Zr_4(OH)_{14}]^{2+}$, and $[Fe_3O(OCOCH_3)_6]^+$ [12]. On the other hand, the following anions are candidates as pillaring ions for LDH: $[PV_4W_8O_{40}]^{7-}$, $[PV_3W_9O_{40}]^{6-}$, $[PV_2W_9O_{40}]^{5-}$, $[Fe(CN)_6]^{4-}$, $[PW_{12}O_{40}]^{3-}$ and $[B_4O_5(OH)_4]^{2-}$ [62–64].

The PLCs synthesized by pillaring these intercalating ions have 1–2 nm micropores, limited by the sizes of the intercalating polyanions. This difficulty was overcome by Yamanaka et al. [65] who used fine sol particles of TiO_2 and intercalated these particles into the interlayers of montmorillonite in acidic solution. This succeeded because TiO_2 particles have a positive surface charge in acidic conditions. These researchers successfully prepared porous materials of 2–4 nm pore size. We have also synthesized a TiO_2/montmorillonite PLC using a similar method [42]. By changing the synthesis temperature from 50 to 80°C, the pore size can be controlled to be between 2 (microporous) and 20 nm (mesoporous). The formation processes are shown schematically in Fig. 10.10. The TiO_2 sol particles intercalated into the interlayers of montmorillonite are found from the XRD peak shift to be much smaller than expected. At longer reaction times and higher reaction temperatures, the residual TiO_2 sol in the solution increases in size and is sorbed on the surface of the primary composite particles to form an aggregated structure. The microporous sample is thought to be similar to a primary composite-type or panoscopic-type nanostructure, while the mesoporous sample has a card-house type

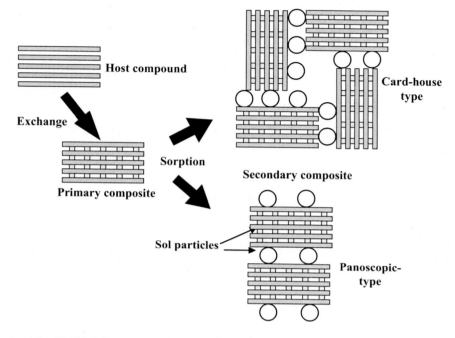

Fig. 10.10. Schematic illustration of the formation process of a new type of PLC.

nanostructure. The maximum specific surface area of the present composites was about 400 m²/g, much higher than previously reported [65]. The mesoporous samples showed higher adsorption for large molecules because of their large mesopores while the microporous samples showed excellent photocatalytic performance in the decomposition of organic matter. This synergy results from the sorption of the target organic molecules by the montmorillonite, followed by photocatalytic decomposition by the anatase particles in the interlayers and external surfaces of the PLC. It is therefore very effective to introduce nanoporosity into TiO$_2$–montmorillonite PLC composites by pillaring with TiO$_2$ sol particles which enhance the photocatalytic properties.

10.2.4 Porous materials by selective leaching

It is well known that the solubility of each component of a mineral changes with the pH of the solution in relation to the solubility products. Under ideal conditions, all activities can be taken as unity. The relationship between the concentration of each component and the solution pH is represented by the following equations:

$$Me(OH)_{n(S)} = Me^{n+}(aq) + nOH^-_{(aq)} \qquad (10.1)$$

$$K_{SP} = [Me^{n+}][OH]^n = C \times (nC)^n = n^n C^{n+1} \qquad (10.2)$$

$$\log[Me^{n+}] = (\log(K_{SP}) + 14n) - n \times pH \qquad (10.3)$$

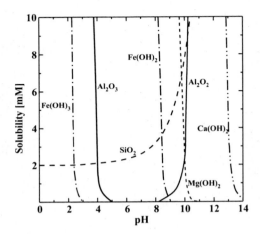

Fig. 10.11. Changes in solubility of various components generally occurring in clay minerals as a function of the solution pH.

where K_{SP} is the solubility product and C is the solubility. The changes in solubility for various components as a function of solution pH are shown in Fig. 10.11. From this figure, it is clear that all the components except SiO_2 are highly soluble in strongly acidic solution. Therefore, selective leaching of silicate compounds occurs in acid solution, the residual silica forming a skeleton structure, i.e., a nanoporous structure. Acid treatment of minerals, especially clay minerals, has been used to prepare activated clays and porous silicas for many years [23]. Although selective leaching is a simple technique, there are many experimental parameters to be optimized to achieve products with good porous properties. The important experimental parameters are the sample/solution ratio, acid concentration, temperature and duration of the leaching, type of acid, etc. The sample/solution ratio has a relatively strong influence on the porous properties of the products in some cases. For example, Kavitratna and Pinnavaia [66] reported a very low specific surface area of 18 m^2/g for acid-treated phlogopite (a mica-group mineral) using a sample/solution ratio of 1.54 g/ml, representing a very high solid concentration. By contrast, Okada et al. [67] successfully prepared porous products with specific surface areas of 532 m^2/g using a sample/solution ratio of 0.02 g/ml. This may be because selective leaching is increasingly suppressed as the concentrations of the various cations approach saturation at excessively high solid concentrations.

The structure and chemical composition of the samples are also important factors in selective leaching. Clay minerals have structures classified as 1:1 type (kaolinite, antigorite, etc), 2:1 type (smectites, vermiculites, chlorites, talc, pyrophyllite, etc), modified 2:1 type (sepiolite and palygorskite), and amorphous type (allophane, imogolite, etc). The crystallinity and particle size of the samples are other influential factors. In terms of the chemical composition, the most important factor is the chemistry of the tetrahedral sheet, i.e., whether or not there is Al substitution for tetrahedral SiO_4, since most of the particle surfaces consist of tetrahedral sheets (on one side in 1:1-type minerals and on both sides in 2:1-type minerals). Acid attack occurs from the surfaces of these sheets and edges. Kavitratna and Pinnavaia [66] divided the selective leaching mechanism into gallery access and edge attack mechanisms. Since the surface area of the sheets is much larger than the edges in a platy particle such as a clay mineral, gallery access is a very important factor both in selective leaching and in determining the porous properties of the resulting products. The leaching ratios of the Al_2O_3 component (occupying tetrahedral and/or octahedral sites) and the MgO component (occupying octahedral sites in various types of clay minerals) are shown in Fig. 10.12 as a function of leaching time [68]. The decrease in the ratio of Al_2O_3 during selective leaching is very different for the different clay minerals, and follows the order: montmorillonite (2:1 type with a SiO_4 sheet), metakaolinite (heat-treated kaolinite; 1:1 type with a SiO_4 sheet) < phlogopite (2:1 type with a $(Si,Al)O_4$ sheet) ≪ vermiculite (2:1 type with a $(Si,Al)O_4$ sheet) < saponite (2:1 type with a $(Si,Al)O_4$ sheet). It is therefore clear that clay minerals whose tetrahedral sheets have Al substitution show higher leaching rates than those without Al substitution. This strongly suggests the operation of a gallery access mechanism in addition to edge attack in the tetrahedral Al-substituted clay minerals, due to the dissolution of AlO_4 from the tetrahedral sheets. By contrast, only edge attack occurs in the unsubstituted clay minerals because of their surface anti-acidic SiO_4 tetrahedral sheets. These leaching mechanisms also influence the dissolution of the octahedral sheets, with talc (2:1 type with a SiO_4 sheet) and montmorillonite showing higher resistance to acid attack than the other clay minerals (2:1 type with $(Si,Al)O_4$ sheets) as shown in Fig. 10.12(b).

Differences in the leaching mechanisms of clay minerals with and without Al-substituted SiO_4 tetrahedral sheets also influence the porous properties of the resulting products.

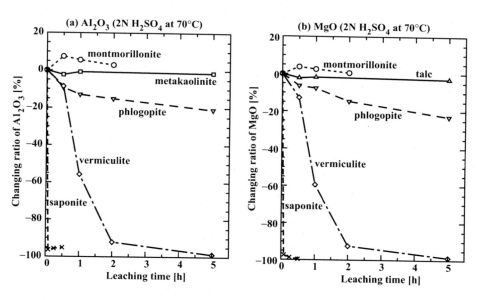

Fig. 10.12. Ratios of Al_2O_3 (a) and MgO (b) components in various clay minerals subjected to selective leaching treatment.

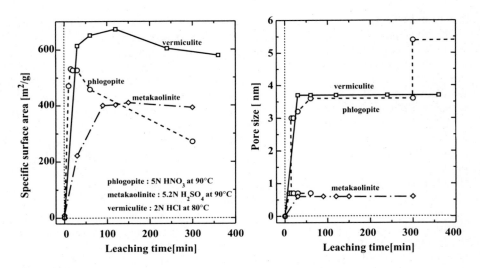

Fig. 10.13. Change in the specific surface areas (a) and pore sizes (b) of products from phlogopite, vermiculite, and metakaolinite as a function of leaching time.

Changes in the specific surface area and pore size of these products are shown in Fig. 10.13 as a function of leaching time. The specific surface areas of the products from both phlogopite and vermiculite (with Al-substituted SiO_4 tetrahedral sheets) show a clear maximum at an optimum leaching time, the specific surface area then decreasing with longer leaching times. This decrease is thought to occur by rearrangement of the SiO_4 tetrahedra fragmented by selective dissolution of the AlO_4 in the tetrahedral sheets. This suggestion is compatible with the pore size change in these samples (Fig. 10.13(b)). The micropores formed in the initial stage of leaching are thought to correspond to the pores formed by selective leaching of the octahedral sheets. However, the micropores change to mesopores within a short leaching time by re-arrangement of the SiO_4 tetrahedra as discussed above. By contrast, the specific surface area of the products from metakaolinite reach a maximum after about 90 min of leaching, but maintain this surface area even after prolonged treatment. The micropores formed in this sample also retain this pore size because the nanoporous structure in this material is stable due to the lack of Al substitution in the SiO_4 tetrahedral sheets as shown schematically in Fig. 10.14, which also shows the porous silica product obtained from antigorite ($Mg_3Si_2O_5(OH)_4$) [46].

As described above, the selective leaching method is a very simple process for obtaining porous materials. Although the process is simple, a variety of nanoporous materials can be obtained by using different types of starting material. Differences in solubility of silica and the other components with solution pH can be effectively used in most cases. This process can also make use of differences in the dissolution rates of crystalline and amorphous phases. For example, both the SiO_2 and Al_2O_3 components are highly soluble in alkaline solution (Fig. 10.11). However, if there is a difference in their dissolution rates, we can selectively leach either component. When kaolinite is heated at about 900–1000°C, phase separation and crystallization occur [23,24], producing a nanotexture consisting of a silica matrix with dispersed nanocrystalline γ-Al_2O_3 grains

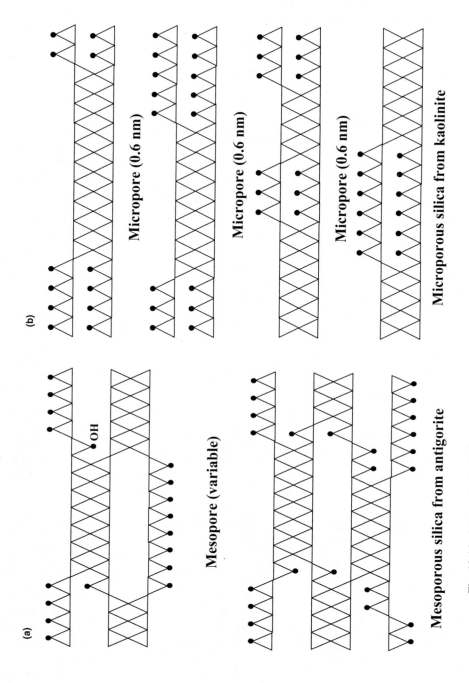

Fig. 10.14. Schematically depicted structure models for products from antigorite (a) and kaolinite (b).

(several nanometers in size) in the pseudomorphic platy particles of the kaolinite. This texture is similar to the bimodal texture in a phase-separated system. Since SiO_2 is amorphous and Al_2O_3 is crystalline in this nanostructure, we would expect a difference in their dissolution rates in alkaline solution. KOH treatment at 90°C can selectively leach out SiO_2, leaving an Al_2O_3-rich product retaining the original platy particle shape. Thus, the spaces left by SiO_2 dissolution form mesopores inside the particles. Since the bimodal texture is of very uniform distribution (even in its random state) and in the size of the nanocrystalline grains, the mesopores generated are of uniform size of several nanometers, and can be controlled in the size range of 4–6 nm by controlling the experimental conditions. The nanostructure of this mesoporous γ-Al_2O_3 is unique compared with conventionally prepared γ-Al_2O_3 because the crystalline γ-Al_2O_3 grains are connected by residual amorphous silica rather than by direct particle contact as in conventional porous γ-Al_2O_3. Because of this unique nanostructure, selectively leached mesoporous γ-Al_2O_3 has excellent thermal stability and humidity control capability [23].

10.2.5 Shaping and coating of porous materials

The previous sections describe the various important factors for the preparation of nanoporous materials and the determination of their porous properties and other characteristics. Nanoporous materials are generally synthesized as relatively fine powders. The various applications of these nanoporous materials, as shown in Fig. 10.1, make it necessary to shape them into pellets, films, membranes, monoliths, etc. Up till now, pellet shapes have commonly been used for catalysts, supports, sorbents, etc. Considerable amounts of binders, forming, and sintering additives are necessary to form pellets by conventional methods, but we have developed a new preparation method for zeolite pellets requiring only a small amount of glass fibers as the forming additive. This is achieved by direct conversion of a wet precursor gel pellet to a zeolite pellet under mild hydrothermal conditions [69]. Monolithic zeolite honeycombs were prepared by Kato [70] by a conventional sintering method using zeolite powder with a considerable amount of additional fibers and binders. It is however more effective to form porous films coated on to substrates such as honeycombs, nets, pipes, etc, to enhance the catalytic and sorption reactions.

The coating of porous materials has been most actively investigated using zeolite films. The proposed methods reported in the literature involve conventional dry gel conversion [71], in-situ crystallization [72], and in-situ reaction [73]. In the conventional method, the substrate is dipped in the precursor solution and reacted under hydrothermal conditions. This is a simple method but the choice of substrate is restricted to materials which are resistant to degradation by the strong alkaline solution. Another disadvantage is the poor adherence strength between the zeolite coating and the substrate. The dry gel conversion method addresses these disadvantages by separating the sample from the solution in the reaction vessel [71]. This method is especially suitable for preparing filter materials by filling the pores of the substrate with zeolite grains grown directly in the pores. However, this method needs a special arrangement to separate the sample and solution in the vessel, and requires more space than a simple autoclave. In the in-situ crystallization method [72], a wet precursor gel is coated on to a substrate and directly autoclaved to produce the zeolite film using only the gel moisture. Since the wet precursor gel films are converted to

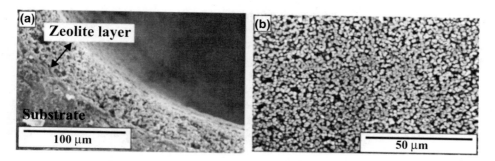

Fig. 10.15. SEM photographs of ZSM-5 zeolite film coated on cordierite honeycomb by the in-situ crystallization method. (a) Honeycomb cell corner with ZSM-5 film, and (b) surface of the ZSM-5 zeolite particles.

zeolite films directly on the substrate, the resulting films form a macroporous texture due to the evaporation of the gel water and shrinkage of gel into crystalline zeolite particles. SEM photographs of ZSM-5 zeolite film prepared by in-situ crystallization are shown in Fig. 10.15. Large macropores are seen not only on the surface but also inside the film. Such macropores are preferable for the speedy transport of target molecules into the interior of the film. This works most effectively in filtration applications. Since a degree of chemical reaction occurs in the interlayer between the coated zeolite film and substrate during autoclaving, especially in the case of silicate substrates, the adherence strength of the coated films is much better than in those prepared by the conventional method.

The in-situ reaction method is a unique technique for preparing zeolite coatings on substrates by partial dissolution of Al_2O_3 and SiO_2 components from the surface of substrate and in-situ reaction with Na^+ from the solution [73]. Commercial glass fibers were used as the substrates on which zeolite Na–A [73] and zeolite X [74] were successfully synthesized. Since the dissolution rate of glass fibers is slower than the commonly used precursor gel, the size of the synthesized zeolite particles is larger than those formed by other coating methods using reactive precursors. In the coating of zeolite on glass fibers, the fibers must be completely covered by the solution as in the conventional coating method, resulting in the degradation of the fibers by the alkaline solution. Where zeolite is coated on to a glass plate substrate, degradation of the substrate can be avoided by floating it on the solution to form the zeolite film on only one side of the substrate without significantly degrading it.

10.3 Practical/Future Applications Related to Various Properties

Increasing global environmental problems are leading to an increased interest in saving energy and protecting the environment. Nanoporous materials are one of the key materials for overcoming these problems. The various applications of nanoporous materials are discussed in this chapter in relation to their properties. Possible future applications are also described.

10.3.1 Solid acidity and applications for catalysts

Since catalytic reactions proceed on the catalyst surfaces and supports, all the porous properties, including the specific surface area, pore size, and pore volume are important parameters. Pore size is especially important. In addition to these geometrical factors, the solid acidity of nanoporous materials is also an important factor for catalyst applications. Solid acidity can be generated by several mechanisms, giving the well-known Brønsted acid and Lewis acid sites [7]. One of the mechanisms for generating solid acidity is by chemical adsorption of H^+ in sites which are locally deficient in formal charge, by substitution of lower valence cations. In nanoporous materials, solid acidity is mostly introduced by substitution of Al^{3+} for Si^{4+} in the tetrahedral sites. Relationships between the amount of solid acidity and the pore size of various nanoporous materials are summarized in Fig. 10.16. A general trend in the solid acidity of these nanoporous materials follows the order: acid-clays < pillared-acid-clay $\leq$ PCHs < K-10 < MPSs < Al–Si gel < zeolites. By contrast, the pore size of these nanoporous materials increases in the following order: zeolites < PCHs < MPSs $\leq$ Al–Si gel $\leq$ K-10 $\leq$ pillared-acid-clay $\leq$ acid-clays. Thus, the trend is towards increasing solid acidity with decreasing pore size. The solid acidity in nanoporous materials is also known to change with cation exchange treatment. In the case of smectites, solid acidity is generated by the H^+ formed by polarization of hydration water associated with the exchangeable cations in the interlayers [82]. The strength of the solid acidity should therefore correlate with the hydration enthalpy of those cations. Onaka et al. [83]

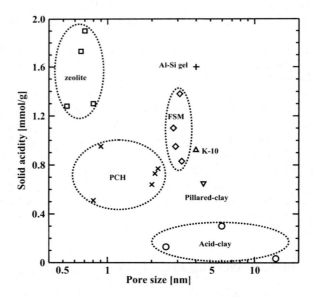

Fig. 10.16. Relationships between solid acidity and pore size of various nanoporous materials: zeolites [74,75], MPSs [76,77], PCHs [78], Al–Si gel [79], K-10 [74], pillared-acid-clay [80], and acid-clays [81].

reported a good correlation between the hydration enthalpy and completion time of silylation of 1-decanol with alkyltrimethylsilane, i.e., higher activity (stronger acid) results in shorter times. This is seen for a series of cation-exchanged montmorillonites which follow the order: Ni- ≤ Zn- < Cu- < H- < Al- < K-10- < Fe- < Sn-montmorillonites. Nanoporous PCHs and PLCs formed by using clay minerals as host layer compounds should show similar trends when cation exchanged. Such treatment of PCHs and PLCs should result in new types of catalysts.

Since the solid acidity in nanoporous materials is generated by substitution of Al for tetrahedral SiO_4, the solid acidity is correlated with the Al/Si ratio. The relationship between the Al/Si ratio and the solid acidity of MPSs has been studied [84,85], giving data which are more similar to SiO_2–Al_2O_3 gels than to zeolites. This may reflect a similarity in the amorphous structures of MPSs and gels. Thus, it may be more difficult to generate solid acidity in MPSs than in zeolites.

The advantages of MPSs and PCHs compared with other nanoporous materials are their pore sizes, which are larger than in zeolites and more uniform in their arrangement than in gels and acid-clays, which should show a similar width of pore size distribution. These advantages of MPSs are illustrated by the synthesis of tetraarylporphyrin (TTP) from aliphatic aldehydes and pyrrole, as reported by Shinoda et al. [30]. Since this aromatic compound is relatively large (1.8 nm), the pore size of the catalyst is an important condition for the reaction. These researchers examined the conversion yield of TTP using three FSMs with pore sizes of 2.0, 2.8, and 3.4 nm and found that the FSM with the pore size of 2.8 nm performed best. This result strongly suggests an optimum pore size for this catalytic reaction. Commercially available K-10, prepared by sulfuric acid treatment of montmorillonite, is also an effective catalyst, showing conversion yields similar to FSM, probably due to the similarity of its pore size (2.7 nm) to that of FSM (2.8 nm). However, the pores in K-10 are formed by a card-house-like aggregation of silica-rich gel particles which is considered to be less stable than the pore structure in FSM. Therefore, the conversion yield of TTP by K-10 decreased to 0% after only two cycles of use while FSM maintained its conversion yield after repeated use. It is clear that MPSs and PCHs have a good potential as catalysts for reactions of relatively large molecules such as aromatic compounds. Many of the practical applications of MPSs as catalysts have been reviewed in detail [8,47,48].

The aforementioned PCHs and MPSs have the advantage that the surfaces of their nanopores can be tailored by manipulating the surface Si-OH groups. Many studies have been reported on organic–inorganic hybrid catalysts using MPSs [86]. Another breakthrough for MPSs was the success in crystallizing the wall matrix by introducing phenylene groups into the silica by the use of organic–inorganic compounds [87]. The nanostructures thus synthesized were of three different types, hexagonal with 3.8 nm pores, cubic with three-dimensionally aligned mesopores, and hexagonal with both larger mesopores (6–12.4 nm) and micropores. Although these MPSs have poor thermal stability because of the organic groups included in their structure, they have many advantages of structural regularity and variety of pore structure, compared with purely inorganic MPSs. For example, these materials provide good support for proteins and enzymes in biotechnological applications.

10.3.2 Sorption properties and applications to sorbents

Most of the nanoporous materials except activated carbon and high-silica zeolites have hydrophilic surface properties, making them highly adsorbing for polar molecules such as water vapor and alcohols. Water vapor adsorption isotherms (25°C) of hydrophilic zeolite (Na-X), hydrophobic zeolite (silicalite), and FSM are shown in Fig. 10.17. Large differences are observed in these isotherms. Zeolite Na-X has a relatively low Si/Al ratio and shows good adsorption in the low P/P_0 range due to its hydrophilic micropores while silicalite, with a high Si/Al ratio shows very low adsorption due to its hydrophobic surface character. FSM has larger mesopores than the zeolites and shows very steeply increased adsorption at $P/P_0 \approx 0.6$ due to capillary condensation which results in a maximum adsorption of about 1200 ml/g. A distinctly large hysteresis is observed in the desorption isotherm, due to the delay of vapor evaporation from the mesopores. This hysteresis is typical of that seen in mesoporous materials [88] but the very large hysteresis in this sample is due to the formation of silanol groups (Si-OH) after the first adsorption experiment. These groups change the surface properties from hydrophobic to hydrophilic prior to the adsorption experiment because they are dehydrated by the pre-treatment for the sorption experiment. Since the silanol groups remain in the surface of the mesopores after the desorption experiment, the second isotherm is quite different from the first. The characteristics of the water vapor sorption isotherms of the various nanoporous materials illustrate why hydrophilic zeolites are used as dehydration materials while MPSs are candidates as humidity control materials.

The adsorption properties of various nanoporous materials for non-polar molecules such as volatile organic compounds (VOCs) have been examined because of their relevance to the 'sick-house' syndrome, especially in Japan. Relationships between the adsorption of CCl_4 (an equi-axial-shaped molecule $0.58 \times 0.58 \times 0.58$ nm^3 in size) and benzene (a plate-shaped molecule $0.66 \times 0.75 \times 0.32$ nm^3 in size) and the pore sizes of nanoporous

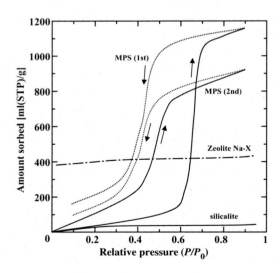

Fig. 10.17. Water vapor adsorption–desorption isotherms of zeolites and MPS.

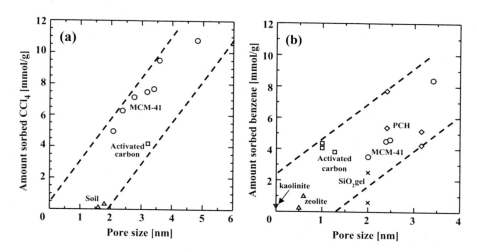

Fig. 10.18. Relationship between the amount of CCl_4 (a) and benzene (b) adsorption and pore size of various nanoporous materials: MCM-41 samples [89,90], activated carbon [91] and soil [92] in CCl_4, and MCM-41 [93], PCH [94], kaolinite [95], zeolite [96], and silica gel [37] in benzene.

materials are shown in Figs. 10.18(a) and (b), respectively. All these adsorbents show similarly increased adsorption with increasing pore size. This indicates that the pore size is the most important factor in the adsorption of these molecules. The maximum amount of CCl_4 adsorption (10.8 mmol/g) was achieved with MCM-41 prepared using a long chain $C_{18}TMA^+$ as the pore former (pore size 5.3 nm). By contrast, the maximum amount of benzene adsorption (8.4 mmol/g) was achieved with MCM-41 prepared using $C_{16}TMA^+$ as the pore former (pore size 3.4 nm). Since MPSs with larger mesopores than MCM-41 have been synthesized, even higher adsorption is expected by these mesoporous MPSs. MPS is also effective in adsorbing gases such as CO_2 (1.27), CH_4 (0.57), and C_2H_5 (1.11 mmol/g), these sorbed amounts being much higher than obtained with acid-clays [97]. Thus, nanoporous materials of the MPS- and PCH-type are very good sorbent materials for various organic molecules, both in solution and in the atmosphere, and will be more widely used as high performance sorbents in the near future.

The adsorption properties of MPSs are further enhanced by surface modification using various silanes. Inumaru et al. [98] introduced octylsilanes into the mesopores of MCM-41 and investigated the effect of the octylsilane chain length on the uptake performance for nonylphenol, an endocrine disrupter. Although the specific surface areas and pore volumes of the octylsilane-grafted MCM-41s became lower than in MCM-41 by the introduction of the octylsilanes, the amount of nonylphenol sorption by C_8- and C_{12}-grafted MCM-41 was 7 and 3 times higher than by MCM-41, respectively. Grafting of the silane with an appropriate chain length gives the surface of the mesopores a hydrophobic character, maintaining sufficient space for sorption and playing an important role in the adsorption of some unusual organic molecules. In this way, MPSs and PCHs are candidates for the highly selective adsorption of harmful organic matter. Their performance is expected to be further enhanced by the introduction of functional groups with high selectivity for unusual organic molecules.

Although activated carbon is most widely used as sorbents for various adsorbates at present, two new developments are expected to become important in the future, namely, (1) high-performance absorption by nanoporous materials such as MPSs, PCHs, and PLCs, with high sorption capacity and/or target selectivity, and (2) cost-effective sorbents produced by activation of widely available template minerals.

10.3.3 Miscellaneous properties and expected applications

As repeatedly stated, one of the advantages of MPSs such as MCMs, FSMs, and PCHs is their ordered nanostructure with uniformly aligned mesopores. These pores are generally utilized as the spaces for catalytic and sorption reactions. Another idea for using these mesopores more effectively is to grow nanoparticles in the pores. The size of the mesopores thus produced in these nanoporous materials is in the range which should show a 'quantum effect.' Abe *et al.* [99] synthesized various sizes of Fe_2O_3 nanoparticles in MPSs and found a variation in the band gap energy of the Fe_2O_3 from 2.14 eV (similar to the bulk sample) to 4.6 eV as the particle size decreased from about 5–2 nm. Tai *et al.* [100] prepared composites of polymer/MCM-41 using the meospores as conducting paths by infiltrating electrolytes. Other proposed applications are electron transfer photosensitizers [101], sensing devices [102], semiconductors [103], polymer wires [104], non-linear optic materials [105], etc. However, these new approaches require more detailed study before real applications emerge.

10.4 Summary

The coupling of inorganic and organic compounds has resulted in the development of new types of porous materials with a variety of porous properties. Much work is still going on in this area. Less attention has been paid to the uniqueness of mineral templates in the synthesis of nanoporous materials compared with the greater interest shown until now in organic templates. However, it is worthwhile to pay more attention to the use of mineral templates, as exemplified by the microporous silica prepared by heating and acid-selective leaching of kaolinite. This microporous silica shows a unique adsorption property attributed to the presence of slit-shaped micropores with hydrophilic surface properties. This was also found to be a very good starting material for the synthesis of MFI zeolites with a well-developed (010) plane; this is the most important plane for catalytic and sorption reactions because the cylindrical micropores run perpendicular to this plane. Kaolinite is also useful for synthesizing mesoporous γ-Al_2O_3 with unique properties.

Mineral templates play an important role as host layer compounds in the syntheses of PCHs and PLCs. Until now, smectites have been used more commonly as templates, but swellable and exchangeable synthetic micas may be more functional as host layer compounds because of their tunable layer charge, higher crystallinity, and higher purity compared with the smectites. More precise control of the interlayers using synthetic micas has the potential for developing new types of porous materials in the future.

Finally, we would emphasize the importance of forming these nanoporous materials into shapes which effectively utilize the functional nanopores. The development of highly functional nanoporous materials is only the first step, and effective texturing of these nanoporous materials is an important next step towards making optimal use of these materials. The panoscopic nanotextures observed in some plants are a good model for the development of highly functional artificial materials.

References

[1] T. Tomita, Ph.D. Thesis, Tokyo Institute of Technology 2004 (in Japanese).

[2] F. Rouquerol, J. Rouquerol and K. Sing, Adsorption by Powders and Porous Solids, Academic Press: San Diego, USA, 1999.

[3] K. Okada, T. Tomita, Y. Kameshima, A. Yasumori, T. Yano and K.J.D. MacKenzie, Microp. Mesop. Mater. **37** (2000) 355.

[4] S.S. Kistler, Nature **127** (1931) 741.

[5] J. Mrowiec-Bialon, A.B. Jarzebski, A.I. Lachowski, J.J. Malinowaski and Y.I. Aristov, Chem. Mater. **9** (1997) 2486.

[6] T. Okubo, Expect. Mater. Future **5** (2005) 38 (in Japanese).

[7] Y. Ono and T. Yashima, Eds., Science and Engineering of Zeolite, Kodansha: Tokyo, Japan, 2000 (in Japanese).

[8] A. Sayari, Stud. Surf. Sci. Catal. **102** (1996) 1.

[9] C.T. Kresge, M.E. Leonowicz, W.J. Roth, J.C. Vartuli and J.S. Beck, Nature **359** (1992) 710.

[10] T. Yanagisawa, T. Shimizu, K. Kuroda and C. Kato, Bull. Chem. Soc. Japan **63** (1990) 988.

[11] K. Ishizaki, S. Komarneni and M. Nanko, Porous Materials, Kulwer Academic Pub.: Dordrecht, Netherlands, 1998, pp. 1–11.

[12] M.S. Whittingham and A.J. Jacobson, Eds., Intercalation Chemistry, Academic Press: New York, USA, 1982.

[13] J.W. Akitt, N.N. Greenwood, B.L. Khandelwal and G.D. Lester, J. Chem. Soc. Dalton Trans. **1972** (1972) 604.

[14] T. Yamaguchi, Y. Sakai and K. Kitajima, J. Mater. Sci. **34** (1999) 5771.

[15] A. Galarneau, A. Barodawalla and T.J. Pinnavaia, Nature **374** (1995) 529.

[16] J. Ahenach, P. Cool and E.F. Vansant, Phys. Chem. Chem. Phys. **2** (2000) 5750.

[17] D. Lozano-Castello, M.A. Lillo-Rodenas, D. Cazorla-Amoros and A. Linares-Solano, Carbon **39** (2001) 741.

[18] K. Okada, N. Yamamoto, Y. Kameshima and A. Yasumori, J. Colloid Interf. Sci. **262** (2003) 179.

[19] K. Okada, Y. Shimizu, Y. Kameshima and A. Nakajima, J. Porous Mater. **12** (2005) 289.

[20] Y. Kobayashi, K. Iwasaki, T. Kyotani and A. Tomita, J. Mater. Sci. **31** (1996) 6185.

[21] T. Kondo and K. Uosaki, Funct. Mater. **14** (1994) 5 (in Japanese).

[22] C. Breen, J. Madejova and P. Komadel, Appl. Clay Sci. **10** (1995) 219.

[23] K. Okada, Int. J. Soc. Mater. Eng. Resour. **8** (2000) 50.

[24] K. Okada, H. Kawashima, Y. Saito, S. Hayashi and A. Yasumori, J. Mater. Chem. **5** (1995) 1241.

[25] S. Wang, T. Dou, Y. Li, Y. Zhang, X. Li and Z. Yan, J. Solid State Chem. **177** (2004) 4800.

[26] Y.Z. Yao and S. Kawi, J. Porous Mater. **6** (1999) 77.

[27] S. Kawi, Mater. Lett. **38** (1999) 351.

[28] Y. Liu and T.J. Pinnavaia, J. Mater. Chem. **14** (2004) 3416.

[29] C.D. Madhusoodana, Y. Kameshima, A. Nakajima, K. Okada, T. Kogure and K.J.D. MacKenzie, J. Colloid Interf. Sci. (2006) in press.

[30] T. Shinoda, Y. Izumi and M. Onaka, J. Chem. Soc. Chem. Commun. **1995** (1995) 1801.

[31] T. Kimura, K. Kuroda, Y. Sugahara and K. Kuroda, J. Porous Mater. **5** (1998) 127.

[32] T. Linssen, P. Cool, M. Baroudi, K. Cassiers, E.F. Vansant, O. Lebedev and J. Van Landuyt, J. Phys. Chem. B **106** (2002) 4470.

[33] L. Mercier and T.J. Pinnavaia, Microp. Mesop. Mater. **20** (1998) 101.

[34] M. Benjelloun, P. Cool, T. Linssen and E.F. Vansant, Microp. Mesop. Mater. **49** (2001) 83.

[35] M. Pichowicz and R. Mokaya, Chem. Commun. **2001** (2001) 2100.

[36] M. Polverejan, T.R. Pauly and T.J. Pinnavaia, Chem. Mater. **12** (2000) 2698.

[37] J. Pires, A.C. Araujo, A.P. Carvalho, M.L. Pinto, J.M. Gonzalez-Calbet and J. Ramirez-Castellanos, Microp. Mesop. Mater. **73** (2004) 175.

[38] L. Zhu, S. Tian and Y. Shi, Clays Clay Miner. **53** (2005) 123.

[39] R. Ishii, M. Nakatsuji and K. Ooi, Microp. Mesop. Mater. **79** (2005) 111.

[40] R. Mokaya and W. Jones, J. Catal. **153** (1995) 76.

[41] T. Yamaguchi, A. Shirai, S. Taruta and K. Kitajima, Ceram. Silikaty **45** (2001) 43.

[42] Y. Kameshima, Y. Tamura, A. Nakajima and K. Okada, Proc. of 13th Inter. Clay Conf. (2005), p. 76.

[43] J.L. Venaruzzo, C. Volzone, M.L. Rueda and J. Ortiga, Microp. Mesop. Mater. **56** (2002) 73.

[44] K. Okada, A. Shimai, S. Hayashi and A. Yasumori, Microp. Mesop. Mater. **21** (1998) 289.

[45] J. Temuujin, K. Okada and K.J.D. MacKenzie, Appl. Clay Sci. **22** (2003) 187.

[46] K. Kosuge, K. Shimada and A. Tsunashima, **7** (1995) 2241.

[47] G.J.A.A. Soler-Illia, C. Sanchez, B. Lebeau and J. Patarin, Chem. Rev. **102** (2002) 4093.

[48] K.J.C. van Bommel, A. Friggeri and S. Shinkai, Chem. Int. Ed. **42** (2003) 980.

[49] I.S. Paulino and U. Schuchardt, Stud. Surf. Sci. Catal. **141** (2002) 93.

[50] S. Namba, A. Mochizuki and M. Kito, Chem. Lett. **1998** (1998) 569.

[51] M.L. Occelli and S. Biz, J. Molecul. Catal. A **151** (2000) 225.

[52] Y. Xia, R. Mokaya and J.J. Titman, J. Phys. Chem. B **108** (2004) 11361.

[53] P. Kumar, N. Mal, Y. Oumi, T. Sato and K. Yamana, Stud. Surf. Sci. Catal. **141** (2002) 159.

[54] S. Inagaki, Y. Fukushima and K. Kuroda, Chem. Comm. **1993** (1993) 680.

[55] L. Chu, M.I. Tejedor-Tejedor and M.A. Anderson, Microp. Mater. **8** (1997) 207.

[56] S. Tomura, M. Maeda, R. Miyawaki, H. Mizuta, Y. Shibasaki and N. Tone, Kobutsugaku-Zasshi **19** (1990) 331 (in Japanese).

[57] R.K. Iler, The Chemistry of Silica, John Wiley Sons: New York, 1979.

[58] M. Ogura, Y. Kawazu, H. Takahashi and T. Okubo, Chem. Mater. **15** (2003) 2661.

[59] B.P. Feuston and J.B. Higgins, J. Phys. Chem. **98** (1994) 4459.

[60] H. Kopka, K. Beneke and G. Lagaly, J. Colloid Interf. Sci. **123** (1988) 427.

[61] S. Miyata and T. Hirose, Clays Clay Miner. **26** (1978) 441.

[62] K. Okada, F. Matsushita, S. Hayashi and A. Yasumori, Clay Sci. **10** (1996) 1.

[63] J. Wnag, Y. Tian, R.-C. Wang and A. Clearfield, Chem. Mater. **4** (1992) 1276.

[64] L. Li, S. Ma, X. Liu, Y. Yue, J. Hui, R. Xu, Y. Bao and J. Rocha, Chem Mater. **8** (1996) 204.

[65] S. Yamanaka, T. Nishihara, M. Hattori and Y. Suzuki, Mater. Chem. Phys. **17** (1987) 87.

[66] H. Kavitratna and T.J. Pinnavaia, Clays Clay Miner. **42** (1994) 717.

[67] . Okada, N. Nakazawa, Y. Kameshima, A. Yasumori, J. Temuujin, K.J.D. MacKenzie and M.E. Smith, Clays Clay Miner. **50** (2002) 624.

[68] K. Okada and K.J.D. MacKenzie, Current Topics Colloid Interf. Sci. (2006) in press.

[69] C.D. Madhusoodana, Y. Kameshima, A. Yasumori and K. Okada, J. Mater. Sci. Lett. **22** (2003) 553.

[70] T. Kato, Mater. Integr. **13** (2000) 1331 (in Japanese).

[71] R. Aiello, F. Crea, F. Testa and A.S. Gattuso, Stud. Surf. Sci. Catal. **105** (1999) 29.

[72] K. Okada, Y. Kameshima, C.D. Madhusoodana and R.N. Das, Sci. Tech. Adv. Mater. **5** (2004) 479.

[73] K. Okada, H. Shinkawa, T. Takei, S. Hayashi and A. Yasumori, J. Porous Mater. **5** (1998) 163.

[74] T. Fukushima, H. Miyazaki and S. Asano, J. TOSOH Res. **33** (1989) 155.

[75] B. Thomas, S. Prathapan and S. Sugunan, Appl. Catal. **277** (2004) 247.

[76] S. Inagaki, Y. Fukushima, A. Okada, T. Kurauchi, K. Kuroda and C. Kato, Proc. 9th Inter. Zeolite Conf., Reed Publishing, USA (1993), p. 305.

[77] T. Linssen, F. Meers, K. Cassiers, P. Cool and E.F. Vansant, J. Phys. Chem. B **107** (2003) 8599.

[78] J. Ahenach, P. Cool and E.F. Vansant, Phys. Chem. Chem. Phys. **2** (2000) 5750.

[79] K. Okada, T. Tomita, Y. Kameshima, A. Yasumori and K.J.D. MacKenzie, J. Colloid Sci. Interf. **219** (1999) 195.

[80] R. Mokaya and W. Jones, J. Catal. **153** (1995) 76.

[81] K. Okada, N. Arimitsu, Y. Kameshima, A. Nakajima and K.J.D. MacKenzie, Appl. Clay Sci. **31** (2006) 185.

[82] M. Frenkel, Clays Clay Miner. **22** (1974) 435.

[83] M. Onaka, Y. Hosokawa, K. Higuchi and Y. Izumi, Tetrahedron Lett. **7** (1993) 1171.

[84] S. Inagaki, Y. Yamada and Y. Fukushima, Prog. Zeolite Microp. Mater. Stud. Sci. Catal. **105** (1997) 109.

[85] Y. Cesteros and G.L. Haller, Microp. Mesop. Mater. **43** (2001) 171.

[86] A.P. Wight and M.E. Davis, Chem. Rev. **102** (2002) 3589.

[87] Y. Goto and S. Inagaki, Expect. Mater. Future **5** (2005) 32 (in Japanese).

[88] K.S.W. Sing, D.H. Everett, R.A.W. Haul, L. Moscou, R.A. Pierotti, J. Rouquerol and T. Siemiemewska, Pure Appl. Chem. **57** (1985) 603.

[89] P.J. Branton, K.S.W. Sing and J.W. White, J. Chem. Soc. Farad. Trans. **93** (1997) 2337.

[90] M. Hakuman and H. Naono, J. Colloid. Interf. Sci. **241** (2001) 127.

[91] J.W. Lee, W.G. Shim, M.S. Yang and H. Moon, J. Chem. Eng. Data **49** (2004) 502.

[92] N.H. Dural, C.-H. Chen and R.K. Puri, Chem. Eng. Commun. **162** (1997) 75.

[93] J.X. Liu, M. Dong, Z.L. Sun, Z.F. Qin and J.G. Wang, Adsorption **10** (2004) 205.

[94] L. Zhu, S. Tian and Y. Shi, Clays Clay Miner. **53** (2005) 123.

[95] Yu.I. Tarasevich, V.E. Polyakov, S.V. Bondarenko and E.V. Aksenenko, Colloid J. **66** (2004) 584.

[96] M.A. Hermandez, L. Corona, A.I. Gonzalez, F. Rojas, V.H. Lara and F. Silva, Ind. Eng. Chem. Res. **44** (2005) 2908.

[97] P. Pendleton, J. Colloid. Interf. Sci. **227** (2000) 227.

[98] K. Inumaru, J. Kiyoto and S. Ymamanaka, Chem. Commun. **2000** (2000) 903.

[99] T. Abe, Y. Tachibana, T. Uematsu and M. Iwamoto, J. Chem. Soc., Chem. Commun. **1995** (1995) 1617.

[100] X. Tai, G. Wu, Y. Tominaga, S. Asai and M. Sumita, J. Polym. Sci. B. Polym. Phys. **43** (2005) 184.

[101] A. Corma, V. Fomes, H. Garcia, M.A. Miranda and M.J. Sabater, J. Am. Chem. Soc. **116** (1994) 9767.

[102] D.H. Olson, G.D. Stucky and J.C. Vartuli, US Patent No.5 364 797 (1994).

[103] R. Leon, D. Margolese, G.D. Stucky and P.M. Petroff, Phys. Rev. B **52** (1995) R2285.

[104] C.G. Wu and T. Bein, Stud. Surf. Sci. Catal. **84** (1994) 2269.

[105] J.S. Beck, G.H. Kuehl, D.H. Olson, J.L. Schlenker, G.D. Stucky and J.C. Vautuli, US Patent No. 5 348 687 (1994).

CHAPTER 11

Enhancement of Thermoelectric Figure of Merit through Nanostructural Control on Intermetallic Semiconductors toward High-Temperature Applications

Yoshinao Mishima, Yoshisato Kimura, and Sung Wng Kim

Abstract

Practical applications of thermoelectric modules have been most common in Peltier cooling and heating in various systems at around room temperature. Recently, recovery of electricity from heat wasted in various situations has been expected for effective utilization of energy resources. One of the key technologies for this purpose is to utilize Seebeck electric generation by finding thermoelectric materials exhibiting a high dimensionless figure of merit over 1.0 at such temperatures around 500–800 K. In the present chapter, the efforts so far attempted by various investigators on improving thermoelectric properties are summarized with a focus on the so-called Phonon Glass Electron Crystal (PGEC) materials including such compounds as skutterudite and half-Heusler. Design strategy to increase the dimensionless figure of merit in these materials has been to reduce lattice thermal conductivity through phonon scattering, enhanced by the rattling effect of alloying elements and to optimize carrier density by doping elements. Some recent results by the group of the present authors are also presented. Finally, the updated results on the substantial enhancement in thermoelectric properties in nanostructured materials which could exhibit the dimensionless figure of merit of over 2.0 are presented.

Keywords: **thermoelectric materials, intermetallic compounds, semiconductor, electrical resistivity, Seebeck coefficient, thermal conductivity, figure of merit, skutterudite, half-Heusler, nano-structure.**

11.1 Background and Principles

11.1.1 History

The physical principles of thermoelectric generation and refrigeration are based on the findings in the early 1800s, although commercial thermoelectric modules were not available until almost 1960. The first important discovery relating to thermoelectricity was made in 1821 when a German scientist, Thomas Seebeck, found that electric current continuously flows in a closed circuit made up of two dissimilar metals, provided that the junctions of the metals are maintained at two different temperatures. In 1834, a French physicist, Jean Peltier, while investigating the Seebeck effect, found that there exists an opposite phenomenon whereby thermal energy is absorbed at one dissimilar metal junction and discharged at the other junction when electric current flows within the closed circuit. Twenty years later, William Thomson comprehensively explained the Seebeck and Peltier effects and described their interrelationship. At that time, however, these phenomena were still considered to be mere laboratory curiosities and without practical applications. In the 1930s, Russian scientists began some of the earliest thermoelectric studies in an effort to construct power generators for use at remote locations throughout the country. In 1949, Ioffe developed a theory of semiconductor thermoelements, and in 1954, Goldsmid and Douglas demonstrated that cooling from a typical ambient temperature down to below 273 K is possible [1].

Figure 11.1 shows the operating principles of thermoelectric generation and refrigeration. By applying a low voltage DC power source to a thermoelectric module, heat is transferred through the module from one side to the other. One module face, therefore, is cooled, while the opposite face is simultaneously heated. It is important to note that this phenomenon may be reversed whereby a change in the polarity of the applied DC voltage

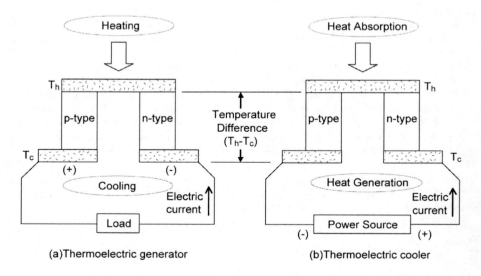

Fig. 11.1. Operating principles of thermoelectric generation and refrigeration.

11.1.2 Thermoelectric figure of merit

The maximum conversion efficiency of a thermoelectric material, ϕ_{max}, is expressed as in [2]

$$\phi_{max} = \frac{T_h - T_c}{T_h} \frac{(1 + Z\overline{T})^{1/2} - 1}{(1 + Z\overline{T})^{1/2} + T_c/T_h}, \quad \overline{T} = \frac{T_c + T_h}{2}. \quad (11.1)$$

The quantity of Z, which determines the maximum conversion efficiency at the operation temperature, is termed as the thermoelectric figure of merit and consists of three transport parameters.

$$Z = \frac{\alpha^2}{\rho \cdot \kappa}, \quad (11.2)$$

where α is the Seebeck coefficient, or thermoelectric power, which is the thermoelectric electromotive force per degree K, ρ the electrical resistivity, and κ the thermal conductivity. A more convenient quantity than Z is the dimensionless figure of merit ZT.

$$ZT = \frac{\alpha^2 \cdot T}{\rho \cdot \kappa} \quad (11.3)$$

Here, T is the temperature in K. From Eq. (11.2), a material will have a high Z, if it has a large Seebeck coefficient, a small electrical resistivity, and a small thermal conductivity. In a semiconducting thermoelectric material, the total thermal conductivity is a combination of the lattice or phonon thermal conductivity κ_{lat} and the electronic thermal conductivity κ_{el}:

$$\kappa = \kappa_{lat} + \kappa_{el}. \quad (11.4)$$

In most heavily doped semiconductors, the total thermal conductivity is dominated by the lattice thermal conductivity contribution. Thus, the problems of producing a high-ZT thermoelectric are maximizing the power factor α^2/ρ and minimizing the lattice thermal conductivity κ_{lat}. As shown in Fig. 11.2, these three transport parameters functionally depend on the carrier concentration; therefore, modifying one parameter affects the others as well. According to the Wiedemann–Franz law, the electronic thermal conductivity in Eq. (11.4) is

$$\kappa_{el} = \frac{L_0 \cdot T}{\rho}, \quad (11.5)$$

where $L_0 = (\pi^2/3) \cdot (k_B/e)^2$ is the Lorenz number, ρ the electrical resistivity, and T the absolute temperature in K.

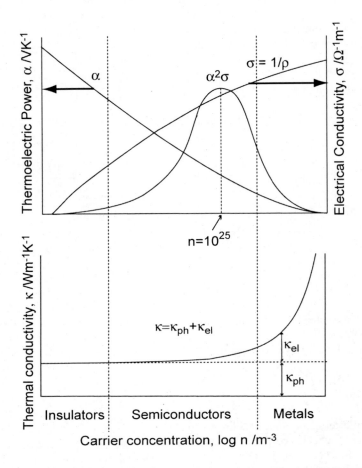

Fig. 11.2. Schematic illustrations of Seebeck coefficient (thermoelectric power) α, electrical resistivity ρ, and thermal conductivity κ, as functions of carrier concentration. Note that the thermal conductivity consists of the lattice vibration κ_{ph} and electronic contribution κ_{el}. The term α^2/ρ is the so-called power factor.

From Eq. (11.5), when the electrical conductivity increases, the electronic contribution of the thermal conductivity increases. Eventually, there is no net increase in the figure of merit Z. Recently, a research study on thermoelectric materials, which is based on lowering the lattice thermal conductivity without decreasing the power factor, has been conducted.

11.1.3 Thermoelectric transport parameters

The Seebeck coefficient and electrical conductivity strongly depend on the Fermi level, which in turn depends on the carrier concentration, carrier effective mass, and temperature. In theoretical formulations, it is convenient to express the thermoelectric transport coefficients in terms of the Fermi energy. Since the thermal conductivity only weakly depends on the carrier concentration, the general effect of an increase in carrier concentration manifests itself in the figure of merit, through the power factor.

In the case of a single band with a typical parabolic density of states and assuming that the carriers obey classical statistics, the Seebeck coefficient can be expressed as

$$\alpha = \pm \frac{k_B}{e}(5/2 + s - \eta^*). \tag{11.6}$$

The + sign is for the valence band, while the − sign is for the conduction band. η^* is the reduced Fermi level, $\eta^* = \eta/k_B T$, k_B Boltzmann's constant, T the temperature in K, and s the scattering parameter. It is assumed that the carrier relaxation time can be expressed in terms of the carrier energy in a simple manner, which is proportional to E^s.

The electrical resistivity is given by

$$\frac{1}{\rho} = n \cdot e \cdot \mu \tag{11.7}$$

where e is the electronic charge, n the carrier concentration, and μ the carrier mobility. The carrier concentration is related to the reduced Fermi energy, as shown by

$$n = 4\pi \left(\frac{2m^* k_B T}{h^2}\right)^{3/2} \exp(\eta^*). \tag{11.8}$$

The optimum reduced Fermi energy is obtained using $dZ/d\eta^* = 0$. This solution gives the η^*_{opt} values of 0 and 2 in the acoustic phonon and ionized impurity scatterings, respectively. Obviously, the thermoelectric material has to be doped heavily to obtain Fermi levels well within the bands. Corresponding optimum carrier concentrations can be estimated from Eq. (11.8).

The use of classical statistics in describing the behavior of the carriers is justified only in the limit of low carrier concentration. Because the thermoelectric material has to be doped heavily, Fermi–Dirac statistics should be employed. The Seebeck coefficient can be expressed as

$$\alpha = \pm \frac{k_B}{e}\left(\frac{(2+s)F_{1+s}(\eta^*)}{(1+s)F_s(\eta^*)} - \eta^*\right), \tag{11.9}$$

$$F_n = \int_0^\infty \frac{x^n}{1 + \exp(x - \eta^*)} dx, \tag{11.10}$$

where η^* is the reduced Fermi level, $\eta^* = \eta/k_B T$ and F_n is a Fermi integral of the order x [3–7]. The + sign is for the valence band, while the − sign is for the conduction band. The parameter s gives the exponent of the energy dependence of the charge carrier mean free path. For the pure phonon or lattice vibration scattering of the charge carriers, $s = 0$. When the carriers are scattered primarily by charged impurities (ionized donors and acceptors), $s = 2$; the intermediate value $s = 1$ represents mixed scattering. The scattering of electrical carrier transport is represented mainly by two scattering mechanisms, namely, the acoustic phonon and ionized impurity scattering mechanisms. The scattering mechanism that affects electrical transport can be determined from the temperature dependence of carrier mobility. The acoustic phonon and ionized impurity scattering mechanisms depend on $T^{-3/2}$ and $T^{1/2}$, respectively.

Because metallic materials have a relatively high thermal conductivity and insulators show a very high electrical resistivity, the most favorable materials for thermoelectrics are usually semiconductors having a Fermi level near the band gap. There are several theoretical research studies on optimizing the band gap for the performance of thermoelectrics. Mahan reported that the optimal band gap size is 10 $k_B T$, 0.25 eV [8].

Generally, n-type materials have much larger Seebeck coefficients than p-type materials at a given carrier concentration due to higher effective mass of electrons than that of holes [4]. The Seebeck coefficient decreases with an increase in carrier concentration, which can be expressed as

$$n = 4\pi \left(\frac{2m^* k_B T}{h^2} \right)^{3/2} F_{1/2}(\eta^*), \tag{11.11}$$

where m^* is the effective mass and T is the temperature in K [6,7]. The reduced Fermi level can be calculated from the experimentally determined Seebeck coefficient using Eq. (11.9), which is then used to calculate the effective masses from Eq. (11.11) together with the measured carrier concentrations. The thermal conductivity is written as a sum of the lattice and electronic contributions as Eq. (11.4). In this case, similar to the Seebeck coefficient, the Lorenz number in the case that acoustic phonon scattering is predominant can be expressed as [9]

$$L_0 = \frac{k_B^2}{e^2} \frac{3 F_0(\eta^*) \cdot F_2(\eta^*) - 4 F_1^2(\eta^*)}{F_0^2(\eta^*)} \tag{11.12}$$

11.1.4 Practical applications

Utilization of thermoelectric materials for a solid state energy conversion system has attracted considerable attention due to such advantages as compactness, quietness (no moving part), and environmental friendliness. Such an energy conversion system also provides localized heating or cooling and a long-term stability with a lifetime on the order of 20 years or more. To utilize thermoelectric materials for practical applications, the development of materials having a large ZT and fabrication of a module from p- and n-type the temperature dependence of which was previously investigated for various p- and n-type thermoelectric materials. It is noted that the parabolic curve in the Fig. 11.3, which corresponds to $ZT = 1$, seems thus far to depict the upper limit of the performance of both the p- and n-type thermoelectric materials. It can be seen that a

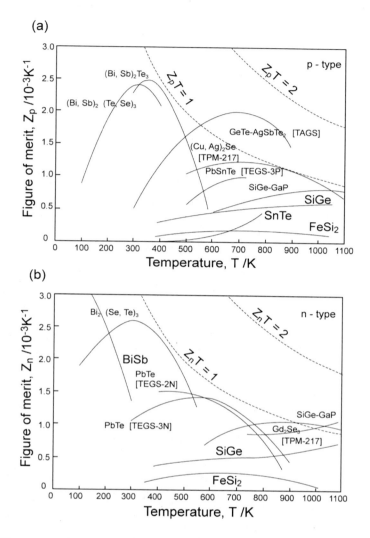

Fig. 11.3. Temperature dependence of thermoelectric figure of merit ZT of typical thermoelectric materials, namely, (a) p-type and (b) n-type semiconductors.

group of intermetallic compounds including Bi-Te solid solution alloys and BiSb, exhibit the highest ZT at around room temperature, and Si–Ge solid solution alloys at higher temperatures [10–12]. It is well known that the materials based on BiTe have been commercially and most extensively utilized for both refrigeration and power generation. A cooling module for a central processing unit of personal computers and a wine cooler are examples of the former use, while a wrist watch operating according to differences in ambient and body temperatures is an example of the latter use. The Si–Ge thermoelectric materials are well known for generating electric power necessary for transmitting various data to and from the earth using a radioisotope as a heat source in the Voyager spacecraft.

Recently, the requirements in the development of materials have been focused on those that contribute to energy conservation and environmental protection. The technologies for conversion of fossil or other fuel sources into electricity are generally accompanied by heat waste. A more efficient energy conversion system, such as a combined cycle electricity generation systems must be designed and developed for more effective utilization of energy resources and lower emission of CO_2 and NO_X; however, the technologies of recovering the heat wasted in these systems and converting it into electricity are in process. According to Kjikawa and Ninno [13], the amount of heat wasted from an industrial process in Japan is estimated to be about 148×10^{12} kcal per year. The potential thermal resource from gasoline engine vehicles and diesel engine trucks is about 390×10^{12} kcal per year. Moreover, the potential energy resource from the combustion heat of solid waste is about 67.5×10^{12} kcal per year. On the assumptions that the ZT of the state-of-the-art material is 0.77 and that of the future material is 2.0, the total contribution to recover the energy consumptions by thermoelectric generation is 4.5 Mkl (equivalent to oil) per year for $ZT = 0.77$ and 8.6 Mkl per year for $ZT = 2.0$. Although these values are about few percent of the total energy consumption in Japan, they are markedly larger than that due to the photovoltaic power (1.13 Mkl per year at 2010) estimated by the Japanese government. Hence, the efficient use of waste heat sources is very important for energy conservation and the reduction in carbon dioxide emission level. It is consequently necessary to develop new thermoelectric materials exhibiting $ZT = 1$ or higher at operating temperatures of 500–800 K.

11.2 Bulk Intermetallic Semiconductors for Future High-Temperature Applications

11.2.1 Phonon glass and electron crystal (PGEC) materials

In the last couple of years, the demand for finding a thermoelectric material to be used at high temperatures between 500 and 800 K prompted many researchers to investigate a new class of materials instead of modifying existing materials. One of the most interesting ideas proposed by Slack [14] is about the 'phonon glass and electron crystal (PGEC).' The essence of this idea of PGEC is to synthesize a material in which phonon mean free paths are as short as possible and in which electron mean free paths are as long as possible. The objective of the research on PGEC materials is to obtain as low a lattice thermal conductivity as possible. If the minimum thermal conductivity of most other heavy element semiconductors, i.e., 0.1–0.5 W/mK, should be obtained and those materials represent moderate electrical properties, the figure of merit will be more than 1 at room temperature [14]. Several PGEC thermoelectric materials, such as skutterudite [15–18], clathrate [19–22], and chevrel [23,24] have been investigated by many researchers. These materials have a complex unit cell, and are composed of heavy elements, thus a possibility of having a very low lattice thermal conductivity exists.

Skutterudite among PGEC materials has been researched owing to its unique structure and electrical properties. The most remarkable electrical transport property of such a material is the extremely high Hall mobility, which is one of the requirements of a PGEC material [14]. Moreover, skutterudite shows a moderate Seebeck coefficient over an entire temperature range. Due to these reasons, a skutterudite system including $IrSb_3$ and $CoSb_3$ has been researched as a promising material. The most distinct property of skutterudite is also characterized by weakly bonded atoms in an oversized atomic cage. Such atoms undergo large local anharmonic vibrations, independent of the other atoms in the crystal, and hence are referred to as rattlers. In insulating crystals, localized rattlers (or Einstein oscillators) can, in some cases, markedly decrease the thermal conductivity to values comparable to the heat conducted by a glass with the same composition [25]. Although a somewhat complex unit cell structure with a large number of atoms is typically associated with a low thermal conductivity, unfortunately skutterudite as a thermoelectric material has a rather high room-temperature thermal conductivity, $\kappa > 10$ W/mK [26]. Due to the high lattice thermal conductivity, most research studies involving skutterudite materials are focused on the decrease in lattice thermal conductivity by filling vacancies in n-type $CoSb_3$-based compounds by adding a proper ternary element, in most cases a rare-earth element. However, the thermoelectric properties of p-type $IrSb_3$-based skutterudite have not been extensively investigated.

A ternary half-Heusler compound has been considered to be the other promising PGEC thermoelectric materials. This half-Heusler compound was reported to exhibit unconventional transport properties, suggesting that it has a gap at the Fermi level [32]. It was found that half-Heusler compounds also exhibit large Seebeck coefficients of more than -200 μV/K, making them promising candidates for thermoelectric material. These large Seebeck coefficients can be attributed to the narrow band gaps of 0.1–0.5 eV [33–35]. From the fact that the lattice thermal conductivity of a given class of solids tends to be lower when the mean atomic weight is higher, a lower lattice thermal conductivity can also be achieved when ternary compounds are used rather than binary ones. Among the half-Heusler systems, ZrNiSn and HfNiSn have been researched by many groups. To reduce the relatively high thermal conductivity of these compounds, the effect on the thermal conductivity of (Zr,Hf)NiSn solid solution has also been performed at room temperature. However, the solid solution of (Ti,Hf)NiSn has not been investigated.

11.2.2 Skutterudite compounds

The unit cell of a binary skutterudite, for example, $IrSb_3$, contains 32 atoms as well as two structural vacancies (space group, $Im3$). These atoms form a cubic sublattice partially filled with six almost square pnicogen (Sb) rings oriented according to the cubic direction. An interesting characteristic of the skutterudite structure is the existence of two structural vacancies in its unit cell. It is known that these vacancies can be filled with lanthanide and other elements. If these vacancies can be filled, the general formula is RAB_3, where R is the rare-earth element, A is Co, Ir, or Rh, and B is As, P, or Sb. The metal A atoms form a simple cubic sublattice and the R atoms are located in the two structural vacancies in the unfilled AB_3 skutterudite structure. Figure 11.4 shows the unit cell of the filled skutterudite structure. Jeitschko and Braun first discovered filled skutterudite structure materials of the $LaFe_4P_{12}$ compound in 1977 [36]. Structural investigations of filled skutterudite materials based on $LnFe_4P_{12}$ (Ln = lanthanide elements) have been conducted by many groups [28,37–40].

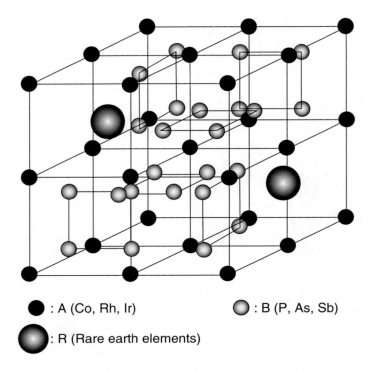

● : A (Co, Rh, Ir)   ◯ : B (P, As, Sb)
● : R (Rare earth elements)

Fig. 11.4. Unit cell of the filled skutterudite, RAB_3.

As described in the preceding section, binary antimonide skutterudites exhibit excellent electrical transport properties, such as a high Hall mobility. However, unfortunately, a high thermal conductivity (more than 10 W/mK) inhibits the use of these materials in thermoelectric applications [26]. One of the distinguishing characteristics of filled skutterudites, i.e., the significant reduction in lattice thermal conductivity due to the rattling of filler atoms positioned in oversized Sb-dodecahedron voids, has been reported [27]. It is believed that the R atoms rattle in the vacancies of the structure and therefore interact with a broad spectrum of lattice phonons, substantially reducing the mean free paths of unfilled skutterudites. Thus, the rattling of the R atoms in the vacancies of filled skutterudites should significantly enhance phonon scattering and reduce the lattice thermal conductivity.

Since the publication of the original articles of Jeitschko and Braun, many other variations of materials with this structure have been synthesized [41–48]. Since it is difficult to synthesize such materials, the intrinsic properties of filled skutterudite compounds are not clarified in detail. Thus far, various atoms, including Ba, La, Ce, Nd, Sm, Eu, Er, and Yb, have been used to fill the vacancies of $CoSb_3$-based skutterudite compounds [27–31], and it has been known that the filler atoms also greatly affect the electrical transport and parameters such as the carrier concentration, carrier mobility, and carrier effective mass. The lanthanide elements placed at the vacancies, usually end up as positively charged ions in the skutterudite structure; therefore, they donate electrons to the conduction or valence bands of semiconductor skutterudite. According to previous studies, filled $CoSb_3$-based skutterudites exhibit a larger effective mass and a lower smaller mobility than unfilled $CoSb_3$-based ones [57], the lattice thermal conductivity of the filled $CoSb_3$-based skutterudites is also tremendously reduced compared to that of the unfilled ones [27–30,49,51]. The striking thermoelectric properties of filled $CoSb_3$-based skutterudites determined indicated that these materials have a high dimensionless figure of merit, ZT, of up to 1–1.4 at 700–900 K [28–30,51].

The following section shows an example of a recent research study performed by the present authors to examine the effect of filling vacancies on the thermoelectric properties of $IrSb_3$-based compounds.

Effect of La filling in $IrSb_3$-based skutterudite compounds [50]. King has explained that a binary skutterudite compound, the donated valence electron count (VEC) of which per formula unit is equal to 72, is a semiconductor [55]. One of the attractive features of binary skutterudite compounds with VEC of 72 is an extremely high Hall mobility, which is closely associated with their unique electronic structure [55,56]. However, since rare-earth atoms usually end up as positively charged ions in the skutterudite structure, their electrons are donated to the conduction bands of these semiconductors. The excellent electrical transport properties of binary compounds degrade with positively charged rare-earth ions in filled skutterudites. Therefore, the objective of research on skutterudite materials is to minimize the thermal conductivity by preparing filled skutterudites without degrading the excellent electrical properties of binary compounds [27–29,51–54].

Compounds based on $IrSb_3$ skutterudite are prepared by a powder metallurgy (PM) technique. The additions of La and Ge are systematically performed according to the formulas $La_X Ir_4 Sb_{12}$ ($X = 0$, 0.2, and 0.3) and $La_Y Ir_4 Ge_{3Y} Sb_{12-3Y}$ ($Y = 0$, 0.1, 0.3, and 1). Figure 11.5 shows the X-ray diffraction patterns of hot-pressed (a) $La_X Ir_4 Sb_{12}$ and (b) $La_Y Ir_4 Ge_{3Y} Sb_{12-3Y}$ compounds. The X-ray diffraction pattern of hot-pressed $La_Y Ir_4 Ge_{3Y} Sb_{12-3Y}$ is different from that of hot-pressed $La_X Ir_4 Sb_{12}$. According to the

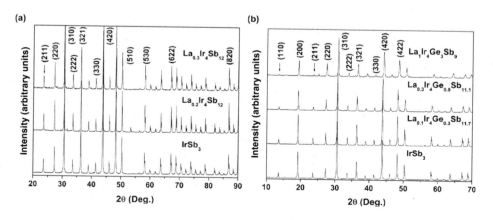

Fig. 11.5. X-ray diffraction patterns of hot-pressed (a) $La_X Ir_4 Sb_{12}$ and (b) $La_Y Ir_4 Ge_{3Y} Sb_{12-3Y}$ compounds.

Table 11.1. Summary of data and refinement parameters for Rietveld analysis.

	Lattice parameter (Å)	Occupation (La)	Thermal parameter (La)
$IrSb_3$	9.2472 ± 0.0001	0	—
$La_{0.1}Ir_4Ge_{0.3}Sb_{11.7}$	9.2396 ± 0.0001	0.08	1.0
$La_{0.3}Ir_4Ge_{0.9}Sb_{11.1}$	9.2241 ± 0.0002	0.21	1.5
$La_1Ir_4Ge_3Sb_9$	9.2016 ± 0.0001	0.53	1.5

report by Nolas et al., the (100) and (211) reflections of a binary skutterudite compound disappear in the case of complete filling of cages by rare-earth or other elements [51]. As shown in Fig. 11.5(a), the relative intensity of the (211) reflection of $La_XIr_4Sb_{12}$ compounds remains nearly constant with the La amount. As contrasted with $La_XIr_4Sb_{12}$ compounds, the intensities of the (110) and (210–211) reflections decrease with the La and Ge amounts of $La_YIr_4Ge_{3Y}Sb_{12-3Y}$ compounds. On the basis of these experimental findings, it is suggested that La filling is not only achieved by La additions, but also by Ge charge compensation in $La_YIr_4Ge_{3Y}Sb_{12-3Y}$ compounds.

To investigate the changes in the lattice parameter and filling fraction of the La element, a Rietveld X-ray analysis is performed in $La_YIr_4Ge_{3Y}Sb_{12-3Y}$ compounds [58]. This Rietveld analysis yields reliable results within a 10% difference between the observed and calculated profiles. Table 11.1 shows a summary of the results of the Rietveld analysis of $La_YIr_4Ge_{3Y}Sb_{12-3Y}$ compounds. The lattice parameters of the La-filled and Ge-compensated compounds are smaller than that of binary $IrSb_3$. The lattice parameter model calculation of unfilled $Ir_4Ge_3Sb_9$ indicates that the lattice parameter is 9.0021 Å, as reported by Nolas et al. [51]. The reductions in the lattice parameters of the La-filled and Ge-compensated skutterudite compounds compared to that of $IrSb_3$ are therefore presumably due to the compensated Ge atom that is smaller than the Sb atom. However, it is notable that the La ions in the cages produce a lattice expansion compared to the reported unfilled $Ir_4Ge_3Sb_9$. The obtained La filling fraction at the (0,0,0) position is smaller than the La amount of nominal compositions; this smaller La filling fraction is due to the formation of undesired phases, such as $LaSb_2$. The large thermal parameter of the La atom indicates that La is poorly bound in the cage and can move about its equilibrium position. Moreover, the larger thermal parameter of La than Ir, Sb, or Ge supports the possibility of La rattling in the cages [28,40,62].

Figure 11.6 shows the temperature dependences of the (a) Seebeck coefficient and (b) electrical resistivity for hot-pressed $La_XIr_4Sb_{12}$ compounds. A previous study showed that the unfilled and Ru- and Ge-doped ternary $IrSb_3$ compounds are p-type; however, $La_XIr_4Sb_{12}$ compounds show n-type properties. The n-type properties of these compounds may be attributed to the fact that the positively charged La^{3+} ions donate their electrons. As contrasted with the semiconductor skutterudite with VEC of 72, $La_XIr_4Sb_{12}$ compounds exhibit a high electrical resistivity that decreases with temperature, indicating the hole–electron interaction induced by donor impurities.

In order for the La-filled skutterudite compounds to maintain the excellent semiconducting properties of binary $IrSb_3$, the adjustment of VEC of 72 is performed by additional alloying to compensate the effect of rare-earth ions in $La_XIr_4Sb_{12}$ skutterudite compounds. For this purpose, the addition of Ge is attempted according to the formula

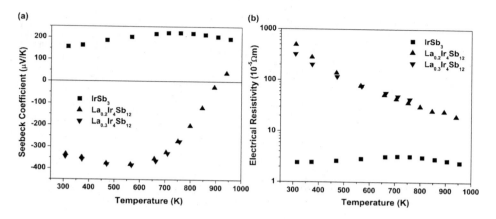

Fig. 11.6. Temperature dependence of the (a) Seebeck coefficient and (b) electrical resistivity for hot-pressed $La_X Ir_4 Sb_{12}$ compounds.

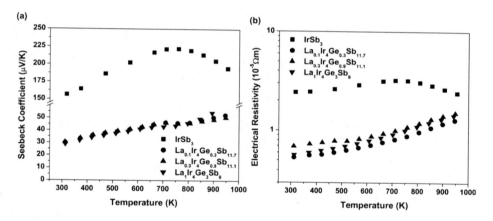

Fig. 11.7. Temperature dependences of the (a) Seebeck coefficient and (b) electrical resistivity for La-filled and Ge-charge-compensated $La_Y Ir_4 Ge_{3Y} Sb_{12-3Y}$ skutterudite compounds.

$La_Y Ir_4 Ge_{3Y} Sb_{12-3Y}$. Figure 11.7 shows the temperature dependences of the (a) Seebeck coefficient and (b) electrical resistivity for the La-filled and Ge-charge-compensated $La_Y Ir_4 Ge_{3Y} Sb_{12-3Y}$ skutterudite compounds. Both the Seebeck coefficient and electrical resistivity markedly decrease in the La-filled and Ge-charge-compensated compounds. The Seebeck coefficient increases slightly with temperature and is very small. The electrical resistivity also increases with temperature. The reduced Seebeck coefficient and electrical resistivity may be attributed to the increase in carrier concentration. In contrast to the expectation of charge compensation to adjust VEC of 72 in binary compounds, $La_Y Ir_4 Ge_{3Y} Sb_{12-3Y}$ compounds exhibit metallic properties. On the basis of the experimental results of structural characterization and electrical properties, it is speculated that a complete La filling of $La_Y Ir_4 Ge_{3Y} Sb_{12-3Y}$ compounds will not be achieved. These results

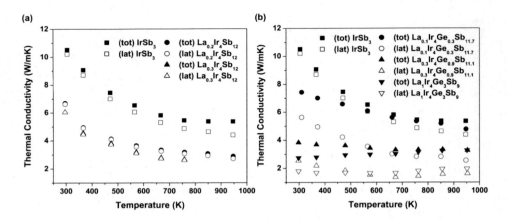

Fig. 11.8. Temperature dependences of the total thermal conductivity (tot) and lattice thermal conductivity (lat) for hot-pressed (a) $La_X Ir_4 Sb_{12}$ and (b) $La_Y Ir_4 Ge_{3Y} Sb_{12-3Y}$ compounds.

indicate that the amount of nominal composition of La in $La_Y Ir_4 Ge_{3Y} Sb_{12-3Y}$ compounds is not present in cages, implicating that the amount of doped Ge is excessive for the adjustment of VEC of 72. Therefore, excessive Ge is the cause of the metallic properties of $La_Y Ir_4 Ge_{3Y} Sb_{12-3Y}$ compounds.

Figure 11.8 shows the temperature dependences of the total thermal conductivity (tot) and lattice thermal conductivity (lat) for the hot-pressed (a) $La_X Ir_4 Sb_{12}$ and (b) La-filled and Ge-charge-compensated $La_Y Ir_4 Ge_{3Y} Sb_{12-3Y}$ skutterudite compounds. As can be seen in Fig. 11.7(a) to 11.8(a), the thermal conductivity of $La_X Ir_4 Sb_{12}$ compounds exhibits a temperature dependence similar to that of binary compounds, showing the decrease behavior with the increase in temperature. The reduced thermal conductivities of $La_{0.2} Ir_4 Sb_{12}$ and $La_{0.3} Ir_4 Sb_{12}$ may be attributed to the degradation of phase stability, such as the formation of an undesired $LaSb_2$ phase or La filling in the cages. It is necessary to investigate the effect of phase stability on thermal conductivity in further studies. It is shown that the thermal conductivities of the La-filled and Ge-charge-compensated compounds are effectively reduced when compared to that of binary $IrSb_3$ owing to the reduction in lattice thermal conductivity. The thermal conductivity of $La_1 Ir_4 Ge_3 Sb_9$ markedly decreases to 2.7 W/mK at room temperature. The lattice thermal conductivity of $La_1 Ir_4 Ge_3 Sb_9$ shows an 83% decrease (from 10.2 to 1.8 W/mK) of binary compounds at room temperature. However, the amount of Ge addition for the charge compensation is in excess of that theoretically required. This results in an increase in electronic thermal conductivity, indicating the different temperature dependence of total thermal conductivity in $La_1 Ir_4 Ge_3 Sb_9$. The contribution of electronic thermal conductivity increases from 2% of binary compounds to 34% of $La_1 Ir_4 Ge_3 Sb_9$ at room temperature and increases with temperature. Although the electronic thermal conductivity increases, the total thermal conductivity decreases owing to the significantly decreased lattice thermal conductivity. From our Rietveld analysis results, it is concluded that the tremendously decreased lattice thermal conductivity is attributed to the enhanced phonon scattering by the rattling of La ions in the cages.

The other reason for the marked reduction in thermal conductivity with the rattling of La ions is the point defect scattering between the vacant and La-filled cages that can be theoretically predicted. The theory for this point defect scattering was originally introduced by Klemens, and Callaway and von Baeyer, and later applied to the solid-solution problem of Abeles [49,59,60]. In the model of Rayleigh-type point defect scattering, the scattering parameter $\Gamma = \alpha(1 - \alpha)(\Delta M/M_{av})^2$ is inversely proportional to the lattice thermal conductivity, where α is the relative concentration of impurity (La-filled cage) and ΔM is the difference between the mass of the impurity and that of the host (vacant cage). The partial La filling gives another freedom to reduce the lattice thermal conductivity and it is maximized at approximately $\alpha = 0.5$ [61]. Actually, it is verified that the 31%-La-filled $CoSb_3$-based skutterudite compounds exhibit a lower lattice thermal conductivity than the 90%-La-filled compounds [18]. Therefore, it is speculated that the 53% La filling in the $La_1Ir_4Ge_3Sb_9$ compound is effective in reducing the lattice thermal conductivity.

11.2.3 Half-Heusler compounds

Several PGEC thermoelectric materials including skutterudite, clathrate, and chevrel have been researched due to their promising thermoelectric properties, particularly the low lattice thermal conductivity. Generally, a low lattice thermal conductivity is expected when polycrystalline samples are used instead of single crystals. The lattice thermal conductivity of a given class of solids also tends to be low when its mean atomic weight is high. In this sense, a low lattice thermal conductivity can also be achieved when ternary compounds are used. Recently, ternary semiconductor compounds, half-Heusler compounds, have received attention as new PGEC thermoelectric materials.

The half-Heusler ternary intermetallic compound ABX (space group, $F\bar{4}3m$) is derived from the normal Heusler structure AB_2X (space group, $Fm3m$), where A is a transition metal of the left-hand side of the periodic table (titanium or vanadium group elements), B is a transition metal of the right-hand side of the periodic table (iron, cobalt, or nickel group elements), and X is one of the main group elements, Ga, Sn, or Sb. In the early years of research, Dwight et al. [63], Jeitschko [64], and some other research groups [65–67] reported on the physical properties of several ternary half-Heusler stannides and antimonides. More stannides and antimonides have been reported since then [68–70]. An updated list of the ternary intermetallic compounds ABX, stannides, and antimonides is given in Table 11.2.

Figure 11.9 shows the unit cell of ABX with a cubic MgAgAs structure, the position of which is shown by A, B, and X atoms. One unit cell holds four formula units, with A atoms in the 4b (1/2, 1/2, 1/2) position, B atoms in the 4c (1/4, 1/4, 1/4) position, and X atoms in the 4a (0, 0, 0) position. This structure may be regarded as four interpenetrating fcc lattices: a lattice of A atoms and a lattice of X atoms, together forming a rock-salt structure and a lattice of B atoms occupying the center of every other cube. The centers of the remaining cubes are empty, thereby forming an fcc lattice of vacancies. When the lattice of vacancies is filled up with additional four B atoms, one obtains AB_2X, a Heusler compound with a cubic $MbCu_2Al$ structure.

Table 11.2. Ternary half-Heusler compounds.

Antimonide	TiFeSb TiCoSb TiNiSb TiRuSb TiRhSb
	VFeSb VCoSb VNiSb VRuSb
	NbFeSb NbCoSb NbRuSb NbRhSb
	ZrCoSb ZrRhSb ZrRuSb
	HfCoSb HfRuSb HfRhSb
	MnPdSb MnPtSb
	TaCoSb TaRuSb
	HoPdSb ErPdSb TmNiSb YbNiSb YNiSb
Stannide	TiCoSn TiNiSn TiRhSn TiIrSn TiPtSn
	NbCoSn NbRhSn
	ZrNiSn ZrPdSn ZrPtSn
	HfNiSn HfPdSn
	MnIrSn MnPtSn
Others	ZrCoBi ZrNiBi GdPtBi

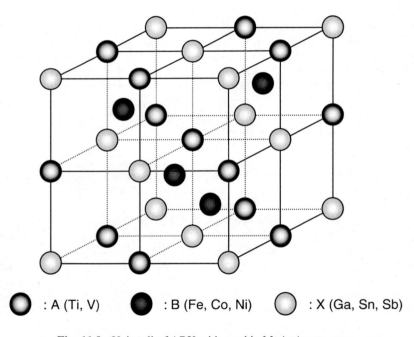

: A (Ti, V)   : B (Fe, Co, Ni)   : X (Ga, Sn, Sb)

Fig. 11.9. Unit cell of ABX with a cubic MgAgAs structure.

In 1983, de Groot *et al.* [71] reported half-Heusler compound as a new class materials, the so-called semimetallic ferromagnetics, experimentally confirmed later [72]. Several nonmagnetic transition-metal-based half-Heusler compounds produced a narrow band gap near the Fermi level, which is required for obtaining promising thermoelectric materials [33]. Recent interest is concentrated on the transition-metal-based stannides and antimonides and other half-Heusler compounds due to their promising thermoelectric properties [73–79] and interesting electronic and magnetic structure properties [35,80,81].

Tobola et al. demonstrated in band structure studies that the transition-metal-based half-Heusler compounds have a high propensity for forming a band gap [32]. They reported the detailed experimental and theoretical studies of the electronic phase diagrams of ordered half-Heusler systems as a function VEC: TiFeSb and TiCoSn with VEC = 17, TiNiSn, ZrNiSn, TiCoSb, NbCoSn with VEC = 18, VCoSb, TiNiSb, NbCoSb with VEC = 19, MnCoSb with VEC = 21 and MnNiSb with VEC = 22, as well as their solid solutions. The generality of band gap formation in these phases can be attributed to the small coordination number of the d-band metals in the MgAgAs structure. They also determined the conducting state of these compounds from the position of the Fermi level E_F with respect to the gap. If the position of E_F is changed as a function of VEC, the thermoelectric properties can be considerably affected. Many types of half-Heusler compound have been investigated as new thermoelectric materials; however, a systematic study of high-temperature thermoelectric properties for half-Heusler compounds as a function of VEC has not yet been reported.

An extensive investigation of the effect of annealing on the thermoelectric properties of the ZrNiSn, HfNiSn, and (Zr,Hf)NiSn solid solutions was conducted by Uher et al. [82]. They found that both the electrical and thermal properties strongly depend on the annealing time. This indicates that the physical properties of these alloys are very sensitive to a particular structural arrangement, and that any kind of disorder strongly influences the transport behavior. Despite this important study, the optimum condition of annealing for a high thermoelectric performance level was not determined in the ZrNiSn and HfNiSn half-Heusler compounds.

The TiNiSn half-Heusler compound exhibits a high Seebeck coefficient and a low electrical resistivity, which yield a relatively large power factor. The largest power factor of half-Heusler compounds was reported to be 6.9 mW/mK^2, which makes this material very attractive as a potential thermoelectric material [83]. The TiNiSn half-Heusler compound was reported as a semiconductor with a narrow band gap of 0.12 eV, which is markedly different from the theoretically calculated band gap of 0.51 eV [33,35].

The following section shows an example of a recent research study performed by the present authors to enhance the effect of the dimensionless figure of merit by alloying and doping, on the thermoelectric properties of TiNiSn-based half-Heusler compounds.

Improvement of ZT in TiNiSn-based half-Heusler compounds [84]. Among the half-Heusler compounds, MNiSn (M = Ti, Zr, Hf) has been reported to be a narrow-band-gap semiconductor with a 0.1–0.5 eV indirect band gap near the Fermi level and exhibits a high Seebeck coefficient and a low electrical resistivity [33,34,64,76], which would yield a relatively large power factor. Although the large power factor has been reported in many studies [79,83], a relatively high thermal conductivity has been measured to be around 10 W/mK at room temperature, which is unfavorable for achieving a high dimensionless figure of merit, ZT, of more than 0.5. Uher et al. [74] investigated the effect of annealing on ZrNiSn, HfNiSn, and (Zr,Hf)NiSn systems, and found that a low thermal conductivity of 4.3 W/mK, at room temperature is obtained for the $Zr_{0.5}Hf_{0.5}NiSn$ sample due to the mass-defect scattering between $Zr(M_{Zr} = 91)$ and $Hf(M_{Hf} = 178)$, where M_x is the atomic mass of the element x. Moreover, even with a small amount of Sb (less than 0.1 at%) doping on the Sn site of the TiNiSn system, a marked effect on the transport properties has been observed by Bhattacharya et al. [83]. It has also been noted that one of the effective methods of reducing the thermal conductivity is the material preparation by the powder metallurgy (PM) technique [85]. Shen et al. [86] reported the lowest thermal conductivity of 2 W/mK, which was obtained by 50% Pd substitution on the Ni site of the ZrNiSn-based half-Heusler system prepared by the PM technique.

Recently, the present authors have systematically investigated the high-temperature thermoelectric properties of TiNiSn-based half-Heusler compounds [84]. Efforts have been made on the reduction in the thermal conductivity of the compound by Hf alloying on the Ti site by introducing a stronger mass-defect scattering between $Ti(M_{Ti} = 48)$ and $Hf(M_{Hf} = 178)$ than that between $Zr(M_{Zr} = 91)$ and $Hf(M_{Hf} = 178)$ of the (Zr,Hf)NiSn system together with the utilization of the PM technique. Such efforts also involve the optimum amount of Sb doping on the Sn site for enhancing the power factor. Polycrystalline compounds were prepared by two fabrication techniques, namely casting (CA) and PM.

The nominal compositions of the compounds prepared are listed in Table 11.3 with the fabrication processes employed. In CA, to ensure homogeneity, stoichiometric amounts of the constituent elements were arc-melted several times on a water-cooled copper hearth in an argon atmosphere. The resulting buttons were wrapped in a Ta foil and sealed in an evacuated quartz tube for annealing at 1073 K for 2 weeks. In PM, the buttons prepared by arc melting were ground using mortar, and those of below 45 μm diameter were separated. Then, the powders were hot pressed at 1073 K for 5 h under 35 MPa. The hot-pressed samples were annealed at 1073 K for 2 weeks in the evacuated quartz tube.

Table 11.3. Nominal composition and fabrication process of TiNiSn-based half-Heusler compounds examined.

Alloys	Fabrication process
TiNiSn	Arc HT and HP HT
$Ti_{0.95}Hf_{0.05}NiSn_{1-X_{Sb}}X$ $(0 \leq X \leq 0.05)$	Arc HT
$Ti_{0.95}Hf_{0.05}NiSn_{0.99}Sb_{0.01}$	HP HT
$Ti_{0.8}Hf_{0.2}NiSn_{0.99}Sb_{0.01}$	HP HT
$Ti_{0.95}Hf_{0.05}Ni_{0.8}Pd_{0.2}Sn_{0.99}Sb_{0.01}$	HP HT
$Ti_{0.5}Hf_{0.5}Ni_{0.5}Pd_{0.5}Sn_{0.99}Sb_{0.01}$	HP HT
$Ti_{0.95}Hf_{0.05}Ni_{0.95}Pt_{0.05}Sn_{0.99}Sb_{0.01}$	HP HT

Arc: Arc melting; HP: Hot pressed at 1073 K for 5 h under 35 Mpa; HT: Annealed at 1073 K for 2 weeks.

The results of our X-ray powder diffraction and microstructure observation show that the compounds prepared by CA consist of TiNiSn with small volume fractions of Ti_6Sn_5 and Sn solid solution phases. However, the compounds prepared by PM are nearly single phase with a minimal volume fraction of the Ti_6Sn_5 phase.

The Seebeck coefficient α and electrical resistivity ρ were simultaneously measured by the standard four-probe DC method in a temperature range from room temperature to 1000 K for consistency in an He atmosphere. The Hall measurement performed to deduce the carrier concentration was also carried out at room temperature using the van der Pauw technique. The thermal conductivity was calculated from the heat capacity and thermal diffusivity measured by the laser flash method in the temperature range from room temperature to 1000 K under vacuum.

Figure 11.10 shows the temperature dependence of the electrical resistivity for the TiNiSn-based half-Heusler compounds prepared by CA and PM techniques. The electrical resistivity of cast $Ti_{0.95}Hf_{0.05}NiSn$ decreases as the temperature increases, showing a semiconductor-like temperature dependence similar to that observed for the cast ternary TiNiSn. The most distinctive effect of Sb doping on the electrical resistivity is the change in temperature dependence like that of a metal, showing a slightly positive temperature dependence. The electrical resistivity of Sb-doped compounds revealed by the measurement of the Hall effect (0.5–0.9×10^{-5} Ωm) are lower than those of the cast TiNiSn (16.5×10^{-5} Ωm) and $Ti_{0.95}Hf_{0.05}NiSn$ (11×10^{-5} Ωm) by 12–35 factors at room temperature. The reduction and change in the temperature dependence in electrical resistivity of Sb-doped compounds are ascribed to the increase in carrier concentration.

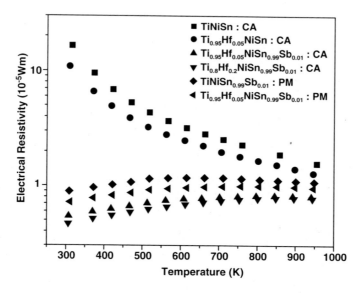

Fig. 11.10. Temperature dependence of electrical resistivity for the TiNiSn-based half-Heusler compounds prepared by casting (CA) and powder metallurgy (PM) technique.

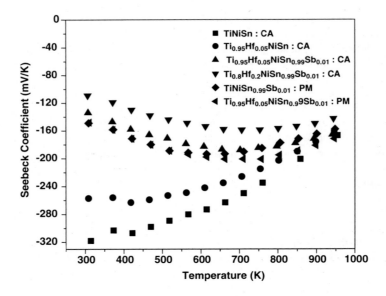

Fig. 11.11. Temperature dependence of the Seebeck coefficient for the TiNiSn-based half-Heusler compounds prepared by casting (CA) and powder metallurgy (PM) technique.

At room temperature, the carrier concentration of the cast TiNiSn is 1.3×10^{19} cm^{-3} and that of the cast Ti$_{0.95}$Hf$_{0.05}$NiSn$_{0.99}$Sb$_{0.01}$ is 4×10^{20} cm^{-3}.

The temperature dependence of the Seebeck coefficient is shown in Fig. 11.11 for the TiNiSn-based half-Heusler compounds prepared by CA and PM techniques. There are two types of temperature dependence, (1) decreasing the behavior of $|\alpha|$ over the entire temperature range in TiNiSn and Ti$_{0.95}$Hf$_{0.05}$NiSn, and (2) first increasing and then decreasing the behavior of $|\alpha|$ in all the compounds doped with Sb. The latter behavior is ascribed to the excitation of electron–hole pairs across the energy gap and the extrinsic conduction by excited carriers from an impurity state.

Both Hf alloying and Sb doping degrade the Seebeck coefficient due to the increase in carrier concentration. The increase in carrier concentration by 35 factors degrades the Seebeck coefficient from -320 μV/K (1.3×10^{19} cm^{-3} of the cast TiNiSn) to -110 μV/K (4.6×10^{20} cm^{-3} of the cast Ti$_{0.8}$Hf$_{0.2}$NiSn$_{0.99}$Sb$_{0.01}$). Hf-alloyed samples show an increased carrier concentration, resulting in decreased properties due to the change in the energy band gap size of TiNiSn. We do not know the exact reason on how Hf alloying increases the carrier concentration. However, the transport properties might be affected by the change in the energy band gap size of TiNiSn. Hf alloying on the Ti site is expected to change the band gap size by affecting the bonding of the Ti and Sn atoms, which is the origin of the band gap of the half-Heusler compound [35]. The effective mass is calculated in several samples with the measured Seebeck coefficient and carrier concentration at room temperature. The effective mass of the cast TiNiSn is about 1.9 m_e and the highest value is about 3.9 m_e for the cast Ti$_{0.95}$Hf$_{0.05}$NiSn$_{0.99}$Sb$_{0.01}$.

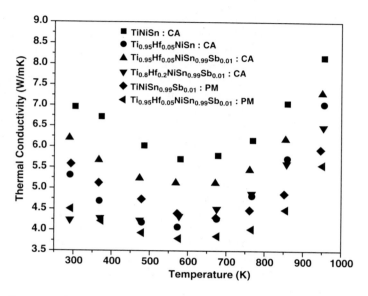

Fig. 11.12. Temperature dependence of the thermal conductivity for the TiNiSn-based half-Heusler compounds prepared by casting (CA) and powder metallurgy (PM) technique.

This high value seems to sustain a relatively high Seebeck coefficient in these materials. As a result, the power factor is markedly increased by Sb doping, and the maximum power factor in this study is 4.5 mW/mK^2 for the cast Ti$_{0.95}$Hf$_{0.05}$NiSn$_{0.99}$Sb$_{0.01}$ sample.

The temperature dependence of the thermal conductivity for the TiNiSn-based half-Heusler compounds prepared by CA and PM techniques is shown in Fig. 11.12. The thermal conductivity of the cast TiNiSn at room temperature is 7 W/mK. Using the Wiedemann–Franz equation, the calculated lattice thermal conductivity of the cast TiNiSn is found out to be 6.9 W/mK at room temperature, where the contribution of the electronic thermal conductivity is found to be less than 2% over the entire temperature range. This indicates that the total thermal conductivity of the cast ternary compound is basically the lattice thermal conductivity. The previous investigations on the effects of the Zr, Hf, and Pt alloying on processes showed that Hf alloying is most effective in reducing the lattice thermal conductivity due to the larger difference in atomic mass between Ti and Hf than those between Ti and Zr, and between Pt and Ni. Hf-alloyed samples exhibit a more reduced value than that of the cast TiNiSn, showing 5.3 W/mK at room temperature for Ti$_{0.95}$Hf$_{0.05}$NiSn. This decrease in total thermal conductivity is attributed to the decrease in lattice thermal conductivity. At room temperature, the lattice thermal conductivity of the Ti$_{0.95}$Hf$_{0.05}$NiSn sample decreases below to 5.2 W/mK (25% decrease) from 6.9 W/mK of the cast TiNiSn, which is due to the strong mass-defect scattering between Ti($M_{Ti} = 48$) and Hf($M_{Hf} = 178$).

Although the effect of substitution of Ni by Pd and Pt is also examined, the thermal conductivity of the compounds is not reduced by the Pd and Pt additions. The temperature dependence of the thermal conductivity for the compounds exhibits a metallic nature, i.e., the thermal conductivity increases with temperature. The reason for this might be the reduction in the stability of the half-Heusler structure and the formation of other intermetallic phases. It is actually observed that there are more of Ti_6Sn_5 phases in the compound with these elements. This in turn reveals that the small amounts of Hf alloying and Sb doping maintain the high structural and phase stabilities. It might be concluded that the structural stability is high when the other intermetallic phases are absent, namely a single-phase compound should be chosen first for a low thermal conductivity as well as excellent electrical properties.

Sb doping, on the other hand, negatively affects the thermal conductivity. Because of the increase in electronic thermal conductivity due to the increase in carrier concentration with Sb doping, the total thermal conductivity of the cast $Ti_{0.95}Hf_{0.05}NiSn_{0.99}Sb_{0.01}$ becomes larger than that of the $Ti_{0.95}Hf_{0.05}NiSn$ sample. However, the thermal conductivities of the cast $Ti_{0.95}Hf_{0.05}NiSn_{0.99}Sb_{0.01}$ and $Ti_{0.8}Hf_{0.2}NiSn_{0.99}Sb_{0.01}$ compounds are still lower than that of the cast TiNiSn owing to the significant Hf effect on reducing the lattice thermal conductivity. In Fig. 11.12, it should be noted that the thermal conductivity of hot-pressed compounds is the lowest. Despite the negative effect by Sb doping, the hot-pressed $TiNiSn_{0.99}Sb_{0.01}$ and $Ti_{0.95}Hf_{0.05}NiSn_{0.99}Sb_{0.01}$ alloys show a lower thermal conductivity than the cast TiNiSn and $Ti_{0.95}Hf_{0.05}NiSn$ compounds. This result is ascribed to the effective reduction in lattice thermal conductivity due to phonon scattering at particle boundaries in the compound fabricated by the PM technique.

Figure 11.13 shows the temperature dependence of the dimensionless figure of merit ZT for the TiNiSn-based half-Heusler compounds prepared by CA and PM techniques.

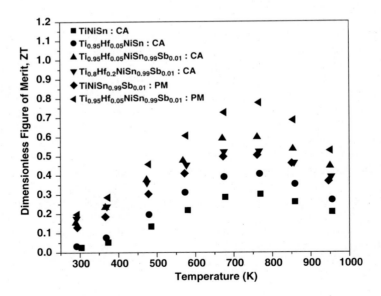

Fig. 11.13. Temperature dependence of the dimensionless figure of merit ZT for TiNiSn-based half-Heusler compounds prepared by casting (CA) and powder metallurgy (PM) technique.

The dimensionless figure of merit ZT of the cast TiNiSn is 0.3 at 750 K. As a result of the effective reduction in thermal conductivity by both Hf alloying and the PM technique, together with the enhancement in power factor by Sb doping, we observe $ZT = 0.78$ at 770 K for the hot-pressed $Ti_{0.95}Hf_{0.05}NiSn_{0.99}Sb_{0.01}$ compounds. This is the highest value for any TiNiSn-based half-Heusler compounds reported thus far.

11.3 A Breakthrough in ZT: Nanostructured Materials

As has been discussed in the preceding sections, large ZT values require a large Seebeck coefficient, a high electrical conductivity, and a low thermal conductivity, since an increase in Seebeck coefficient normally leads to a decrease in electrical conductivity and an increase in electrical conductivity leads to an increase in electronic contribution to the thermal conductivity, as given by the Wiedemann–Franz law. It is therefore very difficult to enhance the ZT values of typical thermoelectric materials. The state-of-the-art thermoelectric material is the Bi–Sb–Te–Se family with $ZT \sim 1$ at room temperature [87]. It is believed that if materials with $ZT \sim 3$ are developed, many practical applications for thermoelectric devices would be realized. However, despite the tremendous efforts after the discovery of the Bi–Sb–Te–Se system with $ZT \sim 1$, a marked enhancement in ZT has not been achieved due principally to the trade-offs among the three thermoelectric parameters in conventional bulk thermoelectric materials (See Fig. 11.14).

The challenge of increasing ZT to more than 1 in bulk materials lies in the fact that the three thermoelectric parameters are interdependent in terms of carrier concentration. The only method of reducing the thermal conductivity without affecting the Seebeck coefficient and electrical conductivity in bulk materials is the use of semiconductors consisting of heavy elements such as a Bi_2Te_3 alloy with Sb or Pb. Although it is

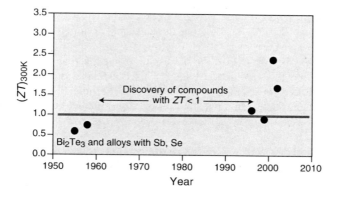

Fig. 11.14. History of dimensionless figure of merit ZT at 300 K. Since the discovery of Bi_2Te_3 and its alloys with Sb and Se in the 1950s, no bulk material with $ZT > 1$ at room temperature has been discovered. There is a huge jump in an extremely short period to $ZT > 2$ using the nanostructured thin film superlattices of Bi_2Te_3/Sb_2Te_3 and PbSeTe/PbTe quantum dot superlattices [88].

possible in principle to achieve $ZT > 3$ in bulk materials, there is no candidate material except the Bi–Sb–Te–Se system.

In the past several years, it was shown that nanostructured materials lead to a marked increase in ZT to more than 2 (see Fig. 11.1). Venkatasubramanian *et al.* reported high room-temperature figures of merit of $ZT \sim 2.4$ of p-type Bi_2Te_3/Sb_2Te_3 superlattices and $ZT \sim 1.4$ for n-type $Bi_2Te_3/Sb_2Te_{2.83}Se_{0.17}$ superlattices [89]. Harman *et al.* demonstrated the improved cooling devices relative to that fabricated with conventional bulk $(Bi,Te)_2(Sb,Te)_3$ materials using PbSeTe/PbTe quantum dot superlattices with ZT in the range of 1.3–1.6 at room temperature [90]. Hsu *et al.* also reported the potential of the nanostructured thermoelectric material $AgPb_mSbTe_{2+m}$ at 800 K for $m = 18$ with $ZT \sim 2.2$, in which the material contains regions 2–4 nm in size that are rich in Ag–Sb and are epitaxially embedded in a matrix that is depleted of Ag and Sb [91].

These achievements using nanostructured materials encourage researchers in thermoelectricity to leverage their efforts for the fabrication of nanostructured thermoelectric materials and devices. Here, we review the recent theoretical and experimental research studies regarding nanostructured thermoelectric materials.

11.3.1 Effect of quantum confinement effect on thermoelectric properties

Theoretical research: two-dimensional (2D) quantum well structure. Early studies on the theoretical evaluations of low-dimensional materials for thermoelectric applications provide a new approach to increasing the dimensionless thermoelectric figure of merit, ZT [92–94]. Hicks and Dresselhaus proposed a theoretical model for a quantum-confined low-dimensional electron gas in 1993, and their calculations implied that a significant enhancement in ZT to more than 6 can be achieved within a two-dimensional (2D) quantum well made of a suitable bulk (3D) thermoelectric material, and that the quantum well thickness is sufficiently small [92]. They emphasized that the marked enhancement in ZT is derived from the increase in Seebeck coefficient due to the enhancement in density of state near E_F without affecting the electrical conductivity.

The calculation was performed on the assumption that the material is of the one-band type (assumed to be the conduction band). This is because one-band materials, such as heavily doped semiconductors, give the highest thermoelectric figure of merit ZT. The only other assumptions are a constant relaxation time and the presence of parabolic bands.

The calculation methods for the Seebeck coefficient, electrical conductivity, and thermal conductivity, have been described elsewhere [95]. Thus, the electronic dispersion relation used is

$$\varepsilon(k_x k_y k_z) = \frac{\hbar^2 k_x^2}{2m_x} + \frac{\hbar^2 k_y^2}{2m_y} + \frac{\hbar^2 k_z^2}{2m_z}. \tag{11.13}$$

The ZT of 3D thermoelectric materials is expressed as

$$Z_{3D}T = \frac{\frac{3}{2}\left(\frac{5F_{3/2}}{3F_{1/2}} - \eta^*\right)^2 F_{1/2}}{\frac{1}{B} + \frac{7}{2}F_{5/2} - \left(\frac{25F_{3/2}^2}{6F_{1/2}}\right)}, \qquad (11.14)$$

where

$$B = \frac{1}{3\pi^2}\left(\frac{2k_B T}{\hbar^2}\right)^{3/2}(m_x m_y m_z)^{1/2}\frac{k_B^2 T \mu_x}{e\kappa_{ph}}, \qquad (11.15)$$

and the Fermi–Dirac function F_i is given by

$$F_n = \int_0^\infty \frac{x^n}{1 + \exp(x - \eta^*)}\, dx, \qquad (11.16)$$

where $\eta^* = \eta/k_B T$ is the reduced chemical potential. Therefore, the ZT for 3D bulk materials is determined by B and the reduced chemical potential, which can be controlled by doping. On the other hand, when the moment of electrons is confined in 2D quantum wells for a suitably fabricated superlattice, the wells are parallel to the x–y plane and the currents flow in the x direction, the electronic dispersion relation used is

$$\varepsilon(k_x k_y) = \frac{\hbar^2 k_x^2}{2m_x} + \frac{\hbar^2 k_y^2}{2m_y} + \frac{\hbar^2 \pi^2}{2m_z a^2}. \qquad (11.17)$$

Then the three calculated thermoelectric properties, ZT for a 2D quantum well is

$$Z_{2D}T = \frac{\left(\frac{2F_1}{F_0} - \eta^*\right)^2 F_0}{\frac{1}{B} + 3F_2 - \frac{4F_1^2}{F_0}}, \qquad (11.18)$$

where

$$B = \frac{1}{2\pi a}\left(\frac{2k_B T}{\hbar^2}\right)^{3/2}(m_x m_y)^{1/2}\frac{k_B^2 T \mu_x}{e\kappa_{ph}}, \qquad (11.19)$$

and the reduced chemical potential is given by

$$\eta^* = \left(\eta - \frac{\hbar^2 \pi^2}{2m_z a^2}\right)\bigg/ k_B T. \qquad (11.20)$$

Note that for a 2D quantum well, the expression for η^* is different from that for the 3D case with the term $\hbar^2\pi^2/2m_z a^2 k_B T$, which is related to the confinement in the quantum well. Thus, the reduced chemical potential may be varied both by doping and by changing the layer thickness a. This leads to the development of a method of increasing $Z_{2D}T$, where one can optimize η^* and make the quantum well layers as thin as possible. The results of calculations are shown in Fig. 11.15.

For the quantum well parallel to the x–y plane, $Z_{2D}T$ is higher than $Z_{3D}T$ for layers thinner than 850 nm. In this study, the highest calculated ZT for bulk Bi_2Te_3 is $Z_{3D}T = 0.5$. As the layers are made even thinner, the increase in $Z_{2D}T$ becomes more significant. To estimate the maximum $Z_{2D}T$ that can be obtained for this layer orientation, $a = 38$ nm, a single layer quantum well, for which $Z_{2D}T = 6.9$, on a 14-fold increase over the bulk value were assumed. After this research, other workers have extended this initial calculation and confirmed that an increase in ZT over the bulk value is possible for a small well thickness [96–100].

Theoretical research: one-dimensional (1D) quantum wire. A greater enhancement in ZT was predicted for a suitable thermoelectric material prepared as a one-dimensional (1D) quantum wire, as shown in Figs. 11.16 and 11.17 [93,101]. The results of calculations suggested that a material having excellent thermoelectric properties in 3D or 2D be expected to exhibit a high ZT value in reduced dimensions within the quantum wire, under an optimum carrier concentration or the most favorable placement of the E_F for a given geometry of a low-dimensional system.

For a 1D quantum wire conductor, the movement of electrons is confined in 1D parallel to the quantum wire, and the electronic dispersion relation used is

$$\varepsilon(k_x) = \frac{\hbar^2 k_x^2}{2m_x} + \frac{\hbar^2 \pi^2}{2m_y a^2} + \frac{\hbar^2 \pi^2}{2m_z a^2}. \quad (11.21)$$

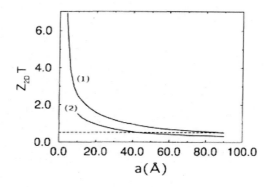

Fig. 11.15. Plot of $Z_{2D}T$ with optimum reduced chemical potential as a function of quantum well thickness a of Bi_2Te_3 quantum well parallel to (1) x–y and (2) x–z planes. The dashed line indicates the highest ZT for 3D bulk Bi_2Te_3 [92].

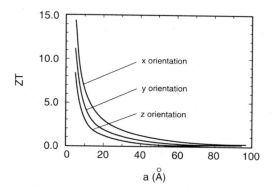

Fig. 11.16. Plot of $Z_{1D}T$ with optimum reduced chemical potential as a function of quantum wire width, a of Bi_2Te_3 quantum wires fabricated along x, y, and z directions [93].

From the three calculated thermoelectric properties, ZT for a 1D quantum well is

$$Z_{1D}T = \frac{\frac{1}{2}\left(\frac{3F_{1/2}}{F_{-1/2}} - \eta^*\right)^2 F_{-1/2}}{\frac{1}{B} + \frac{5}{2}F_{3/2} - \frac{9F_{1/2}^2}{2F_{-1/2}}},$$

where

$$B = \frac{2}{\pi a^2}\left(\frac{2k_B T}{\hbar^2}\right)^{1/2} m_x^{1/2} \frac{k_B^2 T \mu_x}{e\kappa_{ph}},$$

For a 1D quantum wire, it is necessary to maximize B to obtain the maximum $Z_{1D}T$. Therefore, the direction of the highest mobility is expected to give the highest B, and a narrower wire is expected to give a higher B and a higher $Z_{1D}T$.

Figure 11.16 shows that for a 1D quantum wire, the highest calculated figure of merit is $Z_{1D}T = 6$ for a 1-nm-wide wire, and for a 0.5-nm-wide quantum wire, $Z_{1D}T = 14$. This marked increase is mainly due to the increase in density of state, and an additional factor is the decrease in lattice thermal conductivity due to the increased phonon scattering on the surfaces [93].

The reduced dimensionality in 2D and 1D has been considered as one approach to increasing ZT, because of its many advantages. The general advantages of low-dimensional thermoelectric materials include (1) the enhancement of the density of states near E_F, leading to an increase in Seebeck coefficient, (2) opportunities to exploit the anisotropic Fermi surfaces in multivalley semiconductors, and (3) opportunities to increase the boundary scattering of phonons at the barrier well interfaces, without a large increase in the degree of electron scattering.

Proof-of-principle: 2D quantum well superlattice structure. Hicks et al. examined their theoretical predictions by fabricating a 2D $PbTe/Pb_{0.927}Eu_{0.073}Te$ multiquantum

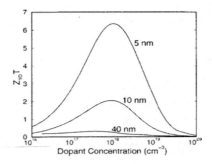

Fig. 11.17. Calculated $Z_{1D}T$ for n-type Bi nanowires at 77 K as a function of carrier concentration for three different wire diameters [101].

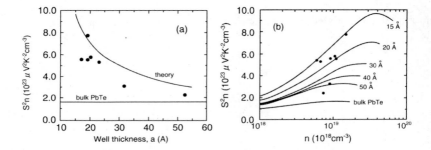

Fig. 11.18. (a) $S^2 n$ results for PbTe/Pb$_{0.927}$Eu$_{0.073}$Te MQWs are denoted by full circles as a function of well thickness a at 300 K. For comparison, the highest experimental bulk PbTe value is also shown. The calculated results for optimum doping are denoted by a solid line. (b) The $S^2 n$ results for the same PbTe/Pb$_{0.927}$Eu$_{0.073}$Te MQW samples as a function of the carrier density n at 300 K are denoted by full circles. The calculated results for different well thicknesses are denoted by solid curves [102].

well (MQW) structure [102]. In the MQW structure, PbTe is the quantum well and Pb$_{0.927}$Eu$_{0.073}$Te is the barrier material. Figure 11.18 shows the $S^2 n$ results for the PbTe/Pb$_{0.927}$Eu$_{0.073}$Te MQW as a function of the (a) well thickness a and (b) carrier density n. One can easily note that ZT increases with $S^2 n$ as the MQW thickness decreases.

They considered only S and n in their experimental studies [102], since it was through these properties that their theory predicted an increase in ZT. However, ZT depends also on the thermal conductivity. If it is assumed that the thermal conductivity of the quantum well is the same as the bulk value, the experimental results indicate that the ZT of the quantum well may be increased by five times the bulk value, giving $ZT = 2.0$ at 300 K, twice the value of state-of-the-art bulk thermoelectric materials. However, it should be noted that this value is for the quantum well alone. The barrier makes the total ZT of the MQW structure significantly lower than that of the quantum well if the thermal conductivity of barrier is the same as that of the quantum well. It is, therefore, important to design barrier materials, which can confine the carrier of the quantum well and have a low thermal conductivity for the enhancement of the ZT of MQW structures.

Another experimental investigation combined with theoretical modeling for the proof-of-principle has been performed within the quantum well of the $Si/Si_{0.7}Ge_{0.3}$ superlattice structure [103]. In this system, superlattices are grown with 15 periods of Si quantum well thickness between 0.1 and 0.5 nm alternating with a barrier $Si_{0.7}Ge_{0.3}$ layer of 30 nm thickness. The experimental result for the Seebeck coefficient follows the theoretical model used reasonably. This system also shows the strong dependence of well thickness on electrical transport, showing that the power factor of a superlattice structure with a 0.1 nm quantum well is four times higher than that of a superlattice structure with a 0.5 nm quantum well at 300 K [103].

Proof-of-principle: 1D quantum wire structure. Heremans *et al.* reported the first experimental observation of a very strong enhancement in the Seebeck coefficient of quantum wires [104]. They successfully fabricated nanocomposites containing bismuth nanowires with diameters of 9 and 15 nm, embedded in porous alumina and porous silica. As shown in Fig. 11.19, Bi nanowires of 200 nm diameter are metallic and show no enhancement in Seebeck coefficient as expected by theoretical calculations. However, a substantial increase in Seebeck coefficient was observed in the nanocomposites. They emphasized that the enhancement is ascribed to the increase in electronic density of states.

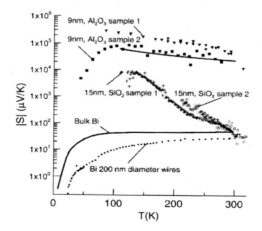

Fig. 11.19. Absolute Seebeck coefficients for two 15 nm Bi/SiO_2 samples and two 9 nm Bi/Al_2O_3 nanocomposite samples. The Seebeck coefficients of bulk Bi, the 200 nm wires, and Bi/SiO_2 composites are negative, while that of the Bi/Al_2O_3 samples is positive [104].

11.3.2 Carrier pocket engineering in superlattice

Koga et al. have discussed another strategy for achieving a higher ZT using a 2D quantum well. They also considered the superlattice structure; however they focused on the evaluation of the entire superlattice structure (3D). The optimization of $Z_{3D}T$ by the theoretical design of the superlattice has been called 'carrier pocket engineering' [105,106], which includes the proper designs of the thickness of the quantum well and barriers, of the chemical composition of the superlattice constituents, and of the strain introduced into the superlattice constituents. In the carrier pocket engineered superlattice sample, the thicknesses for the quantum well and barrier layer are both sufficiently small. Thus, the electronic density of state is basically 2D in both the well and barrier regions. Furthermore, the advantage is to achieve the significant contribution to power factor from barrier. The reduction in thermal conductivity is also expected due to the strong interface phonon scattering at the interfaces; thus, $Z_{3D}T$ would be enhanced in the properly engineered superlattice sample. However, this promising concept for achieving a high ZT in bulk materials awaits experimental validation.

11.3.3 Phonon-blocking electron-transmitting structures

Venkatasubramanian et al. have focused on utilizing superlattice structures to reduce the thermal conductivity by reducing the lattice thermal conductivity, while avoiding solid solution alloying effects on carrier scattering to improve the carrier mobility [89]. The control of the transport of phonons and electrons in superlattice thin films leads to a significant enhancement in $ZT \sim 2.4$ of 1 nm/5 nm Bi_2Te_3/Sb_2Te_3 superlattices (Fig. 11.20). In their ideal superlattice structures, it is possible to obtain a high mobility (383 cm^2/Vs) by transmitting the carrier into the superlattice structures and to reduce the lattice thermal conductivity (2.5 mW/cmK) by the potential reflection effects of phonons at mirror-like superlattice interfaces.

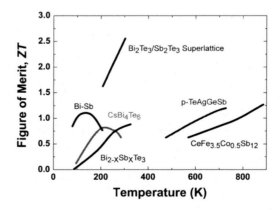

Fig. 11.20. Temperature dependence of ZT of 1 nm/5 nm Bi_2Te_3/Sb_2Te_3 superlattices and several materials [89].

11.3.4 Imbedded quantum dot nanostructured bulk materials

Harman et al. demonstrated the improved cooling devices relative to that fabricated with conventional bulk $(Bi,Te)_2(Sb,Te)_3$ materials using $PbSe_{0.98}Te_{0.02}/PbTe$ quantum dot superlattice thin films with ZT of ~1.6 at room temperature [90]. In the thin films, the pyramidal 'nanodots' of PbSe are formed and surrounded by a larger band gap PbTe matrix. Furthermore, Hsu et al. reported the potential of an imbedded quantum nanodot in bulk materials, $AgPb_mSbTe_{2+m}$ at 800 K for $m = 18$ with $ZT \sim 2.2$, containing regions 2–4 nm in size that are rich in Ag–Sb and are epitaxially embedded in a matrix that is depleted of Ag and Sb [91]. Figure 11.21 shows the TEM images of the nanostructured bulk $AgPb_{18}SbTe_{20}$ material with embedded quantum nanodots.

The origin of high thermoelectric performance of these nanostructured materials is believed to be almost entirely the high density of state, which is derived from the embedded quantum nanodot. In particular, this enhancement in ZT of nanostructured bulk materials containing the quantum nanodot opens the door for exploring the nanostructured bulk materials for various applications.

11.3.5 Remarks

Nanostructured thermoelectric materials have a great potential in improving the thermoelectric performance of cooling devices as well as power generation. Nevertheless, further improvements are anticipated, as both materials and devices are not optimized. The as-expected $ZT > 5$ in low-dimensional systems has not been achieved. Rapid improvement in nanotechnologies, such as the synthesis of materials, and the characteristic physical and/or chemical properties might lead to a tremendous enhancement in thermoelectricity.

It should be noted that a good thermoelectric bulk material is the foundation stone for exploring a good nanostructured bulk and/or thin film thermoelectric material showing

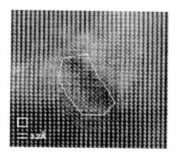

Fig. 11.21. TEM image of $AgPb_{18}SbTe_{20}$ sample showing nanosized region ('nanodot' shown in enclosed area) of crystal structure that is Ag–Sb-rich in composition. The surrounding structure, which is epitaxially related to the nanodot region, is Ag–Sb-deficient in composition close to that of PbTe [91].

a high ZT of more than 5. Therefore, the efforts to discover good thermoelectric bulk materials should be continued.

References

[1] H.J. Goldsmid and R.W. Douglas, J. Appl. Phys. **5** (1954) 386.

[2] P.H. Egli, Thermoelectricity, Wiley: New York, 1960.

[3] J.R. Drabble and H.J. Goldsmid, Thermal Conduction in Semiconductors, Pergamon: Oxford, 1961.

[4] C.M. Bhandari and D.M. Rowe, Thermal Conduction in Semiconductors, Wiley Eastern Ltd.: New Delhi, 1988.

[5] V.A. Johnson, in Progress in Semiconductors, A.F. Gibson, Ed., Heywood: London, 1956.

[6] A.C. Beer, M.N. Chase and P.F. Choquard, Helv. Phys. Acta **28** (1955) 529.

[7] G. Dresselhaus and M.S. Dresselhaus, Phys. Rev. **160** (1967) 649.

[8] G.D. Mahan, J. Appl. Phys. **65** (1989) 1578.

[9] V.I. Fistul, Heavily Doped Semiconductors, Plenum Press: New York, 1969.

[10] A.F. Ioffe, Semiconductor Thermoelements and Thermoelectric Cooling, Inforsearch Ltd.: London, 1957.

[11] I.B. Cadoff and E. Miller, Thermoelectric Materials and Devices, Reinhold Publishing Co.: New York, 1960.

[12] M.J. Smith, R.J. Knight and C.W. Spencer, J. Appl. Phys. **33** (1962) 2186.

[13] T. Kajikawa and M. Niino, Proc. 19th Inter. Conf. on Thermoelectrics, D.M. Rowe, Ed., Barrow Press: Wales, UK, (2000) 51.

[14] G.A. Slack, in CRC Handbook of Thermoelectrics, D.M. Rowe, Ed., 1995, 407.

[15] T. Calliat, A. Borshchevsky and J.-P. Fleurial, in Proc. 13th Inter. Conf. on Thermoelectrics (Kansas City, MO), B. Mathiprakasam and P. Heenan, Eds., (1994) 58.

[16] G.A. Slack and V.G. Tsoukala, J. Appl. Phys. **76** (1994) 1665.

[17] A. Borshchevsky, J.-P. Fleurial, E. Allevato and T. Calliat, Proc. 13th Inter. Conf. on Thermoelectrics (Kansas City, MO), B. Mathiprakasam and P. Heenan, Eds., (1994) 3.

[18] K. Matsubara, T. Iyanaga, T. Tsubouchi, K. Kishimoto and T. Koyanagi, Proc. 13th Inter. Conf. on Thermoelectrics (Kansas City, MO), B. Mathiprakasam and P. Heenan, Eds., (1994) 226.

[19] J.L. Cohn, G.S. Nolas, V. Fessatidis, T.H. Metcalf and G.A. Slack, Phys. Rev. Lett. **82** (1999) 779.

[20] B.C. Sales, B.C. Chakoumakos, R. Jin, J.R. Thompson and D. Mandrus, Phys. Rev. B **63** (2001) 245113.

[21] B.C. Chakoumakos, B.C. Sales, D.G. Mandrus and G.S. Nolas, J. Alloys Compd. **296** (2000) 80.

[22] B.C. Chakoumakos, B.C. Sales and D.G. Mandrus, J. Alloys Compd. **322** (2001) 127.

[23] I. Fisher, Appl. Phys. **1** (1978) 16.

[24] R. Chevrel, M. Sergent and J. Prigent, J. Solid State Chem. **3** (1971) 515.

[25] G.A. Slack, The Thermal Conductivity of Nonmetallic Crystal, in Solid State Physics, Vol. 34, 1979, p. 1. ed. by H. Ehrenreich, F. Seitz, and D. Turnbull, Academic Press, New York.

[26] J.W. Sharp, E.C. Jones, R.K. Williams, P.M. Martin and B.C. Sales, J. Appl. Phys. **78** (1995) 1013.

[27] B.C. Sales, D. Mandrus and R.K. Williams, Science **272** (1996) 1325.

[28] B.C. Sales, D. Mandrus, B.C. Chakoumakos, V. Keppens and J.R. Thompson, Phys. Rev. B **56** (1997) 15081.

[29] B.X. Chen, J.H. Xu, C. Uher, D.T. Morelli, G.P. Meisner, J.-P. Fleurial, T. Caillat and A. Borshchevsky, Phys. Rev. B **55** (1997) 1476.

[30] J.-P. Fleurial, T. Caillat and A. Borshchevsky, Proc. 15th Inter. Conf. of Thermoelectrics, (1996) 92.

[31] X. Tang, L. Chen, T. Goto and T. Hirai, J. Mater. Res. **16** (2001) 837.

[32] J. Tobola and J. Pierre, J. Alloys Comp. **296** (2000) 243.

[33] F.G. Aliev, N.B. Brandt, V.V. Moshchalkov, V.V. Kozyrkov, R.V. Scolozdra and A.I. Belogorokhov, Z. Phys. B: Condens. Matter **75** (1989) 167.

[34] F.G. Aliev, V.V. Kozyrkov, V.V. Moshchalkov, R.V. Scolozdra and K. Durczewski, Z. Phys. B: Condens. Matter **80** (1990) 353.

[35] S. Öğüt and K.M. Rabe, Phys. Rev. B **51** (1995) 10443.

[36] W. Jeitschko and D.J. Braun, Acta Crystallogr. B **33** (1977) 3401.

[37] D.J. Braun and W. Jeitschko, J. Less-Common Met. **72** (1980) 147.

[38] D.J. Braun and W. Jeitschko, J. Less-Common Met. **72** (1980) 33.

[39] D.J. Braun and W. Jeitschko, J. Solid State Chem. **32** (1980) 357.

[40] B.C. Chakoumakos, B.C. Bale, D. Mandrus and V. Keppens, Acta Crystallogr. B **55** (1999) 341.

[41] N.T. Stetson, S.M. Kauzlarich and H. Hope, J. Solid State Chem. **91** (1991) 140.

[42] L.E. DeLong and G.P. Meisner, Solid State Commun. **53** (1985) 119.

[43] D.M. Morelli and G.P. Meisner, J. Appl. Phys. **77** (1995) 3777.

[44] G.P. Meisner, M.S. Torikachvili, K.N. Yang, M.B. Maple and R.P. Guertin, J. Appl. Phys. **57** (1985) 3073.

[45] M.E. Danebrock, C.B.H. Evers and W. Jeitschko, J. Phys. Chem. Solids **57** (1996) 381.

[46] G.P. Meisner, Physica B & C **108B** (1981) 763.

[47] S. Zemi, D. Tranqui, P. Chaudouet, R. Madar and J.P. Senateur, J. Solid State Chem. **65** (1986) 1.

[48] I. Shirotani, T. Adachi, K. Tachi, S. Todo, K. Nozawa, T. Yagi and M. Kinoshita, J. Phys. Chem. Solids **57** (1996) 211.

[49] A. Abeles, Phys. Rev. **131** (1963) 1906.

[50] S.W. Kim, Y. Kimura and Y. Mishima, J. Electronic Mater **32** (2003) 1141.

[51] G.S. Nolas, G.A. Slack, D.T. Morelli, T.M. Tritt and A.C. Ehrlich, J. Appl. Phys. **79** (1996) 4002.

[52] L.D. Chen, T. Kawahara, X.F. Tang, T. Goto, T. Hirai, J.S. Dyck, W. Chen and C. Uher, J. Appl. Phys. **90** (2001) 1864.

[53] G.P. Meisner, D.T. Morelli, S. Hu, J. Yang and C. Uher, Phys. Rev. Lett. **80** (1998) 3551.

[54] T.M. Tritt, G.S. Nolas, G.A. Slack, A.C. Ehrlich, D.J. Gillespie and J.L. Cohn, J. Appl. Phys. **79** (1996) 8412.

[55] R.B. King, Inorg. Chem. **23** (1989) 3048.

[56] D.J. Singh and W.E. Pickett, Phys. Rev. B **50** (1994) 11235.

[57] D.T. Morelli, T. Caillat, J.-P. Fleurial, A. Borshchevsky, J. Vandersande, B. Chen and C. Uher, Phys. Rev. B **51** (1995) 9622.

[58] F. Izumi and T. Ikeda, Mater. Sci. Forum **324** (2000) 198.

[59] P.G. Klemens, Phys. Rev. **119** (1960) 507.

[60] J. Callaway and H.C. von Baeyer, Phys. Rev. **120** (1960) 1149.

[61] G.A. Nolas, J.L. Cohn and G.A. Slack, Phys. Rev. B **58** (1998) 164.

[62] L. Chen, X. Tang, T. Goto and T. Hirai, J. Mater. Res. **15** (2000) 2276.

[63] A.E. Dwight, M.H. Mueller, R.A. Conner Jr., J.W. Downey and H. Knott, Trans. Metall. Soc. AIME **242** (1968) 2075.

[64] W. Jeitschko, Metall. Trans. **1** (1970) 3159.

[65] A.E. Dwight, J. Less-Common Met. **34** (1974) 279.

[66] R. Marazza, R. Ferro and G.J. Rambaldi, J. Less-Common Met. **39** (1975) 341.

[67] Y.V. Stadnyk, L.A. Mykhailiv, V.V. Kuprina and R.V. Skolozdra, Inorg. Mater. **24** (1989) 1196.

[68] R.V. Skolozdra, Y.V. Stadnyk, Y.K. Gorelenko and E.E. Terletskaya, Sov. Phys. Solid State **32** (1990) 1536.

[69] R. Kuentzler, R. Clad, G. Schmerber and Y. Dossmann, J. Magn. Magn. Mater. **104** (1992) 1976.

[70] C.B.H. Evers, C.G. Richter, K. Hartjes and W.J. Jeitschko, J. Alloys Compd. **252** (1997) 93.

[71] R.A. de Groot, F.M. Muller, P.G. Engen and K.H. J. Bushow, Phys. Rev. Lett. **50** (1983) 2024.

[72] P.G. Van Engen, K.H.J. Bushow and R. Jongebreur, Appl. Phys. Lett. **42** (1983) 302.

[73] B.A. Cook, J.L. Harringa, Z.S. Tan and W.A. Jesser, in Proc. ICT'96 15th Inter. Conf. on Thermoelectrics (Piscataway, NJ), IEEE, (1996) 122.

[74] C. Uher, J. Yang, S. Hu, D.T. Morelli and G.P. Meisner, Phys. Rev. B **59** (1999) 8615.

[75] H. Hohl, A.R. Ramirez, W. Kaefer, K. Fess, C.H. Thurner, C.H. Kloc and E. Bucher, Thermoelectric Materials – New Directions and Approaches (MRS Symp. Proc. Vol. 478) T.M. Tritt, M. Kanatzidis, H.B. Lyon Jr. and G.D. Mahan, Eds., 1997, Materials Research Society: Warrendale, PA, p. 109.

[76] H. Hohl, A.R. Ramirez, C. Goldmann, G. Ernst, B. Woelfing and E.J. Bucher, J. Phys.: Condens. Matter **11** (1999) 1697.

[77] V.M. Browning, S.J. Poon, T.M. Tritt, A.L. Pope, S. Bhattacharya, P. Volkov, J.G. Song, V. Ponnambalam and A.C. Ehrlich, Thermoelectric Materials 1998 – The Next Generation Materials for Small-Scale Refrigeration and Power Generation Applications (MRS Symp. Proc. Vol. 545) T.M. Tritt, M. Kanatzidis, H.B. Lyon Jr. and G.D. Mahan, Eds., 1999, Materials Research Society: Warrendale, PA, p. 403.

[78] S. Sportouch, P. Larson, M. Bastea, P. Brazis, J. Ireland, C.R. Kannenwurf, S.D. Mahanti, C. Uher and M.G. Kanatzidis, Thermoelectric Materials 1998 – The Next Generation Materials for Small-Scale Refrigeration and Power Generation Applications (MRS Symp. Proc. Vol. 545) T.M. Tritt, M. Kanatzidis, H.B. Lyon Jr. and G.D. Mahan, Eds., 1999, Materials Research Society: Warrendale, PA, p. 421.

[79] K. Mastronardi, D. Young, C.C. Wang, P. Khalifah, R.J. Cava and A.P. Ramirez, Appl. Phys. Lett. **74** (1999) 1415.

[80] J. Tobola, J. Pierre, S. Kaprzyk, R.V. Skolozdra and M.A. Kouacou, J. Phys.: Condens. Matter **10** (1998) 1013.

[81] K. Kaczmarska, J. Pierre, J. Tobola and R.V. Skolozdra, Phys. Rev. B **60** (1999) 373.

[82] C. Uher, J. Yang, S. Hu, D.T. Morelli and G.P. Meisner, Phys. Rev. B **59** (1999) 8615.

[83] S. Bhattacharya, A.L. Pope, R.T. Littleton IV, M. Tritt, V. Ponnambalam, Y. Xia and S.J. Poon, Appl. Phys. Lett. **77** (2000) 2476.

[84] K. Katayama, S.W. Kim, Y. Kimura and Y. Mishima, J. Electron Mater. **32** (2003) 1160.

[85] S. Bhattacharya, T.M. Tritt, Y. Xia, V. Ponnambalam, S.J. Poon and N. Thadhani, Appl. Phys. Lett. **81** (2002) 43.

[86] Q. Shen, L. Chen, T. Goto, T. Hirai, J. Yang, G.P. Meissner and C. Uher, Appl. Phys. Lett. **79** (2001) 4165.

[87] H.J. Goldsmid, Thermoelectric Refrigeration, Plenum: New York, 1964.

[88] A. Majumdar, Science **303** (2004) 777.

[89] R. Venkatasubramanian, E. Siivola, T. Colpitts and B.O' Quinn, Nature **413** (2001) 597.

[90] C. Harman, P.J. Taylor, M.P. Walsh, and B.E. Laforge, Science **297** (2002) 2229.

[91] K.F. Hsu, S. Loo, F. Guo, W. Chen, J.S. Dyck, C. Uher, T. Hogan, E.K. Polychroniadis, M.G. Kanatzidis, Science **303** (2004) 818.

[92] L.D. Hicks and M.S. Dresselhaus, Phys. Rev. B **47** (1993) 12727.

[93] L.D. Hicks and M.S. Dresselhaus, Phys. Rev. B **47** (1993) 16631.

[94] D. Hicks, T.C. Harman and M.S. Dresselhaus, Appl. Phys. Lett. **63** (1993) 3230.

[95] S. Rittner, J. Appl. Phys. **30** (1959) 702.

[96] O. Sofo and G.D. Mahan, Appl. Phys. Lett. **65** (1994) 2690.

[97] D.A. Broido and T.L. Reinecke, Appl. Phys. Lett. **67** (1995) 100.

[98] D.A. Broido and T.L. Reinecke, Appl. Phys. Lett. **67** (1995) 1170.

[99] D.L. Broido and T.L. Reinecke, Phys. Rev. B **51** (1995) 13797.

[100] D.A. Broido and T.L. Reinecke, Appl. Phys. Lett. **70** (1997) 2834.

[101] Y.M. Lin, X. Sun and M.S. Dresselhaus, Phys. Rev. B **62** (2000) 4610.

[102] L.D. Hicks, T.C. Harman, X. Sun and M.S. Dresselhaus, Phys. Rev. B **53** (1996) R10493.

[103] X. Sun, S.B. Cronin, J. Liu, K.L. Wang, T. Koga, M.S. Dresselhaus and G. Chen, Proc. the 18th Inter. Conf. on Thermoelectrics: ICT Symposium Proceedings, (1999) pp. 652–655.

[104] J.P. Heremans, C.M. Thrush, D.T. Morelli and M.C. Wu, Phys. Rev. Lett. **88** (2002) 216801.

[105] T. Koga, X. Sun, S.B. Cronin and M.S. Dresselhaus, Appl. Phys. Lett. **73** (1998) 2950.

[106] T. Koga, X. Sun, S.B. Cronin and M.S. Dresselhaus, Appl. Phys. Lett. **75** (1999) 2438.

CHAPTER 12

Smart Coatings – Multilayered and Multifunctional in-situ Ultrahigh-temperature Coatings

Hideki Hosoda

Abstract

To develop ultrahigh-temperature structural materials, advanced smart coatings, which provide excellent oxidation resistance, are required. Several functions are also required for practical coatings: self-healing, equivalent thermal expansion, thermal barrier, phase stability, and diffusion barrier. The basic concept in improving oxidation resistance is to reduce the oxygen permeation through the coating to the substrate material. Iridium (Ir) possesses the lowest oxidation permeability of all the materials known. Therefore, Ir is expected to be applicable for the oxygen diffusion barrier (ODB), which is the most important part of ultrahigh-temperature coatings. A serious problem with Ir is that the oxidation resistance itself is not sufficient at high temperature. In order to improve the high-temperature oxidation resistance of Ir, artificial multilayered coatings, such as Ir–Ag and Ir–Al_2O_3, were previously proposed for carbon–carbon composites and were fabricated by deposition techniques. Ag and Al_2O_3 layers act as oxidation-resistant coatings that protect the Ir layer. In such an artificial multilayered coating, if the oxidation-resistant coating is removed accidentally, the Ir coating is rapidly oxidized and evaporated through the formation of gaseous iridium oxide, IrO_3. Hence, instead of such an artificial multilayered coating, an in-situ multilayered smart coating is proposed based on IrAl. When IrAl-based material is deposited onto substrates in an oxidizing environment, the surface of IrAl automatically forms an Al_2O_3 oxidation-resistant coating and Ir becomes the oxygen diffusion barrier due to oxidation. In this review, the basic idea and design of the smart coatings, the fabrication, structure, mechanical properties, oxidation of Ir and IrAl, and effects of ternary additions are described, particularly for ternary Co addition to IrAl.

Keywords: Ultrahigh-temperature coating, oxidation resistance, iridium, IrAl, mechanical properties.

12.1 Introduction

Advanced oxidation-resistant coatings are the key to the practical use of new ultrahigh-temperature structural materials, such as refractory metal alloys (Nb-, Mo-, W-base alloys) and carbon–carbon composites, in the development of higher efficiency generators and advanced hypersonic transport, for example. Although the mechanical properties of these advanced ultrahigh-temperature structural materials have been considerably improved, the oxidation resistance of these materials is still insufficient for practical applications. This is because the oxidation resistance of these materials is intrinsically poor due to the formation of gaseous oxides, thus, excellent coatings are certainly required. Practical high-temperature coatings require (1) excellent oxidation resistance, (2) extremely low oxygen permeability, (3) self-healing capability, (4) similar thermal expansion to that of the substrate, and (5) strong adhesion to the substrate. Low thermal conductivity is often required for the thermal barrier coating (TBC) in order to decrease the temperature of substrate materials. Figure 12.1 shows the oxygen permeability of selected materials [1]. It is clearly seen that the oxides such as ZrO_2 and HfO_2 used for TBC exhibit high oxygen permeability. If these oxides with high oxygen permeability are coated directly onto the substrate, the substrate will still be attacked by oxygen passing through the coatings. Therefore, these oxides used for TBCs cannot be used as oxidation resistant coatings. It is seen in Fig. 12.1 that the oxygen permeabilities

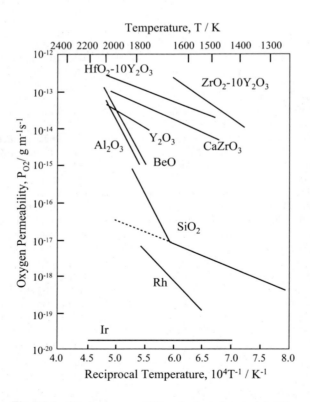

Fig. 12.1. Oxygen permeability of selected materials [1].

of Al_2O_3 and SiO_2 are very low. Thus Al_2O_3 and SiO_2 are major components of current practical oxidation-resistant coatings. In order to form continuous oxide layers at the surface, high-temperature-coating materials such as MCrAlY generally contain large amounts of Al, Si, and/or Cr. However, the oxygen permeability of their oxides becomes greater, making them inapplicable as practical oxygen diffusion barriers (ODBs) for the ultrahigh-temperature materials working above 2000 K. Therefore, other materials having very much lower oxygen diffusivity even at ultrahigh temperature should be developed as ODBs. A promising material is iridium (Ir) because the oxygen permeability of Ir is the lowest of all materials known, and it is 4–7 orders of magnitude lower than those of Al_2O_3 and SiO_2. This means that, from the viewpoint of oxygen diffusivity, the ODB performance of a 1-nm thick Ir layer is comparable or superior to those of 1-μm thick SiO_2 and Al_2O_3 layers above 2000 K. Thus, Ir has a high potential for use as an ODB. Ir is known to exhibit relatively low oxidation rates compared with refractory metals such as Nb and Ta [2]. In addition, Ir has a high melting point of 2720 K [3] and an fcc crystal structure with high crystal symmetry. Because of these excellent high-temperature properties, Ir-based alloys are also expected to be a new type of heat-resistant alloys in place of Ni-based superalloys even though the density of Ir is relatively high (22.5 mg m^{-3}).

On the other hand, the oxidation resistance of Ir itself is poor due to the formation of a gaseous oxide, IrO_3, above 1390 K [2,4]. In order to improve the oxidation resistance and suppress the formation of the Ir oxide, alloying with Al has been evaluated. Lee and Worrell have reported that IrAl containing more than 55mol%Al forms a continuous Al_2O_3 layer and oxidation resistance is greatly improved [5]. In this study, we proposed an effective ODB based on B2 IrAl alloy, and investigated the oxidation behavior of the alloy. IrAl was expected to form a self-healing multifunctional layered structure composed of (1) an Al_2O_3 layer as a protective oxide coating and (2) an Ir layer as an ODB.

In this section, the oxidation resistance of Ir is first described. Second, the design of a smart coating based on Ir is proposed. Then, physical properties, fabrication process, mechanical properties, and oxidation resistance of IrAl are described.

12.2 Oxidation Resistance of Ir

Figure 12.2 shows the oxidation behavior of Ir heated in a pure oxygen (O_2) atmosphere measured by simultaneous thermogravimetry–differential scanning calorimetry (TG–DSC). Bulk Ir of a 5-mm cube was used in this measurement. A large endothermic heat is observed at 1395 K in Fig. 12.2(a), which is in good agreement with the 1393 K for IrO_3 formation in the literature [2]. Corresponding weight changes are obtained in the TG curve in Fig. 12.2(b) – a large mass gain is seen above 1000 K followed by a significant mass loss above 1395 K. These mass changes correspond to the formation of IrO_2 (solid phase) and IrO_3 (gaseous phase), respectively. Thus, it is clear that Ir itself was oxidized above 1000 K, and vaporized above 1400 K in an O_2 environment, resulting in the gaseous oxide IrO_3.

Figure 12.3 shows the isothermal weight change of Ir at 1273, 1473, and 1673 K in an O_2 environment. The Ir samples used were bulk specimens. Based on the results of

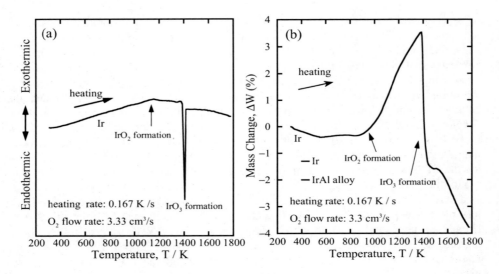

Fig. 12.2. Oxidation behavior of Ir heated in pure O_2 environment: (a) reaction heat measured by differential scanning calorimetry (DSC) and (b) weight change measured by thermogravimetry (TG).

heating oxidation in Fig. 12.2, it is predicted that weight gain must occur at 1273 K, where the solid oxide IrO_2 forms, and that weight loss occurs at 1473 and 1673 K, where the gaseous oxide IrO_3 forms. It is seen in Fig. 12.3 that almost no weight change is recognized at 1273 K. This suggests that the growth rate of IrO_2 is not so high at 1273 K and solid IrO_2 was covered with the surface of the specimen. Further oxidation occurs when oxygen atoms pass through the oxide layer. IrO_2 plays the role of a protective oxide in this temperature range.

On the other hand, when Ir was kept at 1473 and 1673 K, where the gaseous oxide IrO_3 forms, large mass loss was recognized, as predicted. The amount of mass loss increases linearly with time. This is because all the oxide that was formed evaporated, and then the fresh surface that appeared was continuously attacked by oxygen in the environment. Therefore, when Ir is used as the oxygen diffusion barrier at an ultrahigh temperature of around 2000 K, the inhibition of oxide formation is important, and this must be achieved by surface protection of the Ir layer. Some trials of applying a protective coating, such as Ag and Al_2O_3, onto Ir by thermal spraying or deposition were reported previously but were unsuccessful in terms of the duration. The lifetime of the whole coating is equal to that of the protective coating layer in these cases. Moreover, when the outer coating is scratched or partially peeled accidentally, the whole system is vaporized immediately by evaporation due to oxidation.

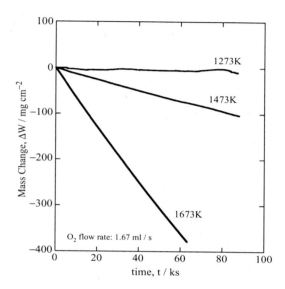

Fig. 12.3. Weight change as a function of time for pure Ir upon isothermal oxidation at 1273, 1473, and 1673 K in an O_2 environment.

12.3 Design of Multifunctional and Multilayered Coating Based on IrAl

For a practical ultrahigh-temperature coating, multilayers are required, where each layer plays its own role such as a thermal barrier, oxidation inhibitor, or oxygen diffusion barrier. In addition, self-healing is an important factor in the reliability and safety of the system. The in-situ multilayered coating made automatically by oxidation is promising. The Ir-based multilayered coating material based on the Ir–Al–O system was designed.

The Ir–Al binary phase diagram is shown in Fig. 12.4 [3]. The B2-type (CsCl) IrAl intermetallic phase exists between 48 and 52mol%Al [3,6]. It is clear that IrAl has a high melting point (2393 K), good crystal symmetry, good oxidation resistance, and relatively low density, owing to the high content of Al. Judging from these characteristics, IrAl alloys are promising new high-temperature smart materials.

Figure 12.5 shows the estimated Ir–Al–O ternary phase equilibrium at a high temperature [4]. Consider the oxidation of Al-rich Ir–Al intermetallics such as Al_5Ir_2 (other than IrAl); Al_5Ir_2 seems to form Al_2O_3 and IrAl, but Ir (Ir solid solution) is not formed, as shown in Fig. 12.5(b). Therefore, Al-rich Ir–Al intermetallics that do not form Ir do not function as oxygen diffusion barriers, although they exhibit good oxidation resistance due to continuous Al_2O_3 formation.

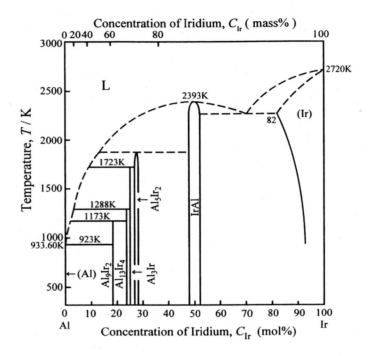

Fig. 12.4. The Ir–Al binary phase diagram [3].

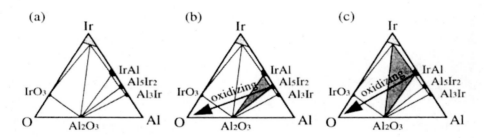

Fig. 12.5. Estimated Ir–Al–O phase diagram and oxidation behavior at high temperature above 1400 K: (a) ternary phase diagram, (b) oxidation of Al_5Ir_2, and (c) oxidation of IrAl. Al_5Ir_2 forms IrAl and Al_2O_3, and IrAl forms Ir and Al_2O_3 upon oxidation.

Second, consider the oxidation of IrAl. IrAl seems to form both Ir and Al_2O_3, while IrO_3 formation may be suppressed, as shown in Fig. 12.5(c). Thus, IrAl will have a high potential to be an oxygen diffusion barrier due to Ir formation under the Al_2O_3 layer, if the Ir layer forms continuously and Al_2O_3 as an oxidation-resistant coating protects the Ir layer. Then, an in-situ multilayered, multifunctional structure can be simply formed by oxidation. An example of an advanced ultrahigh-temperature oxidation-resistant smart coating based on IrAl is demonstrated in Fig. 12.6. It should be noted that the Ir layer was formed through oxidation.

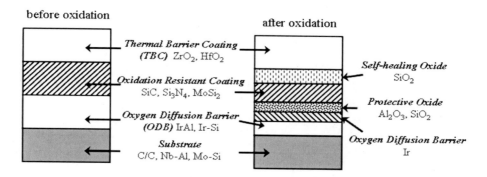

Fig. 12.6. Example of advanced multifunctional, multilayered coating.

12.4 Physical Properties of IrAl Alloys

12.4.1 Microstructures

Figure 12.7 shows the microstructures of several IrAl alloys fabricated by Ar-arc melting (ingot metallurgy, IM), powder sintering (powder metallurgy, PM), and reactive hot pressing (RHP) [7–9]. The chemical compositions of the alloys are listed in Table 12.1 and the thermomechanical treatments used are listed in Table 12.2. PM Ir49Al was made by Ar-arc melting and homogenized, followed by pulverization and sintering by hot pressing in order to obtain the equilibrium state. It should be noted that no phase reaction was observed for any IrAl alloy up to 1773 K in DSC measurements in Ar. Hence, the microstructures obtained are considered to be stable. It is clear that both IM and PM Ir49Al alloys are two-phase alloys containing bright regions. IM Ir49Al shows a eutectic structure and PM Ir49Al contains a small number of voids.

XRD revealed that both IM and PM Ir49Al alloys are composed of fcc Ir and B2 IrAl, as shown in Fig. 12.8. Then, the bright regions of microstructures in Figs. 12.7(a) and (b) correspond to the fcc Ir phase generated through solidification. According to the Ir–Al binary phase diagram in Fig. 12.4, the Ir phase in Ir49Al alloys is not the equilibrium phase, because the chemical composition of Ir49Al lies in the B2 IrAl single-phase region in the phase diagram shown in Fig. 12.4. However, the eutectic Ir does not disappear, even in PM Ir49Al which may be in equilibrium. Then, by taking into account the thermal stability of the eutectic Ir phase, the apparent Ir phase of both Ir49Al alloys may be the equilibrium phase. If this is the case, the phase boundary between IrAl and (Ir + IrAl) might be closer to the stoichiometric composition of Ir-50mol%Al than that in the phase diagram of Fig. 12.4.

Al-rich IrAl alloys were also fabricated by the Ar-arc melting method. However, the chemical compositions shift to the Ir-rich side of stoichiometry, as revealed by inductively coupled plasma (ICP) chemical analysis. The decrease in Al content is considered to be due to the low boiling point of Al compared with the melting point of Ir. Similar eutectic

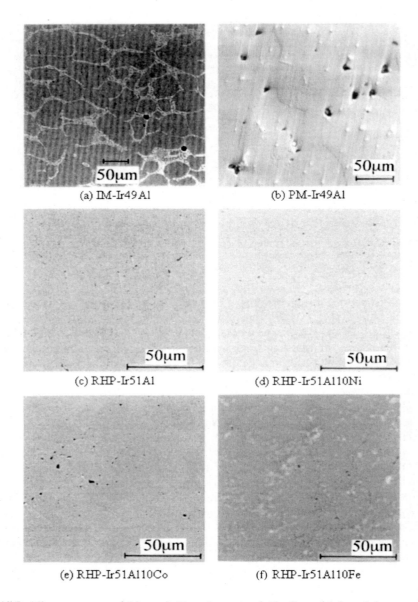

Fig. 12.7. Microstructures of binary IrAl and ternary IrAl alloys fabricated by arc-melting (ingot metallurgy: IM), sintering (powder metallurgy: PM), and reactive hot pressing (RHP): (a) IM-Ir49Al, (b) PM-Ir49Al, (c) RHP-Ir51Al, (d) RHP-Ir51Al10Ni, (e) RHP-Ir51Al10Co, and (f) RHP-Ir51Al10Fe.

Table 12.1. Chemical compositions of the alloys prepared.

Alloy		Ir (mol%)	Al (mol%)	Ni (mol%)	Co (mol%)	Fe (mol%)
Ir49Al alloys	IM Ir49Al*	51.1	48.9	–	–	–
	IM Ir48.3Al*	51.7	48.3			
	IM Ir48Al	52	48.0			
	PM Ir49Al**	51.4	48.6	–	–	–
Ir51Al-based alloys	RHP Ir51Al*	49.0	51.0	–	–	–
	RHP Ir51Al10Ni*	39.0	51.0	10.0	–	–
	RHP Ir51Al10Co*	39.0	51.0	–	10.0	–
	RHP Ir51Al10Fe*	39.0	51.0	–	–	10.0

*By ICP-OES chemical analysis.
**Nominal compositions.

Table 12.2. Alloy fabrication and heat treatment conditions.

Alloy		Fabrication and thermomechanical treatment
Ir49Al alloys	IM Ir49Al	Ar arc melting, 2023 K for 14.4 ks, SC
	PM Ir49Al	Ar arc melting, 2023 K for 14.4 ks, SC, pulverization (<50 μm)*, (HP) 2023 K for 14.4 ks under 70 MPa, 2173 K × 18.0 ks, FC
Ir51Al-based alloys	RHP Ir51Al	(RHP) 2023 K for 18.0 ks under 70 MPa, FC
	RHP Ir51Al10Ni	(RHP) 2023 K for 18.0 ks under 70 MPa, FC
	RHP Ir51Al10Co	(RHP) 2023 K for 18.0 ks under 70 MPa, FC
	RHP Ir51Al10Fe	(RHP) 1523 K for 10.8 ks under 70 MPa, FC

SC: slow cooling (<0.1 K/s), HP: hot pressing, RHP: reactive hot pressing, FC: furnace cooling.
*Pulverization was done mechanically in air.

microstructures were observed in the Al-rich nominal composition when produced by the Ar-arc melting method.

In order to reduce the change in chemical composition, powder metallurgy is advantageous. Figure 12.7(c) shows the microstructure of Al-rich Ir51Al alloy fabricated by the RHP method. In the RHP process, the mixture of Ir powder and Al powder was heated under hot pressing. Then, the reaction from Ir + Al to B2 IrAl occurred due to the large heat of formation. In this case, the IrAl single phase was obtained, even in the Al-rich composition, without significant change in the chemical composition. Some small voids are also seen in the figure, but the fraction of voids is similar to that of IM-Ir49Al. The RHP method is very useful for the fabrication of the IrAl single phase. Ternary IrAl alloys containing Ni, Co, and Fe were fabricated by the RHP method, and their microstructures are also shown in Figs. 12.7(d), (e), and (f), respectively. It was found that the B2 single phase was obtained for the Ni- and Co-added alloys, but that a second phase (bright regions in (e)) and voids along the grain boundaries were formed for the Fe-added alloy. XRD analysis revealed that the Fe-added alloy is composed of B2 IrAl and Al_2O_3. Al_2O_3 is known to form in powder metallurgical methods for materials containing Al. The void formation might be due to the low RHP temperature of 1523 K; synthesis by RHP may not have been complete.

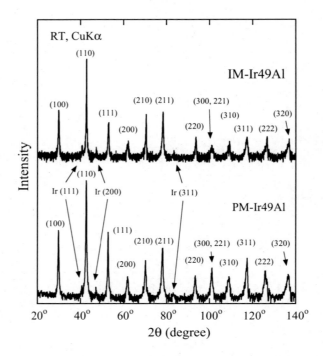

Fig. 12.8. XRD profiles of IM-Ir49Al and Pm-Ir49Al alloys.

12.4.2 Lattice parameter

The lattice parameters of B2 IrAl calculated using the XRD profiles are listed in Table 12.3. The lattice parameter of IrAl is 298.7–298.8 pm for both the Ir49Al alloys. On the other hand, the lattice parameter of Al-rich Ir51Al-based alloys is 298.1–296.9 pm. The value of 298.1 pm for RHP Ir51Al alloy is smaller than those for Ir-rich Ir49Al alloys. The Goldschmidt radius of Al (143 pm) is larger than that of Ir (136 pm); this difference in lattice parameter could not be explained in terms of atomic size but may be explained by the type of compositional defect structures. According to the pseudo-ground-state calculations for the determination of the type of compositional defect structures [10,11], the defect type of Ir-rich IrAl is antistructural (antisite) where excess Ir atoms occupy Al sites,

Table 12.3. Lattice parameters of B2 IrAl phases.

Alloy	Lattice parameter, a/pm
IM Ir49Al	298.8
PM Ir49Al	298.7
RHP Ir51Al	298.1
RHP Ir51Al10Ni	297.2
RHP Ir51Al10Co	296.9
RHP Ir51Al10Fe	296.9

and that of Al-rich IrAl alloy is structural (vacancy type including triple defect) where structural vacancies are formed due to the imbalance of constitutional elements. It is well known that the lattice parameter decreases upon the formation of structural vacancies, as seen in B2 NiAl [12,13]. Detailed investigation of the compositional defect structure is required for the IrAl system. In the case of Ir51Al-based alloys, the lattice parameter of the B2 phase decreases upon adding ternary elements (Ni, Co, and Fe); 297.2–296.9 pm. Since the atomic sizes of Ni, Co, and Fe are smaller than that of Ir, the decrease in lattice parameter in this case can simply be explained by atomic size.

12.4.3 Vickers hardness

The measured Vickers hardness values are listed in Table 12.4 [7,9]. HV values are higher than HV1000 for Ir49Al alloys, and lower than HV1000 for RHP Ir51Al-based alloys. HV values of both types of Ir49Al alloys are scattered depending on the distribution of Ir for IM Ir49Al alloy, and additionally on voids for PM Ir49Al alloy. The HV values of RHP Ir51Al10Fe also are scattered. The scatter of HV values for RHP Ir51Al10Fe may be caused by the presence and distribution of Al_2O_3.

In the case of the binary IrAl alloys, HV values are high on the Ir-rich side and low on the Al-rich side. The existence of structural vacancies usually raises the hardness and yield stress, as is well known for Al-rich NiAl [13]. If it is true that the defect is the structural type (vacancy type) on the Al-rich side and the antistructural on the Ir-rich side of the stoichiometry, the hardness should be high on the Al-rich side and low on the Ir-rich side. Thus, the variation in hardness is contrary to the estimated defect type. In the case of ternary IrAl alloys, ternary additions, Ni and Fe, cause solution hardening, and Co addition causes solution softening. The details of the hardness of IrAl alloys are not sufficiently understood at present.

Morphologies of indentations are shown in Fig. 12.9(a) and (b) for Ir-48.9Al, (c) and (d) for Ir-48.3Al, and (e) and (f) for Ir-48.0Al alloys (in mol%). Figures 12.9(a), (c), and (e) show the indentations on B2 single-phase regions. Although these B2 phases are hard, with HV values of around HV1100, no cracks are introduced by the indenter in alloys containing less than 50mol%Al. No cracks are observed in the eutectic structure even though the Ir phase is relatively large hardness of HV1100. On the other hand, at the grain boundaries of the eutectic structure, cracks are often formed between IrAl and eutectic Ir, as shown in Figs. 12.9(d) and (f). A similar tendency is seen in PM Ir49Al. Crack

Table 12.4. Vickers hardness of IrAl alloys.

Alloy	Average	Scatter
IM Ir49Al	HV1067	HV1000–HV1150
PM Ir49Al	HV1006	HV850–HV1100
RHP Ir51Al	HV884	HV850–HV900
RHP Ir51Al10Ni	HV983	HV950–HV1000
RHP Ir51Al10Co	HV771	HV750–HV800
RHP Ir51Al10Fe	HV932	HV690–HV1050

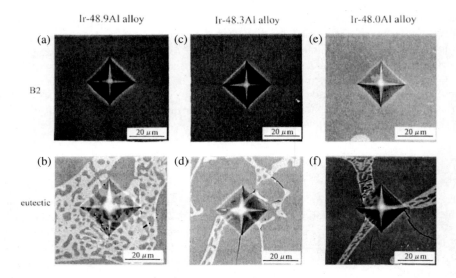

Fig. 12.9. Some examples of indentations observed for Ir-48.9Al ((a) and (b)), Ir-48.3Al ((c) and (d)), and Ir-48.0 Al ((e) and (f)), where these alloys were fabricated by the Ar arc-melting method. It is noted that (a), (c), and (e) show B2 single regions and (b), (d), and (f) show the eutectic part [7].

formation seems to be promoted by the existence of interfaces of Ir domains. Ir-rich IrAl itself seems not to be brittle. Furthermore, cracks are generally seen around the indentation in RHP Ir51Al-base alloys regardless of whether they are binary or ternary alloys. As previously pointed out, the defect structure is estimated to be the vacancy type (structural defect) for an Al-rich composition and the antisite type (substitutional defect) for an Al-poor composition. It is commonly believed that intermetallics become brittle when the vacancy-type defect is formed at nonstoichiometry. Thus, the brittleness of Al-rich IrAl implies the vacancy-type defect.

12.4.4 Compressive mechanical properties

Mechanical properties were evaluated by compression tests between RT and 1873 K. Stress–strain (SS) curves obtained for PM Ir49Al alloys are shown in Fig. 12.10. Ir49Al alloys show brittleness below 1073 K. When compressive ductility appears at a temperature above 1073 K, it is clear that the flow stress shows a positive strain-rate dependence by the strain-rate dip tests: flow stress increases with increasing strain rate. Dislocation motion in IrAl is judged to be a thermal activation process.

The temperature and normalized temperature dependences of 0.2% flow stress, and compressive fracture strength of PM Ir49Al alloy are shown in Fig. 12.11, where the normalized temperatures were calculated using the melting point (T_m) of 2393 K [3]. Tensile data for pure Ir [14] are also shown in the figure. Compressive fracture does not

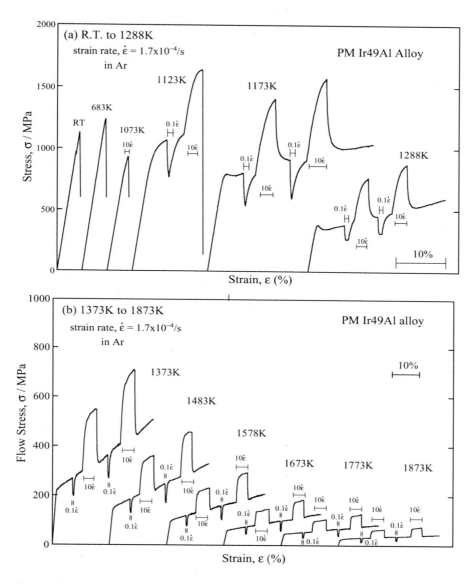

Fig. 12.10. Stress–strain curves of PM Ir49Al alloy: (a) RT-1288 K and (b) 1373–1873 K.

occur above 1100 K (0.45 T_m). At around 0.45 T_m, 0.2% flow stress rapidly decreases with increasing test temperature. These features suggest that the deformation of IrAl is controlled by the Peierls mechanism. It should be mentioned that the yield stress of IrAl is higher than that of Ir throughout the entire temperature range. In addition, both the 0.2% flow stress, as a function of the normalized temperature, and the specific strength calculated from density are higher than those of B2 NiAl [15,16], as shown in Fig. 12.12.

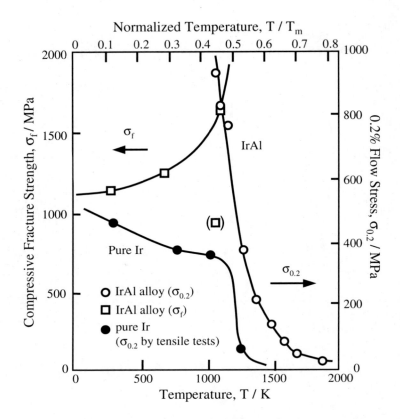

Fig. 12.11. Temperature and normalized temperature dependences of 0.2% flow stress of Ir49Al. Tensile data for pure Ir is also shown [12].

It can be said that IrAl alloys are promising high-temperature structural materials, though Ir is a precious and costly metal.

12.4.5 Thermal expansion

Figure 12.13 shows the thermal expansion of IM Ir49Al alloys up to 2100 K in Ar. It should be noted that no phase transformation was detected either in thermal expansion or in the DSC curve up to 1873 K. Using Fig. 12.13, the coefficient of thermal expansion (CTE) was calculated as a function of temperature and is shown in Fig. 12.14. It is clear that CTE increases with increasing test temperature: 5.7×10^{-6} K^{-1} at 500 K, 7.4×10^{-6} K^{-1} at 1000 K, 8.4×10^{-6} K^{-1} at 1500 K and 8.8×10^{-6} K^{-1} at 2000 K. CTE of IrAl is similar to that of Mo (5.1×10^{-6} K^{-1} at RT and 5.8×10^{-6} K^{-1} at 873 K [17]). Therefore, IrAl-based coatings may be suitable for Mo-based alloys from the viewpoint of thermal expansion.

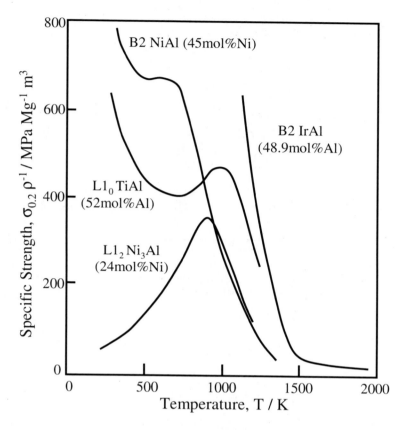

Fig. 12.12. Specific strength of Ir49Al and other intermetallic alloys as a function of temperature. Specific strength was calculated using densities and 0.2% flow stress obtained in this study and from the literature.

12.5 Oxidation Behavior of Ir-rich IrAl

12.5.1 Heating oxidation

Figure 12.15 shows DSC curves of IM-Ir49Al and Ir (shown in Fig. 12.2) in an O_2 environment. No reaction heat was observed for the IrAl alloy. It is suggested that the reaction between IrAl alloy and oxygen is very slow below 1773 K. On the other hand, large endothermic heat was observed for Ir at 1395 K, due to the exothermic IrO_3 formation. Figure 12.16 shows TG curves of IrAl and Ir in an O_2 environment. IrAl alloy did not show significant mass change even though the specimens were powder. On the other hand, for Ir, large mass gain was noticed above 1000 K, followed by significant mass loss above 1395 K. These mass changes are due to the formation of IrO_2 (solid phase) and IrO_3 (gaseous phase). The heat of formation of IrO_2 in IrAl may be very small, which explains the lack of reaction heat observed by DSC below 1370 K.

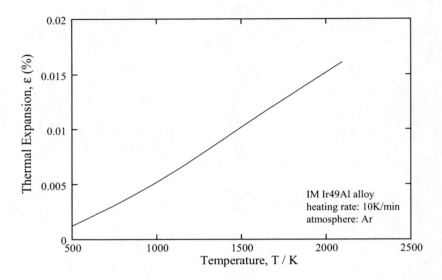

Fig. 12.13. Thermal expansion of IM-Ir49Al alloy.

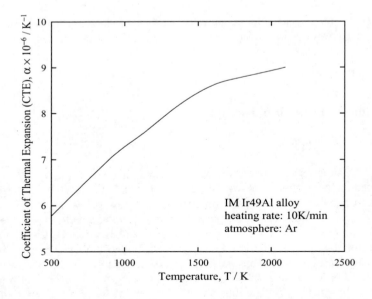

Fig. 12.14. Coefficient of thermal expansion (CTE) of IrAl as a function of temperature.

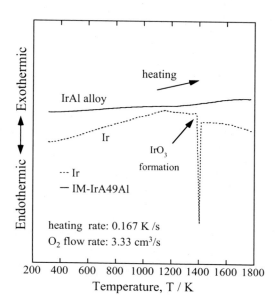

Fig. 12.15. DSC heating curve for IM-Ir49Al in an O_2 environment.

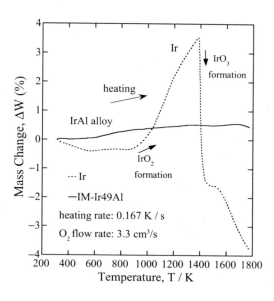

Fig. 12.16. TG heating curve for IM-Ir49Al in an O_2 environment.

It is concluded that formation of iridium oxides, IrO_2 and IrO_3, is suppressed by alloying with Al.

12.5.2 Isothermal oxidation

Figure 12.17 shows TG curves as a function of time for IM-Ir49Al oxidized at 1273, 1473, and 1673 K in an O_2 environment. It should be noted that no significant reaction heat was observed by simultaneous DSC. Little mass change was seen at 1273 K, although mass gain might occur for pure Ir due to IrO_2 formation, as shown in Fig. 12.2. IrO_2 seems to be formed slowly at 1273 K. With increasing test temperature, both materials showed mass loss by oxidation. We compare mass changes of Ir and IrAl after 60 ks, -72 and -360 mg cm^{-2} for Ir and -10 and -115 mg cm^{-2} for IrAl alloy at 1473 and 1673 K, respectively. IrAl alloy has better oxidation resistance than Ir. It is also clear that both the materials exhibit linear oxidation – mass decreased linearly as a function of time above 1473 K. This suggests that the oxidation of both the materials is controlled by a surface or phase boundary process, which involves a steady-state formation of oxide at the metal–oxide interface.

12.5.3 Microstructure after isothermal oxidation

Optical microscopy and XRD revealed that, before oxidation, IM-Ir49Al was composed of primary B2-IrAl and eutectics of B2-IrAl/fcc-Ir. Figure 12.18 shows a cross-sectional micrograph near the surface of IM-Ir49Al after oxidation at 1773 K for 86.4 ks in air.

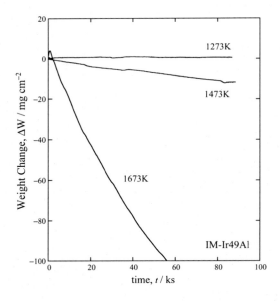

Fig. 12.17. TG curves of IM-Ir49Al at 1273, 1473, and 1673 K in an O_2 environment.

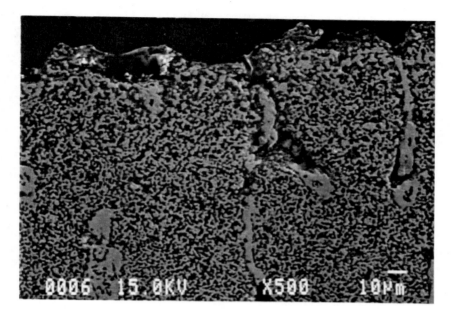

Fig. 12.18. Cross-sectional micrograph near the surface of IM-Ir49Al after oxidation at 1773 K for 86.4 ks in air.

Oxidation of IrAl seems to occur by a grain boundary process, particularly at the eutectics. The existence of voids at grain boundaries is explained by the volatilization of IrO_3.

A cross-sectional micrograph near the substrate of IM-Ir49Al after oxidation at 1773 K for 86.4 ks in air is shown in Fig. 12.19. EDX/WDX revealed that the interface near the substrate was composed of four layers – complicated Al_2O_3 layer containing island-like Ir, continuous Al_2O_3 layer, continuous Ir layer, and substrate. It is clear that the continuous self-healing Al_2O_3 and Ir layers were formed by oxidation, as expected. It is also found that oxygen concentration in the substrate is extremely low, which must be due to the Ir barrier layer. It can be concluded that IrAl alloy is suitable for the oxygen diffusion barrier because of its superior oxidation resistance. However, the complicated Al_2O_3 layer containing island-like Ir is relatively thick and should be improved by alloying. This oxidation was classified to be internal oxidation [18], and a similar oxidized layered structure was reported for (Ru, Ir)Al [19].

12.6 Oxidation Behavior of Al-rich IrAl Alloys

As described above, binary Ir49Al forms a multilayered structure composed of the Ir layer as the ODB and the Al_2O_3 layer as the protective oxide (PO) above the ODB. The problem in the oxidation of binary IM-Ir49Al alloy is the formation of the complex Al_2O_3 layer containing island-like Ir domains and voids. The thickness of the whole

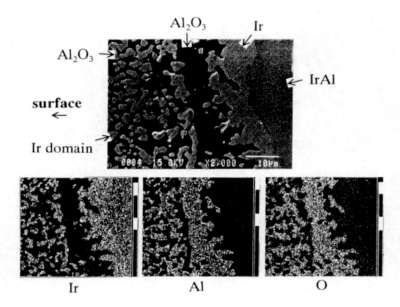

Fig. 12.19. Cross-sectional SEM micrograph near substrate of IM-Ir49Al after oxidation at 1773 K for 86.4 ks in air, and EDS/WDS maps for Ir, Al, and O corresponding to the SEM micrograph. Bright regions of EDS/WDS maps stand for high concentration of each element.

scale is more than 100 μm after oxidation at 1773 K for 86.4 ks – oxidation is not sufficiently suppressed in Ir-rich IrAl with two phases of Ir and IrAl. Al-rich IrAl alloys were then also investigated and discussed in relation with Ir49Al alloys. As shown in Fig. 12.9, the B2 single phase is obtained by Al-enrichment and the RHP method. Ternary additions, Ni, Co, and Fe, were considered to substitute for Ir, which reduces Ir content and stabilizes the B2 phase.

TG curves during isothermal oxidation at 1673 K in an O_2 environment are shown in Fig. 12.20. Weight loss increases with increasing duration in the O_2 environment. The weight change of IrAl is determined by the balance of the reactions $Ir + \frac{3}{2}O_2 = IrO_3$ (gas, mass loss) and $2Al + \frac{3}{2}O_2 = Al_2O_3$ (solid, mass gain). Since Ir is a heavy metal and the atomic weight of an Ir atom is about 12 times larger than that of an O atom, weight loss through IrO_3 formation tends to surpass weight gain through Al_2O_3 formation. It is clear that the weight change is the largest for Ir and that alloying with Al improves the oxidation resistance of Ir. By comparing PM Ir49Al with IM Ir49Al, the oxidation resistance is seen to be slightly improved in the PM Ir49Al alloy because of the removal of the continuous eutectic structure, since the eutectic grain boundary is potentially an easy passage for oxidation from the surface. Although the fabrication processes are different among the three types of binary IrAl alloys, the oxidation resistance of RHP Ir51Al alloy is the worst. This suggests that increasing the Al content not only improves the oxidation resistance in this alloy system, but also influences the oxidation through the formation of the layer. By comparing the four kinds of Ir51Al based alloys, it is easily concluded

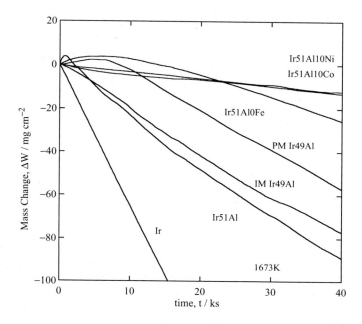

Fig. 12.20. TG curves during isothermal oxidation at 1673 K in an O_2 environment.

that Ni and Co additions significantly improve the oxidation resistance of IrAl alloys, and the effect of Fe addition is moderate.

Figure 12.21 shows cross sections near the interfaces between the oxide layer and the substrate for Ir51Al-based alloys after oxidation. The morphologies of the scales are quite similar for all binary IrAl alloys – Al_2O_3 layer containing Ir islands, continuous Al_2O_3, and continuous Ir layer. It is also clear that continuous, dense Al_2O_3 layers are formed for RHP Ir51Al10Ni and Ir51Al10Co alloys that show the superior oxidation resistance. The growth of Al_2O_3 scales is slow in these two alloys. However, the continuous Ir layer, which is required as the ODB, is not formed in either alloy. Lee and Worrell have reported that Ir–Al alloys containing more than 55mol%Al form a continuous Al_2O_3 layer [5]. Although the test conditions are different from those in Lee and Worrell's experiment, additions of Ni and Co by 10mol% show a similar effect of stabilizing Al_2O_3 without significantly decreasing the melting point. In the case of RHP Ir51Al10Fe, larger scales are formed compared with Ni- and Co-added alloys, while the Ir layer is not formed. It should be noted that oxidation considerably progresses at grain boundaries for RHP Ir51Al10Fe alloy, possibly due to Fe enrichment at the grain boundaries. SEM observations of scale formation and the reduction of the melting point in the Fe-added alloy indicate that Fe addition is not useful, but harmful, in the improvement of oxidation resistance. It could be concluded that the oxidation resistance of IrAl is greatly improved by controlling the scale morphology, which can be achieved by controlling Al and Ni (or Co) concentrations. The effect of Co addition on oxidation will be discussed in the next section.

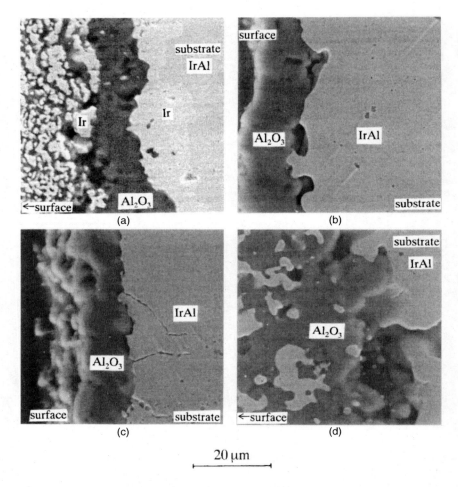

Fig. 12.21. Cross-sectional micrographs near interfaces between oxide scale and substrate for RHP Ir51Al-based alloys after oxidation at 1673 K for 86.4 ks in an O_2 environment: (a) RHP Ir51Al, (b) RHP Ir51Al10Ni, (c) RHP Ir51Al10Co, and (d) RHP Ir51Al10Fe.

12.7 Oxidation Behavior of Co-added IrAl

It was revealed that Ni and Co additions on IrAl are effective in improving oxidation resistance [20,21]. These substitutions are additionally expected to reduce the required amount of the costly and precious Ir. The Ni substitution in Ir sites is also effective for improving the ductility of IrAl [22]. In this section, the details of the oxidation behavior of Co-added IrAl are described. It should be noted that the oxidation behaviors of Co-added and Ni-added IrAl alloys are similar to each other.

12.7.1 Heating oxidation of Co-added IrAl

Figure 12.22 shows (a) TG and (b) DSC curves for binary IrAl and Co-added IrAl in O_2 under the condition of constant heating up to 1863 K. Power samples were used to achieve a clear reaction. Both the alloys show similar tendencies in TG curves – weight gain is very slight up to about 1000 K, weight increases with further increase of temperature up to about 1650 K, and then a small weight loss occurs above 1650 K. The weight gain might be caused by the formation of Al_2O_3 and IrO_2. The weight loss above 1650 K may be caused by the formation of a gaseous oxide, IrO_3. The weight gain is larger for the Co-added IrAl than for the binary IrAl. This is probably because the formation of gaseous iridium oxide, IrO_3, was suppressed by Co addition in comparison with the binary IrAl. Figure 12.22(b) shows simultaneous DSC curves. It was observed for both alloys that the endothermic reaction heat is at around 1150 K and the exothermic heat is at around 1460 K. Additionally, reaction peaks with endothermic heat were seen at around 1595 K for binary IrAl and at 1620 K for the Co-added IrAl. The peak temperature of 1595 K is equal to that of IrO_3 formation in Ir49Al.

Figure 12.23 shows XRD profiles of binary IrAl and Co-added IrAl after heating oxidation. Although the scan was carried out up to 2θ of 158°, the horizontal axis shown in the figure is limited to 2θ of 50° for clear peak separation. XRD revealed that Ir, IrO_2, and Al_2O_3 are formed for both the alloys, and that a complex oxide of $CoAl_2O_4$ was additionally formed for Co-added IrAl but no CoO was detected. It should be noted that no peaks from the B2 phase were recognized. Therefore, all IrAl alloys were oxidized and exhausted under these experimental conditions due to the particle size

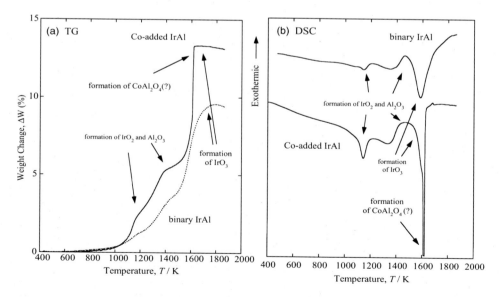

Fig. 12.22. Simultaneous (a) TG and (b) DSC curves of binary IrAl (RHP Ir51Al) and Co-added IrAl (RHP Ir51Al10Co) heated to 1863 K in O_2 environment [17].

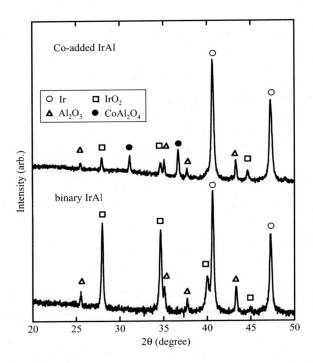

Fig. 12.23. XRD profiles of binary IrAl and Co-added IrAl after heating oxidation.

being smaller than 60 μm (powder specimens used here). Taking TG, DSC, and XRD results into account, the first and second reactions seen in the DSC curves in Fig. 12.22 may correspond to IrO_2, and Al_2O_3 formations, and the third one may correspond to IrO_3 formation for the binary IrAl. Although the reaction temperature of 1620 K seen for Co-added IrAl is not yet clearly understood, it may be related to the formation of $CoAl_2O_4$ detected by XRD.

12.7.2 Isothermal oxidation

TG curves obtained by isothermal oxidations at 1273, 1473, and 1673 K are shown in Fig. 12.24 for Co-added IrAl. It should be noted that no sharp reaction heat was observed in simultaneous DSC measurements and hence the DSC curves are not shown here.

Binary IrAl exhibited linear weight loss above 1273 K, and the weight loss increased with the increasing test temperature due to the formation of gaseous IrO_3. On the other hand, the TG curves for Co-added IrAl were not linear but exhibit small weight gains at 1473 and 1673 K. The weight change as a function of time was complicated for Co-added IrAl. This may be a result of the complicated oxidation mechanism – the total weight change is determined by the sum of the amounts of solid oxides of Al_2O_3 and $CoAl_2O_4$ (weight gain) and gaseous oxides IrO_3 (weight loss). By comparing Fig. 12.24 for RHP Ir51Al10Co and Fig. 12.7 for IM-Ir49Al, it becomes clear that the oxidation rate is dramatically lowered by Co addition, as is particularly easily seen in TG curves obtained for the samples oxidized at 1673 K.

Figure 12.25 shows SEM cross-sectional micrographs of Co-added IrAl oxidized at (a) 1473 K and (b) 1673 K for 86.6 ks. The right side of the figures is the surface side.

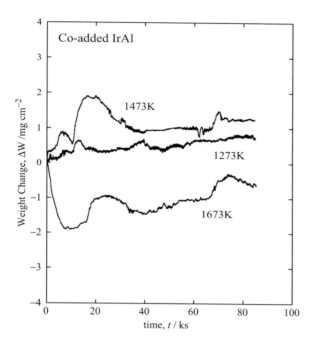

Fig. 12.24. TG curves of Co-added IrAl (RHP Ir51Al10Co) by the isothermal oxidations at 1273, 1473, and 1673 K in an O_2 environment.

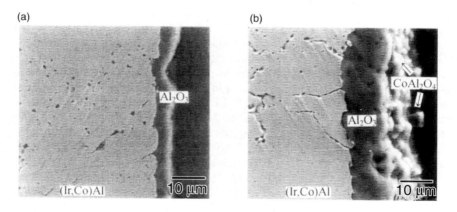

Fig. 12.25. Cross-sectional micrographs of Co-added IrAl (RHP Ir51Al10Co) oxidized at (a) 1473 K and (b) 1673 K for 86.6 ks.

In both the figures, continuous thin Al_2O_3 layers are seen at the surface. The thickness of continuous Al_2O_3 layer is about 3 μm in (a) and 12 μm in (b). The microstructural change is achieved by Co addition. The number of voids is small, contrary to binary IrAl. The low oxidation rate of Co-added IrAl, seen as the TG curves in Fig. 12.24, could be explained by the formation of the continuous thin Al_2O_3. Small particle-like products at the surface must be $CoAl_2O_4$, which was detected by XRD. It should be noted that a smart structure composed of an Ir ODB and Al_2O_3 PO was not clearly seen in the case of 10mol%Co-added IrAl, even though the Ir phase was detected by XRD after heating oxidation up to 1873 K. The oxidation resistance of 10mol%Co-addded IrAl is attributed only to the continuous Al_2O_3 scales. There must exist a moderate Co content between 0 and 10mol% that yields both PO and ODB with excellent oxidation resistance when the Al content is constant at 51mol%.

In order to clarify the formation of the continuous Ir layer in Co-added IrAl, the oxidation behavior of pseudobinary IrAl–CoAl alloys with 50mol%Al has been investigated. CoAl is practically used as an oxidation-resistant coating for Co-based superalloys, particularly used as a part of CoCrAlY. The suitable conditions for the formation of a thin continuous Ir layer has not yet been found, however, a very thin Ir layer was partially formed when Ir-50mol%Al-15mol% Co alloy was oxidized at 1673 K. The formation of a continuous Al_2O_3 layer as well as an Ir layer seems to depend strongly on microstructure, temperature, atmosphere, and chemical composition. It is concluded that IrAl-based multicomponent alloys are highly promising for the new advanced in-situ multilayered, multifunctional smart oxidation-resistant coatings.

12.8 Summary

In order to develop a new, smart oxidation-resistant coating for use at ultrahigh temperature, the fabrication process, phase constitution and stability, mechanical properties, thermal expansion, and oxidation behavior of IrAl alloys were described. The effects of ternary additions of Ni, Co, and Fe on oxidation resistance were also studied. By ingot metallurgy, the eutectic Ir phase is easily introduced, and it is difficult to remove by thermomechanical treatments. The existence of Ir reduces the lower-temperature ductility and oxidation resistance. On the other hand, a powder metallurgical method, such as reactive hot pressing, is promising for fabricating single-phase IrAl. The compositional defect of IrAl is estimated to be the antistructural (antisite) type on the Ir-rich side and the structural (constitutional vacancy) type on the Al-rich side. The composition dependence of hardness is not related to simply the change in lattice parameter and the predicted defect type, i.e., vacancy or substitutional defect. Ir-rich IrAl itself is not brittle in indentation tests. Both the strength determined by temperature normalized by the melting temperature and the specific strength determined from the density are superior to those of NiAl. The coefficient of thermal expansion (CTE) is $5.5–8.8 \times 10^{-6}$ K^{-1} at 500–2000 K; the CTE is similar to that of Mo. Regarding the oxidation behavior, binary IrAl alloys form Ir and Al_2O_3 layers upon oxidation. The Ir layer formed by oxidation must act as the oxygen diffusion barrier (ODB). The oxidation resistance of IrAl alloys is greatly improved by the addition of Ni or Co, although the ODB Ir layer is difficult to be formed when the amount of added ternary elements increases. However, the ternary or multicomponent IrAl alloys with a proper amount of Ni or Co are promising as ultrahigh-temperature in-situ multifunctional coating material in which the functional multilayered structure is formed by oxidation, making the layered structure self-healing.

References

[1] J.M. Criscione, R.A. Mercuri, E.P. Schram, A.W. Smith and H.F. Volk, AFM-TDR-64-174 Part 2, January (1965).

[2] U.R. Evans, The Corrosion and Oxidation of Metals, Edward Arnold Ltd. London (1960), 13.

[3] T.B. Massalski, Ed., Binary Alloy Phase Diagrams (2nd Ed.) Vol. 1 ASM International: OH (1990), 163.

[4] H. Hosoda, T. Kingetsu and S. Hanada, The Third Pacific Rim International Conference on Advanced Materials and Processing (PRICM-3), M.A. Imam, R. DeNale, S. Hanada, Z. Zhong and D.N. Lee, Eds., TMS PA (1998) 2379.

[5] N. Lee and W.L. Worrell, Oxidation of Metals **32** (1989) 357.

[6] P.J. Hill, L.A. Cornish and M.J. Witcomb, J. Alloys and Comp. **280** (1998) 240.

[7] H. Hosoda, T. Takahashi, M. Takehara, T. Kingetsu and H. Masumoto, Mat. Trans. JIM **38** (1997) 871.

[8] H. Hosoda, S. Watanabe and S. Hanada, High-Temperature Ordered Intermetallic Alloys VIII, MRS Symp. Proc. **552** (1999) KK.8.33.1–7.

[9] H. Hosoda, S. Miyazaki and S. Hanada, Intermetallics **8** (2000) 1081–1090.

[10] H. Hosoda, T. Shinoda, T. Suzuki and Y. Mishima, J. Jpn. Inst. Met. **58** (1994) 483.

[11] H. Hosoda, K. Inoue and Y. Mishima, High-Temperature Ordered Intermetallic Alloys VI, J.A. Horton, I. Baker, S. Hanada, R.D. Noebe and D.S. Schwarts, Eds., MRS Symp. Proc. **364** (1995) 437.

[12] A.J. Bradley and A. Taylor, Proc. Roy. Soc. A **159** (1937) 56.

[13] J.H. Westbrook, J. Elect. Soc. **103** (1956) 54.

[14] F.C. Holden, R.W. Douglas and R.I. Jaffee, Symp. Newer Metals, 3rd Pacific Area Meeting Papers ASTM Special Technical Publication PA **272** (1959) 68.

[15] R.T. Pascoe and C.W.A. Newey, Met. Sci. J. **2** (1968) 138.

[16] Y. Mabuchi, H. Hosoda, Y. Tan, Y. Mishima and T. Suzuki, Jpn. Inst. Met. **62** (1998) 912.

[17] Kinzoku Data Book (Second Edition), Jpn. Inst. Met., Maruzen Tokyo (1984) 12.

[18] T.C. Chou, J. Mater. Res. **5** (1996) 378.

[19] P.J. Hill, I.M. Wolff, L.A. Cornish and M.J. Witcomb, Iridium, E.K. Ohriner, R.D. Lanam, P. Panfilov and H. Harada, Eds., TMS (2000) 279.

[20] H. Hosoda, S. Miyazaki, S. Watanabe and S. Hanada, Iridium, E.K. Ohriner, R.D. Lanam, P. Panfilov and H. Harada, Eds., TMS (2000) 271.

[21] H. Hosoda and K. Wakashima, Mat. Sci. Eng. A **352** (2003) 16.

[22] T. Ono, A. Chiba, X.G. Li and S. Takahashi, Mat. Trans. JIM **38** (1997) 171.

Index

Γ–L direction, 268, 270, 273
θ–phase charge-transfer salts, 226–231
   charge order, 231–240
      ^{13}C-NMR line shape, angular dependence, 237, 239, 240
      ^{13}C-NMR spectra, θ-(BEDT-TTF)$_2$RbZn(SCN)$_4$, 234, 235
      correlated insulators, mechanisms, 231, 233
      Knight shift/second moment, 239, 240
      optical conductivity spectrum, 234, 235, 237
      Raman spectrum, temperature dependence, 235, 239
      X-ray oscillation photograph, θ-(BEDT-TTF)$_2$RbZn(SCN)$_4$, 235, 239
   electrical resistivity, temperature dependence,
      hydrostatic pressure, 227, 228
      uniaxial strain, 230, 231
   finite temperature, 250–255
      free energies, charge-ordered patterns, 251, 252
      phase diagrams, 251–254
   fractional charge
      stripe patterns, charge order states, 247, 248
   history, 226, 227
   intermolecular Coulomb repulsion, 240–245
      ground state energies, charge order states, 242, 243
      phase diagrams, 242–246
   issues, 255–257
   molecular arrangement, 225
   overlap integrals, BEDT-TTF molecules, 229, 230
   spin polarization, 248–250
   static magnetic susceptibility, 231, 232
   temperature dependence, lattice constants, 231, 232
   tight-binding band structure, 229
   transfer integrals, 228, 229
   universal phase diagram, 227, 228
Activated carbons, 351
Advanced oxidation-resistant coatings, 420
   selected materials, oxygen permeability, 420
   thermal barrier coating (TBC), 420
Advanced smart coatings, 419–445
   multifunctional and multilayered coating, IrAl based, 423–424
      estimated Ir–Al–O phase diagram, 424
      example, 425
      Ir–Al binary phase diagram, 424
   oxidation resistance, 419
   oxygen diffusion barrier (ODB), 419, 421, 437, 444
   ultrahigh-temperature coating, 419
Ag adatom islands, 276, 277, 290
Ag nanoislands, 277
Ag/GaAs(110), 281, 282, 284
Ag/Si(111), 287
   two-step growth process, 287
   flat films of unique height, 287
Age-hardening phenomenon, 315, 316
   aging time, stages, 318
   curves, 318
Aluminum alloys, nanostructure control, 315–344
   2017 aluminum alloy, 317
   Al-Cu alloys, 316, 324
      Ag-added alloy, 324
      age-hardening curves, 324
   Al–Cu–Mg alloys, 317, 319
   Al–Mg–Cu(-Ag) alloys, 331
      age-hardening curves, 331
   Al–Mg–Si alloys, 319, 334, 337, 339, 340
      properties, 334
      quenched-in vacancies, 334
   Al–Zn–Mg alloys, 333
      direct quenching (DQ), 334
      elongated clusters, 333

Aluminum alloys, nanostructure control *(continued)*
  Al-Zn-Mg alloys *(continued)*
    precipitation sequence, 333
    transformation of small GP-1 zones, 333
    water quenching (WQ), 334
  ductility, 315
  high-strength, 315
  mechanical strength, 315
Annealing, 270, 278, 282, 283, 287, 288

Band dispersion, 268, 269, 270, 273, 277, 295
BEDT-TTF (bis(ethylenedithio)tetrathiafulvalene), organic charge-transfer salt
  crystal structures, 227
  giant nonlinear DC conductance, 225, 226
  molecular structure, 225
Bessel function, 276
Bismuth layer structured dielectrics, size-effect-free characteristics, 107–131
  bismuth layer structured ferroelectric dielectrics (BLD) introduction, 107
    bismuth-oxide layer, 107
    BLDs family layer pervoskite, 108, 109
    crystal structure, illustration, 107, 108
    pseudoperovskite-layer, 107
    simple perovskite, BLD crystal structure, 107
    $SrBi_2Ta_2O_9$ films, ferroelectric behaviour, 108
  BLD electrical properties, orientation dependence, 108
    $Bi_4Ta_3O_{12}$ films, XRD pole figure plots, 110
    $Bi_4Ta_3O_{12}$ films, XRD reciprocal space mappings, 110
    bismuth layer structured dielectrics (BLD), schematic drawing, 114, 115
    BIT single crystal, 113
    electric field function, leakage current density, 113, 114
    expitaxial BLD films and $SrTiO_3$ substrates, crystal growth relationship, 110, 111
    expitaxial films, 108
    oriented $Bi_4Ta_3O_{12}$ films, AFM images, 111, 113
    oriented $SrBi_2Ta_2O_9$ films, TEM images, 110, 112
    P–E hysteresis loops, 113, 115
    relative dielectric constant, 111, 113
    variant BLD film drawings, 110, 111
  BLD films, unique characteristics, 116
  electrode modification, interface property improvement, 125–131
    $CaBi_4Ti_4O_{15}$ films, rocking curve, 129
    $CaBi_4Ti_4O_{15}$ films, thickness dependence, 129, 130
    chemical vapor deposition (CVD), 126
    $LaNiO_3$ bottom electrode, low crystallinity, 125
    $LaNiO_3$ films, crystal quality, 126
    $LaNiO_3/Pt$ substrates, frequency characteristics, 129, 130
    rocking curve FWHM, $SrRuO_3$ 200_c and $LaNiO_3$ 200_c, 126
    XRD patterns and pole figures, 58-nm-thick $CaBi_4Ti_4O_{15}$ films, 128, 129
    XRD patterns, $SrRuO_3$ and $LaNiO_3$ substrates, 128
  epitaxial (001)-oriented SBT film, 116
    simple perovskite and BLD, crystal structure comparison, 116, 117
  single-axis oriented films expansion, 121–125
    30-nm-thick $SrBi_4Ti_4O_{15}$ films, capacitance and dielectric loss, 124
    a–b plane crystal growth, 124
    AFM images, 30-nm-thick $SrBi_4Ti_4O_{15}$ films, 123
    c-axis-oriented BLD films, capacitance stability, 125
    cross-sectional TEM images, 60-nm-thick $SrBi_4Ti_4O_{15}$ films, 123
    diffraction spots, 122
    low lattice mismatch, 122
    non-*c*-axis-oriented $SrBi_4Ti_4O_{15}$ phases, 122
    polycrystalline ceramic substrates, 121
    psi (ψ) scans integrated patterns, 123
    $SrBi_4Ti_4O_{15}$ films, relative dielectric constant and dielectric loss, 125, 126
    $SrBi_4Ti_4O_{15}$ films, X-ray reciprocal space maps, 122
    XRD patterns, 30-nm-thick $SrBi_4Ti_4O_{15}$ films, 121
  $SrBi_2Ta_2O_9$ films, 116–120
    AFM images, 120
    capacitance and applied electric field (C–E) characteristics, 119
    cross-sectional TEM image, 116, 118
    leakage current density, 120
    relative dielectric constant, 119

Bismuth layer structured dielectrics, size-effect-free characteristics *(continued)*
  SrBi$_2$Ta$_2$O$_9$ films *(continued)*
    rocking curve FWHM, film thickness dependence, 116, 118
    XRD patterns, 116, 117
Block copolymer thin films, nanocylinder array structures, 171–223
  blend-polymer-induced orientation, 202, 203
  block copolymers mixture, phase structure in, 187, 188
    TEM images (PS-*b*-PI)α, 188
  common AB block copolymer, experimental studies, 179, 180
  external field, 206–208
    non-microporous films, 207
    POTI domains, 208
    PS-*b*-PMMA, SANS pattern, 208
  liquid-crystalline blocks, 173
    atomic force microscopy (AFM), 174, 192, 193, 197, 200
    classic blocks (C-block), 173
    group transfer polymer (GTR), 175
    PEO, 176, 176, 197
    phase morphology, synthesis, 174
    PMA, 175
    poly(ethylethylene) PEE, 176
    Polymethylmethacrylate (PMMA), 175, 176, 179, 197, 200, 202
    WAXS, 174
  microdomains, self-organization and phase behavior, 173–191
  nanocylindrical-structured, templates, 210–215
    fabrication process and AFM height images, Cr dot arrays, 212
    PEO$_m$*b*-PMA(Az)$_n$ thin film, TEM image, 213, 214, 216
    PEO$_m$*b*-PMA(Az)$_n$, LC block copolymer, 212, 213
    PMA(Az) domains, 215
    PS-based block copolymer, 216
  nanostructured thin films, 171
  nanotemplating, 171, 196, 199, 200, 210, 215
  perpendicular cylinder arrays, template process, 208
    tri-*n*-octylphosphine oxide (TOPO), 210
  phase-segregated nanostructures (block copolymer thin films), 191–194
    poly(styrene-*block*-ethylene oxide) (PS-*b*-PEO), 177
    PS-*b*-PI block copolymer, spin-coated, AFM images, 192
    PS-*b*-PI diblock copolymer, ultrasections, TEM images, 192
    scanning electron microscopy (SEM), 193, 200, 202
    surface morphology, film thickness, function, AFM study, 193, 194
    tapping-mode AFM phase images, SBS films(silicon substrates), 194
    X-ray photoelectron spectroscopy (XPS), 191
  phase-segregated nanostructures (thin films), practical uses, 194–210
    cylinder array structures, orientation, 195, 196
    microphase separation, block copolymer, 194
    perpendicular nanocylinder array structure, 197
    PI oligothiophene-modified side chains (POTI), 198
    poly(ferrocenylsilane) (PFS), 198
    poly(styrene-*block*-butadiene) (PS-*b*-PB), 199
    polyactide (PLA), 198
    polydimethylacryl (PDMA), 198
    polyisoprene (PI), 198
    propylene glycol ether acetate (PGMEA), 198
    reactive ion etching (RIE), 198
  phase transition, 171, 177, 179, 188, 193
  polymer-structure-dependent phase structure, 186, 187
  polymer-surface interaction control, 200
    AFM topographic, phase images (PS-*b*-PEO), 201
    styrene (S) and methylmethacrylate (MMA), copolymers, 200
  restricted microstructures, 203–206
    extreme ultraviolet (EUV), 204
    poly(styrene-*block*-(ethylene-*alt*-propylene) (PS-*b*-PEP), 204
    poly(styrene-*block-tert*-butyl methacrylate) (PS-*b*-P*t*B), 205
  self-organized nanostructure, 171, 177, 192, 194, 206

Block copolymer thin films, nanocylinder array structures *(continued)*
self-organized structure, phase transition, 177
    5-nm-thick triblock film, TEM images, 185
    vector 4111, TEM images, 184
    crew-cut aggregates, multiple morphologies, 181
    cylindrical micelles, chemical structure and TEM image, 184
    diblock copolymer films, TEM images, 186
    diblock copolymers (PS-*b*-PI), TEM images, 183
    hexagonal close packing (*hcp*), 182
    order–disorder transition (ODT), 183
    order–order transition (OOT), 183
    poly (styrene-*block*-acrylic acid) (PS-*b*-PAA), 177
    poly (styrene-*block*-methylmethaacrylate (PS-*b*-PMMA), 177, 202, 205, 206, 211
    poly (styrene-*block*-vinylpyridine) (PS-*b*-PVP), 177, 185, 211
    poly(acrylic acid) (PAA), 179
    poly(ethylene oxide) (PEO), 179, 197, 200, 212
    polybutadiene (PB), 179
    polyisoprene (PI), 179
    polyvinylpyridine (PVP), 179
    PS (A-block), 177, 178, 182
    tetrahydrofuran (THF), 185
small-angle X-ray scattering (SAXS), 172, 174, 175, 177, 183, 186, 191, 197
solvent evaporation, 199, 200
    2-(4-hydroxybenezeazo)benzoic acid (HABA), 199
    field-emission (FE), 220
    tapping-mode AFM phase images, SBS triblock copolymer films, 199
supramolecular self-assembly, phase-structural control, 188–191
    lamellar-within-cylindrical/cylindrical-within-lamellar structures, schematics and TEM images, 190
    salt-added/salt-free sample, schematics and TEM images, 189
synthesis, 172, 173
    anionic polymerization, 172, 185
    atom transfer radical polymerization (ATRP), 173, 176

liquid crystalline (LC) block polymers, 173, 174, 212
poly(styrene-*block*-isoprene) (PS-*b*-P1), 172, 177, 178, 181, 189
polystyrene (PS), 173, 176, 181, 183, 187, 197, 198, 200, 210
radical polymerization, 172
transmission electron microscopy (TEM), 172, 174–177, 181, 191, 192, 197, 212
Bohr-Sommerfeld quantized condition, 269
Boltzmann's constant, 387
Bragg condition/law, 137, 138, 141
Brillouin zone, 137, 138
Bulk intermetallic semiconductors, high-temperature applications, 390–405
    half-Heusler compounds, 397–405
        ABX, unit cell, 398
        band gap, 399
        ternary intermetallic compound ABX, 397, 398
    electron crystal materials, phonon glass and, 390–391
    Rietveld analysis, data summary, 394
    Seebeck coefficient and electrical resistivity, temperature dependences, 394, 395
    skutterudite compounds, 391–397
        binary antimonide skutterudites, 392
        filled skutterudite, unit cell, 392
        IrSb$_3$-based, effect of La filling, 393
    thermal conductivity, 397
    TiNiSn-based half-Heusler compounds, ZT improvement in, 399
        dimensionless figure of merit ZT, temperature dependence, 404
        electrical resistivity, temperature dependence, 401
        nominal composition and fabrication, 400
        Seebeck coefficient, temperature dependence, 402
        thermal conductivity, temperature dependence, 403

Charge density wave state (CDW), 234
Ceramics superplasticity, grain boundary dynamics, 297–310
    future prospects, 310
    physical characteristics, 304–305
        Burger's vector, 304
        grain boundary, atomic structures, 308
        grain compatibility, 304
    superplasticity, definition, 297

Cholesteric liquid crystals (CLC)
    reflection spectra, oblique incidence, 142
    selective reflection, 141–144
        dispersion relation, 143, 144
        emission spectra, dye embedded in, 143, 144
        features, 141
    CLCs, lasing from, 148–159
        anisotropic defect mode, 153–158
            cell structure, 153, 154
            CLC/NLC/CLC, reflectance spectrum, 154, 155
            electrotuning, lasing emission, 158, 159
            far-field pattern, lasing, 157
            polarization characteristics, lasing emissions, 157, 158
            reflection/lasing emission spectra, cells, 155–157
            single CLC, reflectance spectrum, 154, 155
        DFB mode, 148–153
            commercial DCM, structure, 149
            DCM dyes, polar plots of absorption peak, 151, 152
            DCM-doped CLC cells, emission spectra, 149, 150
            DCM-doped CLC cells, reflectance/fluorescence/lasing spectra, 149–153
        lasing threshold, lowering efforts, 158–160
Clusters, 320, 322
    age-hardenable aluminum alloys, 321
    co-clusters, 323, 337
    computer simulation, 328
        Cu/Mg/vacancy complex, 328
        effective nucleation sites, 332
        Mg/vacancy cluster, 328
        Monte Carlo simulation images, 330
        Monte Carlo steps (MCs), 328
        short-range order parameters, 330
    DSC curves, 323
    electrical resistivity changes, aging, 322
    HRTEM images, 325, 326
    micro Vickers hardness, changes, 325
    microalloying elements, cluster formation, 324
    nanoclusters, 237
    phase decomposition, early stage of, 320
        electrical resistivity measurement, 320
        intermediate phase, 321
        precipitation-strengthening, 320
        stable phase, 321
        X-ray diffuse scattering, 320
    three-dimensional atom probe (3DAP), 323, 327, 339
    two-step aging behavior, 332, 333, 337, 339, 340
        mechanism, 340
        pre-aging, 332
        precipitation behavior, 332
        single-aged condition, 339
        single-layer zone, 339
        temperature, 339
        treatment procedure, 338
Confinement potential, 269, 272, 273, 275
    hard-wall confinement potential, 276
Cs salt, energy diagram, 257

Electron quantum confinement, 265, 279, 291
Electron spilling
    interface, 281
    outer surface, 281
Energy band gap (EBG), 137, 138

Fermi-Dirac function, 407
Fermi-Dirac statistics, 387
Fermi energy, 387
Fermi level, 265, 268, 270, 277, 280, 286, 287, 290, 292–295, 386
Fermi wave vectors, 265
Fermi wavelength, 265, 266, 286, 290
Fermi's Golden Rule, 145
Ferroelectric and high permittivity thin films, size effect, 99–101
    actual applications, polycrystalline research, 100
        high /low(dead layer) dielectric constant layer, 100, 119
        less-size-effect approach, 101
    dielectric and ferroelectric properties, size effect research, 100
        $BaTiO_3$ thin film, recent studies, 100
        expitaxial film research, 100
        expitaxial $PbTiO3$ thin films, observations, 100
    dielectric constant size effect, 99
    dynamic random access memories (DRAM), 99
    ferroelectric random access memories (FeRAM), 100
    field effect transistors (FET), 99

Ferroelectric materials, 72–76
   disordered perovskites, 72
      ABO_3 perovskites, 72
      Arrhenius law, 74
      complex dielectric function, 73
      cubic perovskite lattice, 73
      D–E loop, 75
      Debye approximation, 73
      dipolar interactions, Hamiltonian, 74
      dipole nanoregions, 75
      Kramers-Kronig dispersion relations, 74
      polar nanodomains, 75
      polar nanoregions, 75
      polar phonon-mediated interactions, 75
      quantum paraelectric, $SrTiO_3$, 72
      relaxor state, 75
   dielectric properties, 76
      ε' and ε'', frequency dependences, 81
      ferroelectric triglycine sulphate (TGS), 76
      ferroelectrics and relaxors, comparison of properties, 77
      $Pb_{1/3}MgNb_{2/3}O_3$ (PMN), 76
      polar cluster, 78
      STO 18 and SCT, dielectric loss vs. temperature, 80
   dielectric constants, 77
      ferroelectric $SrTi^{18}O_3$ (STO 18), 77
      SCT, dielectric $\varepsilon'(T)$ response, 78
      $Sr_{0.993}Ca_{0.007}TiO_3$ (SCT), 77
      T dependence, Arrhenius plot, 79
      temperature dependence, 78
Film–substrate interface, 281
Flat-top nanofilms, 267
Free-electron-like dispersion, 269, 275
Free-standing film surfaces, 280
Function cultivation, transparent oxides, 3–56
   Clarke number, 4
   oxide ceramics, uses, 4
   oxides purification, techniques, 4
   p-type transparent conductive oxides (TCOs), discovery impact, 5, 7
   research background, 4, 5
   research concept and strategies, 5–7
      artificial periodic nanostructures, 6
      crystal structures, 6
      natural nanaostructures, 6
      oxides, characteristic features, 5
      transparent oxides, 5

Ginzburg–Laudau expansion, 248, 250
GP zones, 319, 320, 322
   crystallography, 321
   Cu-rich layers, 319
   GP(1) zone, 328
      formation, 328
      heterogeneous nucleation, 328
   high-angle annular dark-field STEM (HAADF-STEM), 319
   morphology, 321
   solute clusters, 319
Grain refinement and grain growth suppression, 305–306
   diffusion coefficient, 307
   diffusion enhancement, 306–308
   high-strain-rate superplasticity, 306
   low-temperature superplasticity, 306
   Ostwald ripening, 306
   semi-empirical equation, 305
   shear thickening behaviour, 308
   SiAlON, 308
   silicon nitride (Si_3N_4), 308
   small grain size, 305

Hartree–Fock approximation, 242, 250
Hartree–Fock theory, 241
Highest occupied molecular orbitals (HOMO), 229, 240
Hubbard band gap, 233
Hubbard model, 231, 234, 241, 249

Ion scattering, 270
Interface-induced Friedel oscillations, 281
Ir oxidation resistance, 421
   differential scanning calorimetry (DSC), 422
   thermogravimetry–differential scanning calorimetry (TG–DSC), 421, 422
   thermogravity (TG), 422
IrAl alloys, physical properties, 425–433
   compressive mechanical properties, 430–432
      stress-strain (SS) curves, 430, 431
   lattice parameters, B2 IrAl, 428–429
   microstructures, 425–428
      alloy fabrication, heat treatment conditions and, 427
      alloys, chemical compositions, 427
      hot pressing (HP), 427
      ICP chemical analysis, 425
      IM-Ir49Al and Pm-Ir49Al alloys, XRD profiles, 428
      ingot metallurgy (IM), 425

IrAl alloys, physical properties *(continued)*
  microstructures *(continued)*
    powder metallurgy (PM), 425
    reactive hot pressing (RHP), 425, 427
    slow cooling (SC), 427
  thermal expansion, 432, 433
    specific strength, function of temperature, 433
  Vickers hardness, 429
Ir-rich IrAl, oxidation behavior, 433–437
Al-rich IrAl alloys, oxidation behavior, 437–440
  cross-sectional SEM micrographs, 440
  isothermal oxidation, TG curves, 438, 439
  protective oxide (PO), 437
Co-added IrAl, oxidation behavior, 440–444
  Co-added IrAl (RHP Ir51Al10Co), cross-sectional micrographs, 443
  Co-added IrAl (RHP Ir51Al10Co), TG curves, 442, 443
  heating oxidation, 441
  isothermal oxidation, 442
  XRD profiles, 442
heating oxidation, 433
  coefficient of thermal expansion (CTE), function of temperature, 434, 444
  IM Ir49Al alloy, thermal expansion, 434
  IM-Ir49Al, DSC heating curve, 435
  IM-Ir49Al, TG heating curve, 435
  isothermal oxidation, 436
  microstructure, 436
  TG curve, O_2 environment, 436
Ising model, triangular, 245, 246, 254

Lattice defects, oxides, 62–95
  compositional disorder, 62
  dipolar interaction, 63
Liquid crystal (LC) nanostructures, lasing
  defect mode lasing, 147, 148
    defect mode generation, configuration kinds, 147
    defect structures, kinds, 147, 148
    distributed feedback (DFB) CLC laser, 145–147
    cavity structure, 145, 146
    LC dye laser principle, 144, 145
      density of state (DOS), 145, 146
      simulated transmittance spectrum, 145, 146
Liquid crystal nanostructures, photonic devices, 137–167

Liquid crystals (LCs),
  advantages, 167
  problems, practical applications, 167
  applications, 167
Lithium ion conductivity, oxides, 76–79
  β-eucryptite, 80
    cation distribution, 83
    idealized structure, 84
    $Li_2Ti_3O_7$ ions migration, 86
    $Li_2Ti_3O_7$ ramsdellite structure, 85
    lithium conduction channel sites, 82
    lithium titanate ($Li_2Ti_3O_7$), 81–83
    SiO_4 and AlO_4 tetrahedra layers, 80
    TiO_6 octahedra, 81
  one and quasi-one dimensional lithium ion conductor, 80–83
  two-dimensional lithium ion conductor, 84–88
    β-alumina, ionic size comparison, 86
    A-O-Al spacer units, 84
    Li-β-alumina, 84, 85
    MgI_2O_4 spinel, structure, 84
Low-energy electron microscopy (LEEM), 291

Magic size, 278
Magnetic materials (oxides spin crossover), 63–72
  anisotropic environment, 65
  d^4 to d^8, ground state regions, 66
  D_{4h} elongation, stabilizing effect, 64
  excited Co ions proposed arrangement, 68, 70
    intermediate-spin Co ions, 68, 70
    low-spin Co^{3+}, 68
  high-spin-state Co ions, 68, 69
    experimental data, calculated curves, 69
    fraction, 68
    heat capacity, 68
    magnetic susceptibility, 68
    net excitation energy, 68
  K_2NiF_4 structure, 65
    B′–O, B″–O bonds, 65
    magnetic measurements, 64, 65
    magnetic superexchange couplings, 67
    MO_6 octahedron, 63
    negative cooperative effect, concept, 68
    NiO_6 octahedron, distortion comparisons, 67
  perovskite-type oxides, studies, 63
  stoichiometric $Pr_{0.5}Ca_{0.5}CoO_3$ 68
    Co^{3+} spin crossover, 68
    DC magnetization, 71

Magnetic materials (oxides spin crossover) *(continued)*
  stoichiometric $Pr_{0.5}Ca_{0.5}CoO_3$ *(continued)*
    heat capacities, 71
    lattice constants, unit cell volume, 71
    resistivities, 71
  Tanabe-Sugano diagrams, 63
Man-designed nanostructures, 275
Matrix density, 363
Maxwell equation, 142
Mesoporous silica (MPS), 353, 357–364
Metal-insulator transition, 227
Microalloying, 315
  effects, 316
  elements, 328
Motion and topological evolution, grains, 298, 301
  3D six-grain cluster geometry, 302
  cluster, cage break-up, 303
  conservative motion, grain boundary, 300
  grain boundary sliding, 303
  grain disappearance (T2 process) 300
  grain growth, snapshot of microstructural evolution, 300
  grain rearrangement, 303
  grain switching (T1 process), 300, 301, 302
  Lifshitz sliding, 299
  non-conservative motion, grain boundary, 299
  shear deformation, regular grain array, 299, 303
  soap-froth model, 299
  strain enhanced/dynamic grain growth, 303
  superplasticity topological models, Ashby and Verral, 299
  symmetrical diffusion path, Sphingarn and Nix, 302
Mott insulator, 233, 234

N ion bombardment, 278
Nanocluster assist processing (NACP), 344
Nanoclusters, 315
  Cu-atom clusters, 316
  Cu/Mg/vacancy clusters, 316
  ductility, 342
  elongation, 343
  precipitate-free-zone (PFZ), 342, 343
  proof stress, 343
  quenched-in vacancies, 342
Nanoislands
  Ag islands, 276
    flat-top, 286

Cu islands, 277
  even-layer-height, 286
  highest occupied state (HOS), 286
  lateral size control, 290
    quantum effect, 290
  Pb islands, 284
    flat-top, 285
  odd-layer-height, 286
Nanoporous materials, historical background and development, 349–357
  uses, 349–351
  prous materials, roles, 350
  classification, pores, 351
  mineral templates, synthesis, 357–373
    organic templates, synthesis from, 360
    Paulino and Schuchardt, studies, 360
    properties and characteristics, 359
    schematic models, 358
  preparation methods, 351, 352
    aerogels, 352
    Al_2O_3-SiO_2 xerogels, pore size changes, 353
    historical development, maximum pore sizes, 355
    Keggin-type structure, 356
    pillaring technique, 356
    pore size distributions, 352
    porous clay heterostructures, 356
    processes, build-up, 351, 352
    processes, leaching, 351, 356
    quaternary ammonium hydroxides, synthesizers, 353
    selective leaching, 357
    structure-directing agents (SDA) and zeolites, relationship, 354
    structure-directing agents (SDA), pore formers, 352
Nanoporous materials, practical/future applications, 373–378
  adsorption and pore size, relationship, 376, 377
  catalysts, solid acidity and applications, 374–375
  miscellaneous properties, 378
  MPS and PCH, advantages, 375
  MPS, octylsilane introduction, 377
  solid acidity and pore size, relationship, 374
  sorption properties and applications, 376–378
Nanoscale-controlled microstructures, 317
Nanostructure, 265

Nanostructured materials, breakthrough in ZT, 405–414
  career pocket engineering, superlattice, 412
  dimensionless figure of merit, history at 300 K, 405
  imbedded quantum dot nanostructured bulk materials, 413
    TEM image, 413
  phonon-blocking electron-transmitting structures, 412
  ZT temperature dependence, 412
  quantum confinement effect, effect on thermoelectric properties, 406–414
    1D quantum wire structure, proof-of-principle, 411
    1D quantum wire, theoretical research, 408–409
    2D quantum well structure, theoretical research, 406
    2D quantum well superlattice structure, proof-of-principle, 409–411
    absolute Seebeck coefficients, 411
    calculated $Z_{1D}T$, carrier concentration, 410
    $S2n$ results, multiquantum confinement effect (MQW), 410
    $Z_{1D}T$ plot, optimum reduced chemical potential, 409
Nanostructured materials synthesis, silica sources, 361
  antigorite and kaolinite, structure models, 371
  high specific surface area, reasons for, 363
  ideal cylindrical mesospores, 364
  layered double hydroxides, 365
  microspore size distribution, 363
    Horvath-Kawazoe methods, 363
    Saito-Foley methods, 363
  mineral-templated MPS, 362
  pillared layer compound (PLC), 365–367
    card-house type nanostructure, 366
    host layer compounds, 365
    pillaring ions, 365
    primary composite-/panoscopic-type nanostructure, 366
  porous clay heterostructure (PCH), 364–365
    sources, 365
  porous materials, shaping and coating, 372–373
    dry gel conversion method, 372
  porous materials, selective leaching, 367–372
    leaching ratios and leaching time, 369
  specific surface areas, change in, 370
  surface area calculation, 363
    t-plot method, 363
  various components, solubility changes, 368
  ZSM-5 zeolite film, SEM images, 373
Nematic liquid crystal (NLC) polymers, 154
New functional oxide materials, 94–95

Optical diode (OD)
  cell structures, 160, 161
    three-layered, 162
    two-layered, 161
  experimental results, 164
    transmittance spectra, two-layered device, 164
  principle, 160, 161
  simulations, 162–164
    dispersion curves, 162, 163
    transmittance spectra, three-layered device, 162, 163
  tunability, 164–165
    electrotunability, two-layered device, 165, 166
Organic thyristor, 225, 226, 257, 258
Orowan mechanism, 319

Particle-in-a-box model, 268
Pb/Si(111), 284
Peierls state, 234, 255
Periodic nanostructure encoding, interfering femtosecond pulses, 40–56
  background and history, 40
  embedded holograms, 48
    chemical etching, 50
    chirped pulse, 49
    embedded gratings, 49, 51
    full compressed pulse, 48
    surface-relief gratings, 49, 50
  encoding, progress in, 40
  encoded hologram, example, 46–53
  encoding mechanism, 41
  features, 41
  instrumentation, 42
    experimental setup, 42
    general features, 42
    near infrared (IR) fs laser pulses, 42, 43
    pulse beams, spatial/temporal colliding, 43

Periodic nanostructure encoding, interfering femtosecond pulses *(continued)*
  multi-dimensional periodic structure, 51
    crossed grating encoded, SEM images, 52, 53
    double exposure technique, 51, 52
    resulting encoded structure, 51
  optical device fabrication, application, 54, 55
    DFB laser structure, fabrication, 55
    LiF DFB laser, laser oscillation and measuring set-up, 56
    LiF single crystal, distributed feedback color center laser, 54
    waveguide, optical coupling device in, 54
  surface relief holograms, 46
    infrared reflection spectra, 46
    near-IR fs laser, on silica glass, 46
    silica thin film, 48
    surface-relief gratings, on silica glass, 47
    UV fs laser, on silica glass, 47
    ZnO, gratings recorded in, 47
  temporal coincidence, detecting mechanism, 44–46
    distributed feedback dye laser (DFDL) technique, 46
    optical Kerr gating method, 45
    photodiode-detector (PD), 45
    polarized beam splitter (PBS), 45
    pump and probe technique, 45
    THG intensity, function of collision delay time, 44
Phonon glass and electron crystal (PGEC) materials, 390–391
Photonic band gap (PBG), 138, 139, 143, 145–147, 150, 151, 153–155, 157, 158, 160–162, 164, 165
Photonic crystal (PC), 138, 140
Photonic effect, 137–148
  standing waves, dielectric layers, 138
  superprism effect, 138
Photonic liquid crystal (LC), 139–141
  chiral SmCA* phase, 140, 141
  chiral smectic C (SmC*) phase, 140, 141
  cholesteric LC (CLC) phase, 140, 141
  helical structures, 140
  twist grain boundary (TGB) phase, 140, 141
Photonic structures, 138, 139
Pillared layer compounds (PLCs), 351
Plusiotis batesi, 154

Plusiotis resplendens, 154
Powder metallurgy (PM) alloys, 317
  atomization, 317
PZT thin films, ferroelectricity size effect, 101
  PZT capacitor formation, 101
    capacitance inverse, 104
    leakage current densities, 102
    metal organic chemical vapor deposition (MOCVD) process, 101
    polarization–electric field (P–E) hysteresis loops, 105
    P_r and E_c values, saturation properties, 106
    relative dielectric constant, 104
    XRD pattern change, 102
    XRD reciprocal space mappings, 50-nm-thick PZT film, 102, 103

Quantum confinement, 265
  conduction electrons, 281
  electron quantum confinement, 279
Quantum well states (QWSs), 266, 291
  energy levels, 294
  peaks, 273, 288
  Fermi level crossing, 295
Quantum well width, 269
  dependence, 270

Scanning tunneling spectroscopy (STS), 274
Schrodinger equation, 275
Second-harmonic generation (SHG), 272
  intensity oscillation, 272
Seebeck coefficient, 385
Seebeck effect, 384
Size control, 265
  nanostructures, 278
Size controllable thin films, 293
Shockley-type surface state, 275
Small-thickness region, 280
sp-band electrons, 269
  two-dimensional, 276
Standing-wave patterns, 275, 278
  concentric, 276
Static limit/atomic limit, 241
STM measurements, 277
  dI/dV mapping
Stoner model, 248–250
Structural stability, 293
Superplastic forming (SPF), 309–310
  cavitation, schematic illustration, 309
  sinter forging, 309

Superplastic forming/diffusion bonding (SPF/DB), 309
Surface-parallel/normal direction, 267
Surface-state electrons, 275
Symmetry gap, 273

Thermal disturbances, 266
Thermal energy, 272
Thermoelectric figure of merit enhancement, nanostructural control, 383
  history, 384–385
  practical applications, 388–390
  principles and background, 384–390
  Seebeck coefficient, schematic illustrations of, 386
  thermoelectric generation, operating principles, 384
  thermoelectric figure of merit, 385–386
    temperature dependence, 389
  thermoelectric transport parameters, 386–388
Thin films
  film/interface potential barrier, 294
  nanometer-height, 294
Three-dimensional lithium ion conductor, 85–93
  bottleneck size influence, 91
  glass transition observation, calorimetry, 91
    $La_{0.51}Li_{0.35}TiO_3$, heat capacities, electrical moduli, 91
  Li_4SiO_4 and related compounds, 85
  lithium ion conductivities, 92
    $La_{0.562}Li_{0.16}TiO_{3.0}$, scattering amplitude distribution, 93
    temperature dependence, 92
  lithium ion vacancy concentration, 88–91
    A-sites, 88
    ionic conductivity variation, 90
    ionic conductivity, function of lithium content, 89
    $La_{1/2}Na_{1/2}TiO_3$, 90
    percolation theory, 90
  NASICON-type structure, 85, 87
    Li(1), Li(2) sites, 86
    room temperature ion conductivity, 87
  pervoskite-type oxides, 86, 87
    BX_6 octahedra, 86
    interstitial passageway (bottlenecks), 88
    lanthanum lithium titanate, cross section, 88, 89
    pervoskite structure scheme, 88

TMTTF (tetramethylthetrathiafulvane) salts, 234, 255
Total reflection band, 141
Transparent oxide semiconductors (TCOs), 7–24
  amorphous TOSs, electron mobility, 8
  Cu-based p-type TOSs, 8
  electron transport paths, schematic illustration, 9
  layered oxychalcogenides (LnCuOCh), 17–24
    As-deposited and annealed films, SEM images, 19
    LnCuOCh epitaxial thin film, crystal structure and HRTEM image, 20
    p-type degenerated conduction, 21
    emission spectra, LaCuOS, 22
    DFWM signals, 24
    LnCuOCh growth, 17, 19
    electronic properties, 18, 20
    optical properties, 20, 21
    optical absorption spectra, 21, 22
    optical nonlinearity, exciton-exciton interaction, 23, 24
  layered TOS compounds, natural quantum well structures, 15–24
    epitaxial film growth, reactive solid-phase epitaxy (R-SPE), 17, 18
    examples, 16
    features, layered MQW vs. natural MQW, 17
    LaCuOS, band structure, 16
  new TCOs development, guiding principles, 8
  reactive solid phase epitaxy (R-SPE), 8
  TOS, research frontiers, 9–15
  TOS device, 12–15
    high-performance transparent TFTs, 14, 15
    organic TFTs, 13
    photoluminescence and current-injected luminescence spectra, heterojunction LED, 13
    transparent pn-diode, p-$ZnRh_2O_4$/ n-ZnO, 12
    transparent UV detector, 13
    UV-LED, p-type $SrCu_2O_2$/ n-type ZnO, 12
  TOS epitaxial films, 11, 12
    super-flat ITO epitaxial films, 11
    vanadyl-phthalocyanine (VOPc), lateral epitaxial growth, 12

TOS materials, 9–11
   amorphous TOSs, 10
   Deep-UV (DUV) TOS, 10
   p-type TOSs, 9
   transparent electrochromic material, 10
   valence band maximum (VBM), 8
Transparent nanoporous crystal ($12CaO \cdot 7Al_2O_3$), 24–39
   background and approach, 24
   C12A7, various shapes synthesis, 26, 27
   C12A7 electride, 35
      current-voltage characteristic, 37
      electron emission current, function of temperature, 38
      electron emission, 37
      fabrication via. melting, 36
      optical absorption spectra, 35
      Richardson–Dushman equation, 38
      synthesis, 35
      TFE measurement, experimental setup, 38
   hot-ion implantation, 39
      Ar^+-dose dependance, electrical conductivity, 39
   hydrogen-related species, incorporation, 29–34
   nanoporous crystal, $12CaO \cdot 7Al_2O_3$ (C12A7), 24
      active functions, 25
      anions incorporation, cages, 26
      crystal structure, 25
      UV-light induced electrical conductivity, 26
   oxygen-related species, incorporation, 27–29
      methane, partial oxidation into syngas, 28, 30
      O^- ion beam, generation, 29, 31
      O_2^- radicals in cages, configuration and dynamics, 28, 29
      oxygen radical formation, 27
      partial oxygen pressure and temperature dependences, oxygen radical concentration, 28
   photo-induced insulator-conductor conversion, C12A7, 29–31, 34
      cage conduction band, 31
      embedded cluster approach, 33
      energy band diagram, 33
      photoinduced conversion, mechanism, 31–34
   stable electride, formation, 35–37

Ultraviolet photoelectron spectroscopy, 267

Volatile organic compounds (VOCs), 351

Wiedmann–Franz equation, 403
Wiedmann–Franz law, 385

X-ray photoelectron emission microscopy (XPEEM), 292

Y_2O_3-stabilized tetragonal ZrO_2 polycrystals (Y-TZP), 298, 302–304, 307

Zeolites, 351